TALL BUILDING STRUCTURES

TALL BUILDING STRUCTURES: ANALYSIS AND DESIGN

Bryan Stafford Smith
McGill University
Montreal, Quebec
Canada

Alex Coull
University of Glasgow
Glasgow, Scotland
United Kingdom

A WILEY-INTERSCIENCE PUBLICATION

JOHN WILEY & SONS, INC.
New York • Chichester • Brisbane • Toronto • Singapore

A NOTE TO THE READER:
This book has been electronically reproduced from digital information stored at John Wiley & Sons, Inc. We are pleased that the use of this new technology will enable us to keep works of enduring scholarly value in print as long as there is a reasonable demand for them. The content of this book is identical to previous printings.

In recognition of the importance of preserving what has been written, it is a policy of John Wiley & Sons, Inc., to have books of enduring value published in the United States printed on acid-free paper, and we exert our best efforts to that end.

Copyright © 1991 by John Wiley & Sons, Inc.

All rights reserved. Published simultaneously in Canada.

Reproduction or translation of any part of this work beyond that permitted by Section 107 or 108 of the 1976 United States Copyright Act without the permission of the copyright owner is unlawful. Requests for permission or further information should be addressed to the Permissions Department, John Wiley & Sons, Inc.

Library of Congress Cataloging in Publication Data:
Stafford Smith, Bryan.
   Tall building structures: analysis and design/Bryan Stafford Smith, Alex Coull.
     p. cm.
   "A Wiley-Interscience publication."
   Includes bibliographical references.
   ISBN 0-471—51237-0
   1. Tall buildings-Design and construction. 2. Structural engineering. I. Coull, Alex. II. Title.
TH1611.S59 1991
690—dc20
                                        90-13007
                                            CIP

To Betty and Frances

PREFACE

This book is the indirect outcome of 25 years of research on tall building structures by the two authors. It began with their liaison in the mid-1960s at the University of Southampton, England, and has since continued in their respective Universities, of Surrey, McGill, Strathclyde, and Glasgow.

At the commencement of the period, the evolution of radically new structural forms gave great stimulus to devising appropriate methods of analysis. In the succeeding quarter-century there have been great advances in the design and construction of tall buildings throughout the world, and in the associated development of analytical techniques.

In the early days, approximate techniques were being devised for specific, largely two-dimensional, structural forms, and the analysis of complex three-dimensional systems represented a formidable challenge. Since then, there have been significant advances in both computer hardware and software: the power of computers has increased dramatically, and a large number of comprehensive general purpose analysis programs have been developed, based on the stiffness method of analysis. In principle at least, it is now theoretically possible to analyze accurately virtually any complex elastic structure, the only constraints being the capacity of the available computer, time, and cost.

However, the great power of this analytical facility has to be handled judiciously. Real building structures are so complex that even an elaborate computational model will be a considerable simplification, and the results from an analysis will always be approximate, being at best only as good as the quality of the chosen model and method of analysis. It is thus imperative to be able to devise an analytical model of the real structure that will represent and predict with appropriate accuracy, and as efficiently and economically as possible, the response of the building to the anticipated forces. Models required for the early stages of design will often be of a different, lower level of sophistication than those for checking the final design.

The task of structural modeling is arguably the most difficult one facing the structural analyst, requiring critical judgment and a sound knowledge of the structural behavior of tall building components and assemblies. Also, the resulting data from the analysis must be interpreted and appraised with discernment for use with the real structure, in order to serve as a reasonable basis for making design decisions.

The rapid advances in the past quarter-century have slowed up, and the era is now one of consolidation and utilization of research findings. However, the majority of the research findings still exist only in the form of papers in research journals, which are not generally available or familiar to the design engineer. There is a need to digest and to bring together in a unified and coherent form the main corpus of knowledge that has been accumulated and to disseminate it to the structural engineering profession. This task forms the main objective of this volume.

It is not possible to deal in a comprehensive manner in a single volume with all aspects of tall building design and construction, and attention has been focused on the building structure. Such important related topics as foundation design, construction methods, fire resistance, planning, and economics have had to be omitted. The intention has been to concentrate on the structural aspects that are particularly affected by the quality of tallness; topics that are of equal relevance to low-rise buildings have generally not been considered in any depth.

The major part of the book thus concentrates on the fundamental approaches to the analysis of the behavior of different forms of tall building structures, including frame, shear wall, tubular, core, and outrigger-braced systems. Both accurate computer-based and approximate methods of analyses are included. The latter, although being of value in their own right for the analysis of simplified regular structures, serve also to highlight the most important actions and modes of behavior of components and assemblies, and thus offer guidance to the engineer in devising appropriate models for analytical purposes.

Introductory chapters discuss the forces to which the structure is subjected, the design criteria that are of the greatest relevance and importance to tall buildings, and the various structural forms that have developed over the years since the early skyscrapers were first introduced at the turn of the century. A major chapter is devoted to the modeling of real structures for both preliminary and final analyses. Considerable attention is devoted to the assessment of the stability of the structure, and the significance of creep and shrinkage in tall concrete buildings is discussed. Finally, a chapter is devoted to the dynamic response of structures subjected to wind and earthquake forces, including a discussion of the human response to tall building motions.

In addition to the set of references appropriate to each chapter, a short bibliography has also been presented. This has been designed to serve several purposes: to note historically important papers, to recommend major works that themselves contain large numbers of bibliographic references, and to refer to papers that offer material or information additional to that contained in the different chapters. Space has prevented the production of a comprehensive bibliography, since the literature on the subject is now vast. Apologies are therefore due to the many authors whose work has been omitted due to either the demand for brevity or the oversight of the writers.

In view of the wide variations in practice in different countries, it was decided not to concentrate on a single set of units in the numerical examples presented to illustrate the theory. Thus both SI and US units will be found.

The book is aimed at two different groups. First, as a result of the continuing activity in the design and construction of tall buildings throughout the world, it will be of value to practicing structural engineers. Second, by treating the material in a logical, coherent, and unified form, it is hoped that it can form the basis of an independent academic discipline, serving as a useful text for graduate student courses, and as an introduction to the subject for senior undergraduates.

In writing the book, the authors are conscious of a debt to many sources, to friends, colleagues, and co-workers in the field, and to the stimulating work of those associated with the Council on Tall Buildings and Urban Habitat, the successor to the International Committee for the Planning and Design of Tall Buildings, with whom they have been associated since its inception. A special privilege of working in a university is the opportunity to interact with fresh young minds. Consequently, above all, they acknowledge their indebtedness to the many research students with whom they have worked over the years, who have done so much to assist them in their progress. Many of their names figure in the References and Bibliography, and many are now recognized authorities in this field. The authors owe them much.

Although the subject material has altered considerably over the long period of writing, the authors also wish to acknowledge the helpful discussions with Professor Joseph Schwaighofer of the University of Toronto in the early stages of planning this work.

Finally, the authors wish to express their gratitude to Ann Bless, Regina Gaiotti and Marie José Nollet of McGill University, Andrea Green of Queens University, and June Lawn and Tessa Bryden of Glasgow University, who have contributed greatly to the production of this volume.

B. STAFFORD SMITH
A. COULL

Montreal, Quebec
Glasgow, Scotland
January 1991

CONTENTS

1. Tall Buildings — 1

  1.1 Why Tall Buildings? — 1
  1.2 Factors Affecting Growth, Height, and Structural Form — 2
  1.3 The Tall Building Structure — 4
      1.3.1 The Design Process — 5
  1.4 Philosophy, Scope, and Content — 6
  1.5 Raisons D'Être — 7
  Reference — 8

2. Design Criteria — 9

  2.1 Design Philosophy — 9
  2.2 Loading — 10
      2.2.1 Sequential Loading — 10
  2.3 Strength and Stability — 11
  2.4 Stiffness and Drift Limitations — 11
  2.5 Human Comfort Criteria — 13
  2.6 Creep, Shrinkage, and Temperature Effects — 14
  2.7 Fire — 14
  2.8 Foundation Settlement and Soil-Structure Interaction — 15
  Summary — 16
  References — 17

3. Loading — 18

  3.1 Gravity Loading — 18
      3.1.1 Methods of Live Load Reduction — 19
      3.1.2 Impact Gravity Loading — 20
      3.1.3 Construction Loads — 20
  3.2 Wind Loading — 21
      3.2.1 Simple Static Approach — 22
      3.2.2 Dynamic Methods — 23
  3.3 Earthquake Loading — 25
      3.3.1 Equivalent Lateral Force Procedure — 26
      3.3.2 Modal Analysis Procedure — 29

3.4	Combinations of Loading		29
	3.4.1.	Working Stress Design	30
	3.4.2	Limit States Design	30
	3.4.3	Plastic Design	30
	Summary		31
	References		32

4. Structural Form   34

4.1	Structural Form		37
	4.1.1	Braced-Frame Structures	37
	4.1.2	Rigid-Frame Structures	38
	4.1.3	Infilled-Frame Structures	40
	4.1.4	Flat-Plate and Flat-Slab Structures	41
	4.1.5	Shear Wall Structures	41
	4.1.6	Wall-Frame Structures	42
	4.1.7	Framed-Tube Structures	44
	4.1.8	Outrigger-Braced Structures	49
	4.1.9	Suspended Structures	50
	4.1.10	Core Structures	52
	4.1.11	Space Structures	53
	4.1.12	Hybrid Structures	54
4.2	Floor Systems—Reinforced Concrete		56
	4.2.1	One-Way Slabs on Beams or Walls	57
	4.2.2	One-Way Pan Joists and Beams	57
	4.2.3	One-Way Slab on Beams and Girders	58
	4.2.4	Two-Way Flat Plate	58
	4.2.5	Two-Way Flat Slab	59
	4.2.6	Waffle Flat Slabs	59
	4.2.7	Two-Way Slab and Beam	59
4.3	Floor Systems—Steel Framing		60
	4.3.1	One-Way Beam System	61
	4.3.2	Two-Way Beam System	61
	4.3.3	Three-Way Beam System	62
	4.3.4	Composite Steel–Concrete Floor Systems	62
	Summary		63

5. Modeling for Analysis   65

5.1	Approaches to Analysis		65
	5.1.1	Preliminary Analyses	65
	5.1.2	Intermediate and Final Analysis	66
	5.1.3	Hybrid Approach to Preliminary and Final Analyses	67

5.2	Assumptions		67
	5.2.1	Materials	68
	5.2.2	Participating Components	68
	5.2.3	Floor Slabs	68
	5.2.4	Negligible Stiffnesses	68
	5.2.5	Negligible Deformations	69
	5.2.6	Cracking	69
5.3	High-Rise Behavior		69
5.4	Modeling for Approximate Analyses		70
	5.4.1	Approximate Representation of Bents	71
	5.4.2	Approximate Modeling of Slabs	73
	5.4.3	Modeling for Continuum Analyses	77
5.5	Modeling for Accurate Analysis		78
	5.5.1	Plane Frames	79
	5.5.2	Plane Shear Walls	79
	5.5.3	Three-Dimensional Frame and Wall Structures	83
	5.5.4	P-Delta Effects	86
	5.5.5	The Assembled Model	87
5.6	Reduction Techniques		88
	5.6.1	Symmetry and Antisymmetry	88
	5.6.2	Two-Dimensional Models of Nontwisting Structures	91
	5.6.3	Two-Dimensional Models of Structures That Translate and Twist	94
	5.6.4	Lumping	99
	5.6.5	Wide-Column Deep-Beam Analogies	103
	Summary		104
	References		105

6. Braced Frames — 106

6.1	Types of Bracing		106
6.2	Behavior of Bracing		109
6.3	Behavior of Braced Bents		111
6.4	Methods of Analysis		113
	6.4.1	Member Force Analysis	113
	6.4.2	Drift Analysis	115
	6.4.3	Worked Example for Calculating Drift by Approximate Methods	119
6.5	Use of Large-Scale Bracing		124
	Summary		128
	References		129

7. Rigid-Frame Structures — 130

- 7.1 Rigid-Frame Behavior — 131
- 7.2 Approximate Determination of Member Forces Caused by Gravity Loading — 133
  - 7.2.1 Girder Forces—Code Recommended Values — 133
  - 7.2.2 Two-Cycle Moment Distribution — 133
  - 7.2.3 Column Forces — 138
- 7.3 Approximate Analysis of Member Forces Caused by Horizontal Loading — 138
  - 7.3.1 Allocation of Loading between Bents — 138
  - 7.3.2 Member Force Analysis by Portal Method — 141
  - 7.3.3 Approximate Analysis by Cantilever Method — 146
  - 7.3.4 Approximate Analysis of Rigid Frames with Setbacks — 150
- 7.4 Approximate Analysis for Drift — 150
  - 7.4.1 Components of Drift — 152
  - 7.4.2 Correction of Excessive Drift — 156
  - 7.4.3 Effective Shear Rigidity (GA) — 157
- 7.5 Flat Plate Structure—Analogous Rigid Frame — 158
  - 7.5.1 Worked Example — 159
- 7.6 Computer Analysis of Rigid Frames — 161
- 7.7 Reduction of Rigid Frames for Analysis — 161
  - 7.7.1 Lumped Girder Frame — 161
  - 7.7.2 Single-Bay Substitute Frame — 163
- Summary — 165
- References — 166

8. Infilled-Frame Structures — 168

- 8.1 Behavior of Infilled Frames — 169
- 8.2 Forces in the Infill and Frame — 172
  - 8.2.1 Stresses in the Infill — 172
  - 8.2.2 Forces in the Frame — 174
- 8.3 Development of the Design Procedure — 174
  - 8.3.1 Design of the Infill — 175
  - 8.3.2 Design of the Frame — 177
  - 8.3.3 Horizontal Deflection — 178
- 8.4 Summary of the Design Method — 178
  - 8.4.1 Provisions — 179
  - 8.4.2 Design of the Infill — 179
  - 8.4.3 Design of the Frame — 180
  - 8.4.4 Deflections — 180

	8.5	Worked Example—Infilled Frame	180
		Summary	182
		References	183

9. Shear Wall Structures — 184

- 9.1 Behavior of Shear Wall Structures — 184
- 9.2 Analysis of Proportionate Wall Systems — 186
  - 9.2.1 Proportionate Nontwisting Structures — 186
  - 9.2.2 Proportionate Twisting Structures — 187
- 9.3 Nonproportionate Structures — 190
  - 9.3.1 Nonproportionate Nontwisting Structures — 190
  - 9.3.2 Nonproportionate Twisting Structures — 199
- 9.4 Behavior of Nonproportionate Structures — 199
- 9.5 Effects of Discontinuities at the Base — 202
- 9.6 Stress Analysis of Shear Walls — 206
  - 9.6.1 Membrane Finite Element Analysis — 206
  - 9.6.2 Analogous Frame Analysis — 207
- Summary — 211
- References — 212

10. Coupled Shear Wall Structures — 213

- 10.1 Behavior of Coupled Shear Wall Structures — 213
- 10.2 Methods of Analysis — 215
- 10.3 The Continuous Medium Method — 216
  - 10.3.1 Derivation of the Governing Differential Equations — 216
  - 10.3.2 General Solutions of Governing Equations — 222
  - 10.3.3 Solution for Standard Load Cases — 223
  - 10.3.4 Graphic Design Method — 231
  - 10.3.5 Coupled Shear Walls with Two Symmetrical Bands of Openings — 235
  - 10.3.6 Worked Example of Coupled Shear Wall Structure — 236
  - 10.3.7 Coupled Shear Walls with Different Support Conditions — 243
- 10.4 Computer Analysis by Frame Analogy — 246
  - 10.4.1 Analysis of Analogous Frame — 247
- 10.5 Computer Analysis Using Membrane Finite Elements — 252
- Summary — 253
- References — 254

11. Wall-Frame Structures — 255

- 11.1 Behavior of Symmetric Wall-Frames — 257
- 11.2 Approximate Theory for Wall-Frames — 260
  - 11.2.1 Derivation of the Governing Differential Equation — 260

	11.2.2	Solution for Uniformly Distributed Loading	262
	11.2.3	Forces in the Wall and Frame	264
	11.2.4	Solutions for Alternative Loadings	266
	11.2.5	Determination of Shear Rigidity (GA)	266
11.3	Analysis by the Use of Graphs		268
11.4	Worked Example to Illustrate Approximate Analysis		271
11.5	Computer Analysis		277
11.6	Comments on the Design of Wall–Frame Structures		279
	11.6.1	Optimum Structure	279
	11.6.2	Curtailed or Interrupted Shear Walls	279
	11.6.3	Increased Concentrated Interaction	280
	Summary		281
	References		282

12. Tubular Structures   283

12.1	Structural Behavior of Tubular Structures		283
	12.1.1	Framed-Tube Structures	283
	12.1.2	Bundled-Tube Structures	288
	12.1.3	Braced-Tube Structures	289
12.2	General Three-Dimensional Structural Analysis		296
12.3	Simplified Analytical Models for Symmetrical Tubular Structures		297
	12.3.1	Reduction of Three-Dimensional Framed Tube to an Equivalent Plane Frame	297
	12.3.2	Bundled-Tube Structures	303
	12.3.3	Diagonally Braced Framed-Tube Structures	305
	Summary		306
	References		307

13. Core Structures   308

13.1	Concept of Warping Behavior		310
13.2	Sectorial Properties of Thin-Walled Cores Subjected to Torsion		315
	13.2.1	Sectorial Coordinate ω'	315
	13.2.2	Shear Center	317
	13.2.3	Principal Sectorial Coordinate (ω) Diagram	318
	13.2.4	Sectorial Moment of Inertia I_ω	320
	13.2.5	Shear Torsion Constant J	321
	13.2.6	Calculation of Sectorial Properties: Worked Example	321
13.3	Theory for Restrained Warping of Uniform Cores Subjected to Torsion		323
	13.3.1	Governing Differential Equation	323

	13.3.2	Solution for Uniformly Distributed Torque	325
	13.3.3	Warping Stresses	326
	13.3.4	Elevator Cores with a Partially Closed Section	329
	13.3.5	Forces in Connecting Beams	331
	13.3.6	Solutions for Alternative Loadings	332
13.4	Analysis by the Use of Design Curves		332
13.5	Worked Example to Analyze a Core Using Formulas and Design Curves		333
13.6	Computer Analyses of Core Structures		341
	13.6.1	Membrane Finite Element Model Analysis	341
	13.6.2	Analogous Frame Analysis	344
	13.6.3	Two-Column Analogy	345
	13.6.4	Single Warping-Column Model	349
	Summary		353
	References		354

14. Outrigger-Braced Structures — 355

14.1	Method of Analysis		356
	14.1.1	Assumptions for Analysis	356
	14.1.2	Compatibility Analysis of a Two-Outrigger Structure	358
	14.1.3	Analysis of Forces	362
	14.1.4	Analysis of Horizontal Deflections	362
14.2	Generalized Solutions of Forces and Deflections		363
	14.2.1	Restraining Moments	363
	14.2.2	Horizontal Deflections	364
14.3	Optimum Locations of Outriggers		364
14.4	Performance of Outrigger Structures		365
	14.4.1	Optimum Locations of Outriggers	366
	14.4.2	Effects of Outrigger Flexibility	368
	14.4.3	"Efficiency" of Outrigger Structures	368
	14.4.4	Alternative Loading Conditions	370
	Summary		370
	References		371

15. Generalized Theory — 372

15.1	Coupled Wall Theory		373
15.2	Physical Interpretation of the Deflection Equation		377
15.3	Application to Other Types of Bent		378
	15.3.1	Determination of Rigidity Parameters	379
	15.3.2	Calculation of Deflection	381
15.4	Application to Mixed-Bent Structures		381

15.5	Accuracy of the Method		383
15.6	Numerical Example		384
	Summary		386
	References		387

16. Stability of High-Rise Buildings — 388

16.1	Overall Buckling Analysis of Frames: Approximate Methods		389
	16.1.1	Shear Mode	390
	16.1.2	Flexural Mode	391
	16.1.3	Combined Shear and Flexural Modes	391
16.2	Overall Buckling Analysis of Wall–Frames		392
	16.2.1	Analytical Method	392
	16.2.2	Example: Stability of Wall–Frame Structure	396
16.3	Second-Order Effects of Gravity Loading		398
	16.3.1	The P-Delta Effect	398
	16.3.2	Amplification Factor P-Delta Analysis	399
	16.3.3	Iterative P-Delta Analysis	401
	16.3.4	Iterative Gravity Load P-Delta Analysis	403
	16.3.5	Direct P-Delta Analysis	405
16.4	Simultaneous First-Order and P-Delta Analysis		406
	16.4.1	Development of the Second-Order Matrix	406
	16.4.2	Negative Shear Area Column	408
	16.4.3	Negative Flexural Stiffness Column	410
16.5	Translational-Torsional Instability		411
16.6	Out-of-Plumb Effects		414
16.7	Stiffness of Members in Stability Calculations		414
16.8	Effects of Foundation Rotation		415
	Summary		416
	References		417

17. Dynamic Analysis — 419

17.1	Dynamic Response to Wind Loading		420
	17.1.1	Sensitivity of Structures to Wind Forces	421
	17.1.2	Dynamic Structural Response due to Wind Forces	422
	17.1.3	Along-Wind Response	423
	17.1.4	Cross-Wind Response	429
	17.1.5	Worked Example	430
17.2	Dynamic Response to Earthquake Motions		431
	17.2.1	Response of Tall Buildings to Ground Accelerations	431
	17.2.2	Response Spectrum Analysis	435

	17.2.3	Empirical Relationships for Fundamental Natural Frequency		449
	17.2.4	Structural Damping Ratios		451
17.3	Comfort Criteria: Human Response to Building Motions			452
	17.3.1	Human Perception of Building Motion		452
	17.3.2	Perception Thresholds		453
	17.3.3	Use of Comfort Criteria in Design		457
	Summary			458
	References			459

18. Creep, Shrinkage, and Temperature Effects — 461

18.1	Effects of Differential Movements			461
18.2	Designing for Differential Movement			462
18.3	Creep and Shrinkage Effects			464
	18.3.1	Factors Affecting Creep and Shrinkage Movements in Concrete		464
	18.3.2	Determination of Vertical Shortening of Walls and Columns		468
	18.3.3	Influence of Reinforcement on Column Stresses, Creep, and Shrinkage		471
	18.3.4	Worked Example		472
	18.3.5	Influence of Vertical Shortening on Structural Actions in Horizontal Members		474
18.4	Temperature Effects			475
	Summary			478
	References			478

APPENDIX 1. Formulas and Design Curves for Coupled Shear Walls — 480

	A1.1	Formulas and Design Curves for Alternative Load Cases		480
		A1.1.1	Formulas for Top Concentrated Load and Triangularly Distributed Loading	480
		A1.1.2	Design Curves	482
	A1.2	Formulas for Coupled Shear Walls with Different Flexible Support Conditions		487
	A1.3	Stiffness of Floor Slabs Connecting Shear Walls		489
		A1.3.1	Effective Width of Floor Slab	493
		A1.3.2	Empirical Relationships for Effective Slab Width	495
		A1.3.3	Numerical Examples	497
		References		500

APPENDIX 2. Formulas and Graphs for Wall–Frame and Core Structures   **502**

    A2.1 Formulas and Graphs for Deflections and Forces   502
           A2.1.1 Uniformly Distributed Horizontal Loading   502
           A2.1.2 Triangularly Distributed Horizontal Loading   506
           A2.1.3 Concentrated Horizontal Load at the Top   506

Bibliography   **512**

Index   **527**

CHAPTER 1

Tall Buildings

This book is concerned with tall building structures. Tallness, however, is a relative matter, and tall buildings cannot be defined in specific terms related just to height or to the number of floors. The tallness of a building is a matter of a person's or community's circumstance and their consequent perception; therefore, a measurable definition of a tall building cannot be universally applied. From the structural engineer's point of view, however, a tall building may be defined as one that, because of its height, is affected by lateral forces due to wind or earthquake actions to an extent that they play an important role in the structural design. The influence of these actions must therefore be considered from the very beginning of the design process.

1.1 WHY TALL BUILDINGS?

Tall towers and buildings have fascinated mankind from the beginning of civilization, their construction being initially for defense and subsequently for ecclesiastical purposes. The growth in modern tall building construction, however, which began in the 1880s, has been largely for commercial and residential purposes.

Tall commercial buildings are primarily a response to the demand by business activities to be as close to each other, and to the city center, as possible, thereby putting intense pressure on the available land space. Also, because they form distinctive landmarks, tall commercial buildings are frequently developed in city centers as prestige symbols for corporate organizations. Further, the business and tourist community, with its increasing mobility, has fuelled a need for more, frequently high-rise, city center hotel accommodations.

The rapid growth of the urban population and the consequent pressure on limited space have considerably influenced city residential development. The high cost of land, the desire to avoid a continuous urban sprawl, and the need to preserve important agricultural production have all contributed to drive residential buildings upward. In some cities, for example, Hong Kong and Rio de Janeiro, local topographical restrictions make tall buildings the only feasible solution for housing needs.

1

1.2 FACTORS AFFECTING GROWTH, HEIGHT, AND STRUCTURAL FORM

The feasibility and desirability of high-rise structures have always depended on the available materials, the level of construction technology, and the state of development of the services necessary for the use of the building. As a result, significant advances have occurred from time to time with the advent of a new material, construction facility, or form of service.

Multistory buildings were a feature of ancient Rome: four-story wooden tenement buildings, of post and lintel construction, were common. Those built after the great fire of Nero, however, used the new brick and concrete materials in the form of arch and barrel vault structures. Through the following centuries, the two basic construction materials were timber and masonry. The former lacked strength for buildings of more than about five stories, and always presented a fire hazard. The latter had high compressive strength and fire resistance, but its weight tended to overload the lower supports. With the rapidly increasing number of masonry high-rise buildings in North America toward the end of the nineteenth century, the limits of this form of construction became apparent in 1891 in the 16-story Monadnock Building in Chicago. With the space in its lower floors largely occupied by walls of over 2 m thick, it was the last tall building in the city for which massive load-bearing masonry walls were employed.

The socioeconomic problems that followed industrialization in the nineteenth century, coupled with an increasing demand for space in the growing U.S. cities, created a strong impetus to tall building construction. Yet the ensuing growth could not have been sustained without two major technical innovations that occurred in the middle of that century: the development of higher strength and structurally more efficient materials, wrought iron and subsequently steel, and the introduction of the elevator (cf. Fig. 1.1). Although the elevator had been developed some 20 years earlier, its potential in high-rise buildings was apparently not realized until its incorporation in the Equitable Life Insurance Building in New York in 1870. For the first time, this made the upper stories as attractive to rent as the lower ones, and, consequently, made the taller building financially viable.

The new materials allowed the development of lightweight skeletal structures, permitting buildings of greater height and with larger interior open spaces and windows, although the early wrought-iron frame structures still employed load-bearing masonry facade walls. The first high-rise building totally supported by a metal frame was the 11-story Home Insurance Building in Chicago in 1883, followed in 1889 by the first all-steel frame in the 9-story Rand-McNally Building. Two years later, in the same city, diagonal bracings were introduced in the facade frames of the 20-story Masonic Temple to form vertical trusses, the forerunner of modern shear wall and braced frame construction. It was by then appreciated that at that height wind forces were an important design consideration. Improved design methods and construction techniques allowed the maximum height of steel-frame structures to increase steadily, reaching a height of 60 stories with the construction of the Woolworth Building in New York in 1913. This golden age of

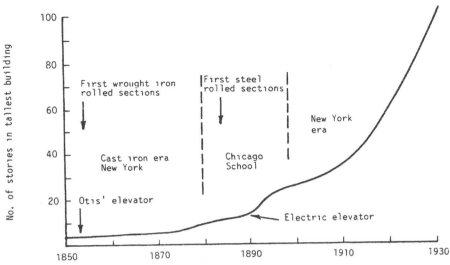

Fig. 1.1 Growth in height of the first great era of American skyscrapers.

American skyscraper construction culminated in 1931 in its crowning glory, the Empire State Building, whose 102-story braced steel frame reached a height of 1250 ft (381 m).

Although reinforced concrete construction began around the turn of the century, it does not appear to have been used for multistory buildings until after the end of World War I. The inherent advantages of the composite material, which could be readily formed to simultaneously satisfy both aesthetic and load-carrying requirements, were not then fully appreciated, and the early systems were purely imitations of their steel counterparts. Progress in reinforced concrete was slow and intermittent, and, at the time the steel-framed Empire State Building was completed, the tallest concrete building, the Exchange Building in Seattle, had attained a height of only 23 stories.

The economic depression of the 1930s put an end to the great skyscraper era, and it was not until some years after the end of World War II that the construction of high-rise buildings recommenced, with radically new structural and architectural solutions. Rather than bringing significant increases in height, however, these modern developments comprised new structural systems, improved material qualities and services, and better design and construction techniques. It was not until 1973 that the Empire State Building was eclipsed in height by the twin towers of the 110-story, 1350 ft (412 m) high World Trade Center in New York, using framed-tube construction, which was followed in 1974 by the 1450 ft (442 m) high bundled-tube Sears Tower in Chicago.

Different structural systems have gradually evolved for residential and office buildings, reflecting their differing functional requirements. In modern office buildings, the need to satisfy the differing requirements of individual clients for floor space arrangements led to the provision of large column-free open areas to allow flexibility in planning. Improved levels of services have frequently necessitated the devotion of entire floors to mechanical plant, but the spaces lost can often be utilized also to accommodate deep girders or trusses connecting the exterior and interior structural systems. The earlier heavy internal partitions and masonry cladding, with their contributions to the reserve of stiffness and strength, have largely given way to light demountable partitions and glass curtain walls, forcing the basic structure alone to provide the required strength and stiffness against both vertical and lateral loads.

Other architectural features of commercial buildings that have influenced structural form are the large entrances and open lobby areas at ground level, the multistory atriums, and the high-level restaurants and viewing galleries that may require more extensive elevator systems and associated sky lobbies.

A residential building's basic functional requirement is the provision of self-contained individual dwelling units, separated by substantial partitions that provide adequate fire and acoustic insulation. Because the partitions are repeated from story to story, modern designs have utilized them in a structural capacity, leading to the shear wall, cross wall, or infilled-frame forms of construction.

The trends to exposed structure and architectural cutouts, and the provision of setbacks at the upper levels to meet daylight requirements, have also been features of modern architecture. The requirement to provide adequately stiff and strong structures, while accommodating these various features, led to radical developments in structural framing, and inspired the new generation of braced frames, framed-tube and hull-core structures, wall-frame systems, and outrigger-braced structures described in Chapter 4. The latest generation of "postmodern" buildings, with their even more varied and irregular external architectural treatment, has led to hybrid double and sometimes triple combinations of the structural monoforms used for modern buildings.

Speed of erection is a vital factor in obtaining a return on the investment involved in such large-scale projects. Most tall buildings are constructed in congested city sites, with difficult access; therefore careful planning and organization of the construction sequence become essential. The story-to-story uniformity of most multistory buildings encourages construction through repetitive operations and prefabrication techniques. Progress in the ability to build tall has gone hand in hand with the development of more efficient equipment and improved methods of construction, such as slip- and flying-formwork, concrete pumping, and the use of tower, climbing, and large mobile cranes.

1.3 THE TALL BUILDING STRUCTURE

Ideally, in the early stages of planning a building, the entire design team, including the architect, structural engineer, and services engineer, should collaborate to agree on a form of structure to satisfy their respective requirements of function, safety and serviceability, and servicing. A compromise between conflicting demands will be almost inevitable. In all but the very tallest structures, however, the structural arrangement will be subservient to the architectural requirements of space arrangement and aesthetics. Often, this will lead to a less-than-ideal structural solution that will tax the ingenuity, and probably the patience, of the structural engineer.

The two primary types of vertical load-resisting elements of tall buildings are columns and walls, the latter acting either independently as shear walls or in assemblies as shear wall cores. The building function will lead naturally to the provision of walls to divide and enclose space, and of cores to contain and convey services such as elevators. Columns will be provided, in otherwise unsupported regions, to transmit gravity loads and, in some types of structure, horizontal loads also. Columns may also serve architecturally as, for example, facade mullions.

The inevitable primary function of the structural elements is to resist the gravity loading from the weight of the building and its contents. Since the loading on different floors tends to be similar, the weight of the floor system per unit floor area is approximately constant, regardless of the building height. Because the gravity load on the columns increases down the height of a building, the weight of columns per unit area increases approximately linearly with the building height.

The highly probable second function of the vertical structural elements is to resist also the parasitic load caused by wind and possibly earthquakes, whose magnitudes will be obtained from National Building Codes or wind tunnel studies. The bending moments on the building caused by these lateral forces increase with at least the square of the height, and their effects will become progressively more important as the building height increases. On the basis of the factors above, the relative quantities of material required in the floors, columns, and wind bracing of a traditional steel frame and the penalty on these due to increasing height are approximately as illustrated in Fig. 4.1.

Because the worst possible effects of lateral forces occur rarely, if ever, in the life of the building, it is imperative to minimize the penalty for height to achieve an optimum design. The constant search for more efficient solutions led to the innovative designs and new structural forms of recent years (cf. Chapter 4). In developing a suitable system for resisting lateral forces, the engineer seeks to devise stiff horizontal interconnections between the various vertical components to form composite assemblies such as coupled walls and rigid frames, which, as demonstrated in later chapters, create a total structural assembly having a lateral stiffness many times greater than the sum of the lateral stiffnesses of the individual vertical components.

1.3.1. The Design Process

Once the functional layout of the structure has been decided, the design process generally follows a well-defined iterative procedure. Preliminary calculations for member sizes are usually based on gravity loading augmented by an arbitrary increment to account for wind forces. The cross-sectional areas of the vertical members will be based on the accumulated loadings from their associated tributary areas, with reductions to account for the probability that not all floors will be subjected simultaneously to their maximum live loading. The initial sizes of beams and slabs are normally based on moments and shears obtained from some simple method of gravity load analysis, such as two-cycle moment distibution, or from codified mid- and end-span values.

A check is then made on the maximum horizontal deflection, and the forces in the major structural members, using some rapid approximate analysis technique. If the deflection is excessive, or some of the members are inadequate, adjustments are made to the member sizes or the structural arrangement. If certain members attract excessive loads, the engineer may reduce their stiffness to redistribute the load to less heavily stressed components. The procedure of preliminary analysis, checking, and adjustment is repeated until a satisfactory solution is obtained.

Invariably, alterations to the initial layout of the building will be required as the client's and architect's ideas of the building evolve. This will call for structural modifications, or perhaps a radical rearrangement, which necessitates a complete review of the structural design. The various preliminary stages may therefore have to be repeated a number of times before a final solution is reached.

A rigorous final analysis, using a more refined analytical model, will then be made to provide a final check on deflections and member strengths. This will usually include the second-order effects of gravity loads on the lateral deflections and member forces (P-Delta effects). A dynamic analysis may also be required if, as a result of wind loading, there is any likelihood of excessive deflections due to oscillations or of comfort criteria being exceeded, or if earthquake loading has to be considered. At some stage in the procedure the deleterious effects of differential movements due to creep, shrinkage, or temperature differentials will also be checked.

In the design process, a thorough knowledge of high-rise structural components and their modes of behavior is a prerequisite to devising an appropriate load-resisting system. Such a system must be efficient, economic, and should minimize the structural penalty for height while maximizing the satisfaction of the basic serviceability requirements. With the increasing availability of general-purpose structural analysis programs, the formation of a concise and properly representative model has become an important part of tall building analysis; this also requires a fundamental knowledge of structural behavior. Modeling for analysis is discussed in Chapter 5.

1.4 PHILOSOPHY, SCOPE, AND CONTENT

The iterative design process described above involves different levels of structural analysis, ranging from relatively crude and approximate techniques for the preliminary stages to sophisticated and accurate methods for the final check. The major part of this book is devoted therefore to a discussion and comparison of the different practical methods of analysis developed for the range of structural forms encountered in tall buildings. The emphasis throughout is on methods particular to tall building structures, with less importance placed on methods for general structural analysis, which are treated comprehensively in other texts. It is thus assumed that the reader is already familiar with the fundamentals of the stiffness matrix method and the finite element method of analysis.

The methods of analysis presented are, almost without exception, static, and assume linear elastic behavior of the structure. Although wind and earthquake forces are transient in nature, it is reasonable and practical to represent them in the majority of design situations by equivalent static force distributions, as described in Chapter 3. Although recognizing that concrete and masonry behave in a nonlinear manner, a linear elastic analysis is still the most important tool for deciding a tall building's structural design. Techniques do exist for the prediction of inelastic behavior, but they are not yet sufficiently well developed to be appropriate for undertaking a detailed analysis of a highly indeterminate tall building structure.

The main emphasis of static linear analysis is applied to both components and assemblies found in tall buildings, ranging from the primary rigid frames, braced and infilled frames, and shear walls, to the more efficient composite systems that include coupled shear walls, wall-frame and framed-tube structures, shear wall cores, and outrigger-braced structures.

Methods suitable for both preliminary and final analyses are described and, where appropriate, detailed worked examples are given to illustrate the steps involved. Although computer-based matrix techniques form the most versatile and accurate methods for practical structural analysis, attention is also devoted to the more limited and approximate continuum techniques. These serve well to provide an understanding of structural behavior and their generalized solutions indicate more clearly and rapidly the influence of changes in structural parameters. Such an understanding can be valuable in selecting a suitable model for computer analysis. The book concludes with a series of Appendices that include useful design formulas and charts, and a selective Bibliography of significant references to the subject matter of the various chapters.

It is impracticable to deal comprehensively in a single volume with all aspects of tall building structures. Important associated topics, therefore, including foundation systems, the detailed treatment of wind and earthquake forces and the associated dynamic structural analysis, and construction procedures, which form major subjects in their own right, have had to be omitted. For a general discussion on all aspects of tall buildings, architectural, social, and technical, the reader is referred to the Reports and Proceedings of the Council on Tall Buildings and Urban Habitat, particularly the five-volume series of definitive Monographs [1.1].

1.5 RAISONS D'ÊTRE

The authors believe that a book devoted to the analysis and design of tall building structures is merited on a number of counts. During the last few decades a large body of knowledge on the subject has accrued from an intensive worldwide research effort. The pace of this research has now abated, but the results are widely dispersed and still generally available only in research journals. Many of the analysis techniques that have been developed are virtually unique to tall buildings, and they form the foundations of an academic discipline that has required the research results to be digested, consolidated, and recorded in a coherent and unified form. Meanwhile high-rise construction continues apace, and there is a continuous demand for information from engineers involved in high-rise design, while structural engineering graduate students are enrolled in courses and conducting further research on tall building structures. This text is aimed to be of value to both the design office and those in the classroom or laboratory.

The object of the book is therefore to offer a coherent and unified treatment of the subject analysis and design of high-rise building structures, for practicing structural engineers concerned with the design of tall buildings, and for senior undergraduate and postgraduate structural engineering students.

REFERENCE

1.1 *Monograph on Planning and Design of Tall Buildings*, Vols. CB, CL, PC, SB, and SC, ASCE, 1980.

CHAPTER 2

Design Criteria

Tall buildings are designed primarily to serve the needs of an intended occupancy, whether residential, commercial, or, in some cases, a combination of the two. The dominant design requirement is therefore the provision of an appropriate internal layout for the building. At the same time, it is essential for the architect to satisfy the client's expectations concerning the aesthetic qualities of the building's exterior. The main design criteria are, therefore, architectural, and it is within these that the engineer is usually constrained to fit his structure. Only in exceptionally tall buildings will structural requirements become a predominant consideration.

The basic layout will be contained within a structural mesh that must be minimally obtrusive to the functional requirements of the building. Simultaneously, there must be an integration of the building structure with the various service systems—heating, ventilating, air-conditioning, water supply and waste disposal, electrical supply, and vertical transportation—which are extensive and complex, and constitute a major part of the cost of a tall building.

Once the functional layout has been established, the engineer must develop a structural system that will satisfy established design criteria as efficiently and economically as possible, while fitting into the architectural layout. The vital structural criteria are an adequate reserve of strength against failure, adequate lateral stiffness, and an efficient performance during the service life of the building.

This chapter provides a brief description of the important criteria that must be considered in the structural design of a tall building. Most of the principles of structural design apply equally to low-rise as to high-rise buildings, and therefore, for brevity, special attention is devoted to only those aspects that have particular consequences for the designers of high-rise buildings.

2.1 DESIGN PHILOSOPHY

Chapter 1 described how radical changes in the structural form of tall buildings occurred in the construction period that followed World War II. Over the same period, a major shift occurred in design philosophy, and the Code formats have progressed from the earlier working stress or ultimate strength deterministic bases to modern more generally accepted probability-based approaches. The probabilistic approach for both structural properties and loading conditions has led to the limit states design philosophy, which is now almost universally accepted. The aim of this approach is to ensure that all structures and their constituent components are designed to resist with reasonable safety the worst loads and deformations that are liable to occur during construction and service, and to have adequate durability during their lifetime.

The entire structure, or any part of it, is considered as having "failed" when it reaches any one of various "limit states," when it no longer meets the prescribed limiting design conditions. Two fundamental types of limit state must be considered: (1) the ultimate limit states corresponding to the loads to cause failure, including instability: since events associated with collapse would be catastrophic, endangering lives and causing serious financial losses, the probability of failure must be very low; and (2) the serviceability limit states, which involve the criteria governing the service life of the building, and which, because the consequences of their failure would not be catastrophic, are permitted a much higher probability of occurrence. These are concerned with the fitness of the building for normal use rather than safety, and are of less critical importance.

A particular limit state may be reached as a result of an adverse combination of random effects. Partial safety factors are employed for different conditions that reflect the probability of certain occurrences or circumstances of the structure and loading existing. The implicit objective of the design calculations is then to ensure that the probability of any particular limit state being reached is maintained below an acceptable value for the type of structure concerned.

The following sections consider the criteria that apply in particular to the design of tall buildings.

2.2 LOADING

The structure must be designed to resist the gravitational and lateral forces, both permanent and transient, that it will be called on to sustain during its construction and subsequent service life. These forces will depend on the size and shape of the building, as well as on its geographic location, and maximum probable values must be established before the design can proceed.

The probable accuracy of estimating the dead and live loads, and the probability of the simultaneous occurrence of different combinations of gravity loading, both dead and live, with either wind or earthquake forces, is included in limit states design through the use of prescribed factors.

The load systems that must be considered are described in Chapter 3.

2.2.1 Sequential Loading

For loads that are applied after completion of the building, such as live, wind, or seismic loading, the analysis is independent of the construction sequence. For dead loads, however, which are applied to the building frame as construction proceeds, the effects of sequential loading should be considered to assess the worst conditions to which any component may be subjected, and also to determine the true behavior of the frame.

In multistory reinforced concrete construction, the usual practice is to shore the freshly placed floor on several previously cast floors. The construction loads in the supporting floors due to the weight of the wet concrete and formwork may appreciably exceed the loads under service conditions. Such loads depend on the sequence and rate of erection.

If column axial deformations are calculated as though the dead loads are applied to the completed structure, bending moments in the horizontal components will result from any differential column shortening that is shown to result. Because of the cumulative effects over the height of the building, the effects are greater in the highest levels of the building. However, the effects of such differential movements would be greatly overestimated because in reality, during the construction sequence, a particular horizontal member is constructed on columns in which the initial axial deformations due to the dead weight of the structure up to that particular level have already taken place. The deformations of that particular floor will then be caused by the loads that are applied subsequent to its construction. Such sequential effects must be considered if an accurate assessment of the structural actions due to dead loads is to be achieved.

2.3 STRENGTH AND STABILITY

For the ultimate limit state, the prime design requirement is that the building structure should have adequate strength to resist, and to remain stable under, the worst probable load actions that may occur during the lifetime of the building, including the period of construction.

This requires an analysis of the forces and stresses that will occur in the members as a result of the most critical possible load combinations, including the augmented moments that may arise from second-order additional deflections (P-Delta effects) (cf. Chapter 16). An adequate reserve of strength, using prescribed load factors, must be present. Particular attention must be paid to critical members, whose failure could prove catastrophic in initiating a progressive collapse of part of or the entire building. Any additional stresses caused by restrained differential movements due to creep, shrinkage, or temperature must be included (cf. Chapter 18).

In addition, a check must be made on the most fundamental condition of equilibrium, to establish that the applied lateral forces will not cause the entire building to topple as a rigid body about one edge of the base. Taking moments about that edge, the resisting moment of the dead weight of the building must be greater than the overturning moment for stability by an acceptable factor of safety.

2.4 STIFFNESS AND DRIFT LIMITATIONS

The provision of adequate stiffness, particularly lateral stiffness, is a major consideration in the design of a tall building for several important reasons. As far as the ultimate limit state is concerned, lateral deflections must be limited to prevent second-order P-Delta effects due to gravity loading being of such a magnitude as to precipitate collapse. In terms of the serviceability limit states, deflections must first be maintained at a sufficiently low level to allow the proper functioning of nonstructural components such as elevators and doors; second, to avoid distress in the structure, to prevent excessive cracking and consequent loss of stiffness, and to avoid any redistribution of load to non-load-bearing partitions, infills, cladding, or glazing; and third, the structure must be sufficiently stiff to prevent dynamic motions becoming large enough to cause discomfort to occupants, prevent delicate work being undertaken, or affect sensitive equipment. In fact, it is in the particular need for concern for the provision of lateral stiffness that the design of a high-rise building largely departs from that of a low-rise building.

One simple parameter that affords an estimate of the lateral stiffness of a building is the drift index, defined as the ratio of the maximum deflection at the top of the building to the total height. In addition, the corresponding value for a single story height, the interstory drift index, gives a measure of possible localized excessive deformation. The control of lateral deflections is of particular importance for modern buildings in which the traditional reserves of stiffness due to heavy internal partitions and outer cladding have largely disappeared. It must be stressed, however, that even if the drift index is kept within traditionally accepted limits, such as $\frac{1}{500}$, it does not necessarily follow that the dynamic comfort criteria will also be satisfactory. Problems may arise, for example, if there is coupling between bending and torsional oscillations that leads to unacceptable complex motions or accelerations. In addition to static deflection calculations, the question of the dynamic response, involving the lateral acceleration, amplitude, and period of oscillation, may also have to be considered.

The establishment of a drift index limit is a major design decision, but, unfortunately, there are no unambiguous or widely accepted values, or even, in some of the National Codes concerned, any firm guidance. The designer is then faced with having to decide on an appropriate value. The figure adopted will reflect the building usage, the type of design criterion employed (for example, working or ultimate load conditions), the form of construction, the materials employed, including any substantial infills or claddings, the wind loads considered, and, in particular, past experience of similar buildings that have performed satisfactorily.

Design drift index limits that have been used in different countries range from 0.001 to 0.005. To put this in perspective, a maximum horizontal top deflection of between 0.1 and 0.5 m (6 to 20 in.) would be allowed in a 33-story, 100-m (330-ft.) high building, or, alternatively, a relative deflection of 3 to 15 mm (0.12 to 0.6 in.) over a story height of 3 m (10 ft). Generally, lower values should be used for hotels or apartment buildings than for office buildings, since noise and movement tend to be more disturbing in the former. Consideration may be given to whether the stiffening effects of any internal partitions, infills, or claddings are included in the deflection calculations.

The consideration of this limit state requires an accurate estimate of the lateral deflections that occur, and involves an assessment of the stiffness of cracked members, the effects of shrinkage and creep and any redistribution of forces that may result, and of any rotational foundation movement. In the design process, the stiffness of joints, particularly in precast or prefabricated structures, must be given special attention to develop adequate lateral stiffness of the structure and to prevent any possible progressive failure. The possibility of torsional deformations must not be overlooked.

In practice, non-load-bearing infills, partitions, external wall panels, and window glazing should be designed with sufficient clearance or with flexible supports to accommodate the calculated movements.

Sound engineering judgment is required when deciding on the drift index limit to be imposed. However, for conventional structures, the preferred acceptable range is 0.0015 to 0.003 (that is, approximately $\frac{1}{650}$ to $\frac{1}{350}$), and sufficient stiffness must be provided to ensure that the top deflection does not exceed this value under extreme load conditions. As the height of the building increases, drift index coefficients should be decreased to the lower end of the range to keep the top story deflection to a suitably low level. Succeeding chapters describe how deflections may be computed.

The drift criteria apply essentially to quasistatic conditons. When extreme force conditions are possible, or where problems involving vortex shedding or other unusual phenomena may occur, a more sophisticated approach involving a dynamic analysis may be required.

If excessive, the drift of a structure can be reduced by changing the geometric configuration to alter the mode of lateral load resistance, increasing the bending stiffness of the horizontal members, adding additional stiffness by the inclusion of stiffer wall or core members, achieving stiffer connections, and even by sloping the exterior columns. In extreme circumstances, it may be necessary to add dampers, which may be of the passive or active type.

2.5 HUMAN COMFORT CRITERIA

If a tall flexible structure is subjected to lateral or torsional deflections under the action of fluctuating wind loads, the resulting oscillatory movements can induce a wide range of responses in the building's occupants, ranging from mild discomfort to acute nausea. Motions that have psychological or physiological effects on the occupants may thus result in an otherwise acceptable structure becoming an undesirable or even unrentable building.

There are as yet no universally accepted international standards for comfort criteria, although they are under consideration, and engineers must base their design criteria on an assessment of published data. It is generally agreed that acceleration is the predominant parameter in determining human response to vibration, but other factors such as period, amplitude, body orientation, visual and acoustic cues, and even past experience can be influential. Threshold curves are available that give various limits for human behavior, ranging from motion perception through work difficulty to ambulatory limits, in terms of acceleration and period.

14 DESIGN CRITERIA

A dynamic analysis is then required to allow the predicted response of the building to be compared with the threshold limits.

The questions of human response to motion, comfort criteria, and their influence on structural design are considered in Chapter 17.

2.6 CREEP, SHRINKAGE, AND TEMPERATURE EFFECTS

In very tall concrete buildings, the cumulative vertical movements due to creep and shrinkage may be sufficiently large to cause distress in nonstructural elements, and to induce significant structural actions in the horizontal elements, especially in the upper regions of the building. In assessing these long-term deformations, the influence of a number of significant factors must be considered, particularly the concrete properties, the loading history and age of the concrete at load application, and the volume–surface ratio and amount of reinforcement in the members concerned. The structural actions in the horizontal elements caused by the resulting relative vertical deflections of their supports can then be estimated. The differential movements due to creep and shrinkage must be considered structurally and accommodated as far as possible in the architectural details at the design stage. However, by attempting to achieve a uniformity of stress in the vertical components, it is possible to reduce as far as possible any relative vertical movement due to creep.

In the construction phase, in addition to creep and shrinkage, elastic shortening will occur in the vertical elements of the lower levels due to the additional loads imposed by the upper stories as they are completed. Any cumulative differential movements will affect the stresses in the subsequent structure, especially in buildings that include both in situ and precast components.

In buildings with partially or fully exposed exterior columns, significant temperature differences may occur between exterior and interior columns, and any restraint to their relative deformations will induce stresses in the members concerned. The analysis of such actions requires a knowledge of the differential temperatures that are likely to occur between the building and its exterior and the temperature gradient through the members. This will allow an evaluation of the free thermal length changes that would occur if no restraint existed, and, hence, using a standard elastic analysis, the resulting thermal stresses and deformations may be determined.

Practical methods for analyzing the effects of creep, shrinkage, and temperature are discussed in Chapter 18.

2.7 FIRE

The design considerations for fire prevention and protection, smoke control, firefighting, and escape are beyond the scope of a book on building structures. However, since fire appears to be by far the most common extreme situation that will cause damage in structures, it must be a primary consideration in the design process.

The characteristic feature of a fire, such as the temperature and duration, can be estimated from a knowledge of the important parameters involved, particularly the quantity and nature of combustible material present, the possibility and extent of ventilation, and the geometric and thermal properties of the fire compartment involved. Once the temperatures at the various surfaces have been determined, from the gas temperature curve, it is possible to estimate the heat flow through the insulation and structural members. A knowledge of the temperature gradient across the member, and the degree of restraint afforded by the supports and surrounding structure, enables the stresses in the member to be evaluated. The mechanical properties of the structural materials, particularly the elastic modulus or stiffness and strength, may deteriorate rapidly as the temperature rises, and the resistance to loads is greatly reduced. For example the yield stress of mild steel at a temperature of 700°C is only some 10-20% of its value at room temperature. Over the same temperature range, the elastic modulus drops by around 40-50%. The critical temperature at which large deflections or collapse occurs will thus depend on the materials used, the nature of the structure, and the loading conditions.

The parameters that govern the approach are stochastic in nature, and the results of any calculation can be given only in probabilistic terms. The aim should be to achieve a homogeneous design in which the risks due to the different extreme situations are comparable.

Designing against fire is, however, a specialist discipline, and the interested reader is referred to the Monograph on Tall Buildings (Vol. CL) [2.1].

2.8 FOUNDATION SETTLEMENT AND SOIL-STRUCTURE INTERACTION

The gravity and lateral forces on the building will be transmitted to the earth through the foundation system, and, as the principles of foundation design are not affected by the quality of tallness of the superstructure, conventional approaches will suffice. The concern of the structural designer is then with the influence of any foundation deformation on the building's structural behavior and on the soil-structure interactive forces.

Because of its height, the loads transmitted by the columns in a tall building can be very heavy. Where the underlying soil is rock or other strong stable subgrade, foundations may be carried down to the stiff load-bearing layers by use of piles, caissons, or deep basements. Problems are not generally encountered with such conditions since large variations in column loadings and spacings can be accommodated with negligible differential settlement. In areas in which soil conditions are poor, loadings on foundation elements must be limited to prevent shearing failures or excessive differential settlements. Relief may be obtained by excavating a weight of soil equal to a significant portion of the gross building weight. Because of the high short-term transient moments and shears that arise from wind loads, particular attention must be given to the design of the foundation system for resisting moments and shears, especially if the precompression due to the dead weight of the building is not sufficient to overcome the highest tensile stresses caused by wind moments, leading to uplift on the foundation.

The major influences of foundation deformations are twofold. First, if the bases of vertical elements yield, a stress redistribution will occur, and the extra loads imposed on other elements may then further increase the deformation there. The influence of the relative displacements on the forces in the horizontal elements must then be assessed. Second, if an overall rotational settlement θ of the entire foundation occurs, the ensuing lateral deflections will be magnified by the height H to give a top deflection of $H\theta$. As well as increasing the maximum drift, the movement will have a destabilising effect on the structure as a whole, by increasing any P-Delta effects that occur (cf. Chapter 16).

Soil–structure interaction involves both static and dynamic behavior. The former is generally treated by simplified models of subgrade behavior, and finite element methods of analysis are usual. When considering dynamic effects, both interactions between soil and structure, and any amplification caused by a coincidence of the natural frequencies of building and foundation, must be included. Severe permanent structural damage may be caused by earthquakes when large deformations occur due to the soil being compacted by the ground vibration, which under certain conditions may result in the development of excess hydrostatic pressures sufficient to produce liquefaction of the soil. These types of soil instability may be prevented or reduced in intensity by appropriate soil investigation and foundation design. On the other hand, the dynamic response of buildings to ground vibrations, which is also affected by soil conditions, cannot be avoided and must be considered in design.

A general discussion of all aspects of the design of foundations for tall buildings is given in Reference 2.2.

SUMMARY

Probability-based limit states concepts form the basis of modern structural design codes. This chapter summarizes the most important limit states involved in the design of tall building structures. Ultimate limit states are concerned with the maximum load and carrying capacity of the structure, where the probability of failure must be very low, whereas serviceability limit states are concerned with actions that occur during the service life of the structure, and are permitted to have a much higher probability of occurrence.

The most important ultimate limit state requirement is that the structure should have adequate strength and remain stable under all probable load combinations that may occur during the construction and subsequent life of the building. When assessing stability, any second-order P-Delta effects in heavily loaded slender members must be considered. Any stresses induced by relative movements caused by creep, shrinkage, and temperature differentials must be included.

One major serviceability limit state criterion lies in the provision of adequate stiffness, particularly lateral stiffness, to avoid excessive cracking in concrete and to avoid any load transfer to non-load-bearing components, to avoid excessive secondary P-Delta moments caused by lateral deflections, and to prevent any dynamic motions that would affect the comfort of the occupants. One measure of the stiffness is the drift of the structure and this should be limited to the range of 0.0015 to 0.003 of the total height. Similar limits should be imposed on the acceptable interstory drift index.

The stresses and loss of stiffness that might result from a building fire must be a major consideration, as this is not a remote possibility. However, designing against fire is a specialist discipline that cannot be covered in any detail here.

Although the principles of foundation design are not affected by the height of a building, the situation for tall buildings is different as a result of the high short-term transient moments and shears that arise from wind loads. The high dead load caused by the height of the building produces large compressive stresses on the foundation, and excessive differential settlements must be avoided. Any lateral deflections caused by rotational settlement will be magnified by the height of the building, and the soil–structure interaction must be considered, particularly under seismic actions.

REFERENCES

2.1 *Tall Building Criteria and Loading.* Vol. CL. *Monograph on Planning and Design of Tall Buildings,* ASCE, 1980, pp. 251–390.

2.2 *Tall Building Systems and Concepts.* Vol. SC. *Monograph on Planning and Design of Tall Buildings,* ASCE, 1980, pp. 259–340.

CHAPTER 3
Loading

Loading on tall buildings differs from loading on low-rise buildings in its accumulation into much larger structural forces, in the increased significance of wind loading, and in the greater importance of dynamic effects. The collection of gravity loading over a large number of stories in a tall building can produce column loads of an order higher than those in low-rise buildings. Wind loading on a tall building acts not only over a very large building surface, but also with greater intensity at the greater heights and with a larger moment arm about the base than on a low-rise building. Although wind loading on a low-rise building usually has an insignificant influence on the design of the structure, wind on a high-rise building can have a dominant influence on its structural arrangement and design. In an extreme case of a very slender or flexible structure, the motion of the building in the wind may have to be considered in assessing the loading applied by the wind.

In earthquake regions, any inertial loads from the shaking of the ground may well exceed the loading due to wind and, therefore, be dominant in influencing the building's structural form, design, and cost. As an inertial problem, the building's dynamic response plays a large part in influencing, and in estimating, the effective loading on the structure.

With the exception of dead loading, the loads on a building cannot be assessed accurately. While maximum gravity live loads can be anticipated approximately from previous field observations, wind and earthquake loadings are random in nature, more difficult to measure from past events, and even more difficult to predict with confidence. The application of probabilistic theory has helped to rationalize, if not in every case to simplify, the approaches to estimating wind and earthquake loading.

It is difficult to discuss approaches to the estimation of loading entirely in generalities because the variety of methods in the different Codes of Practice, although rationally based, tend to be empirical in their presentation. Therefore, in some parts of this chapter, methods from reasonably representative modern Codes are given in detail to illustrate current philosophies and trends.

3.1 GRAVITY LOADING

Although the tributary areas, and therefore the gravity loading, supported by the beams and slabs in a tall building do not differ from those in a low-rise building, the accumulation in the former of many stories of loading by the columns and walls can be very much greater.

As in a low-rise building, dead loading is calculated from the designed member sizes and estimated material densities. This is prone to minor inaccuracies such as differences between the real and the designed sizes, and between the actual and the assumed densities.

Live loading is specified as the intensity of a uniformly distributed floor load, according to the occupancy or use of the space. In certain situations such as in parking areas, offices, and plant rooms, the floors should be considered for the alternative worst possibility of specified concentrated loads.

The magnitudes of live loading specified in the Codes are estimates based on a combination of experience and the results of typical field surveys. The differences between the live load magnitudes in the Codes of different countries (some examples of which are shown in Table 3.1 [3.1]) indicate a lack of unanimity and consistency sufficient to raise questions about their accuracy. Load capacity experiments have shown that even the Code values, which are usually accepted as conservative, may in some circumstances underestimate the maximum possible values.

Pattern distribution of gravity live loading over adjacent and alternate spans should be considered in estimating the local maxima for member forces, while live load reductions may be allowed to account for the improbability of total loading being applied simultaneously over larger areas.

3.1.1 Methods of Live Load Reduction

The philosophy of live load reduction is that although, at some time in the life of a structure, it is probable that a small area may be subjected to the full intensity

TABLE 3.1 Live Load Magnitudes

	United States (ANSI A58.1-1972)		Great Britain (CP3-CH.V PT.1:1967)		Japan (AIJ Standard)		U.S.S.R. (SN and PII-A.11-62)	
	kPa	psf	kPa	psf	kPa	psf	kPa	psf
Office buildings								
Offices	2.4	50	2.5	52	2.9	61	2.0	41
Corridors	3.8	80	2.5	52			2.9	61
Lobbies	4.8	100	2.5	52	2.9	61	2.9	61
Residential								
Apartments	1.9	40	1.5	31	1.8	37	1.5	31
Hotel	·1.9	40	2.0	42	1.8	37	2.0	41
Corridors	3.8	80	a		1.8	37	2.9	61
Public rooms	4.8	100	2.0	42	3.5	74	2.0	41

From Ref. [3.1].

aSame values as for occupancy.

of live load, it is improbable that the whole of a large area or a collection of areas, and the members supporting them, will be subjected simultaneously to the full live load. Consequently, it is reasonable to design the girders and columns supporting a large tributary area for significantly less than the full live loading. The different methods of live load reduction generally allow for the girders, columns, and walls to be designed for a reduced proportion of the full live load with an increased amount of supported area. An upper limit is usually placed on the reduction in order to retain an adequate margin of safety.

The following three examples of methods of live load reduction serve to illustrate how the general philosophy may be applied [3.1].

1. Simple percentages may be specified for the reductions and for the limiting amount. For example, the supporting members may be designed for 100% of the live load on the roof, 85% of that on the top floor, and further reductions of 5% for each successive floor down to a minimum of 50% of the live load.
2. A tributary area formula may be given, allowing a more refined definition of the reduction, with the limit built into the formula. For example, the supporting members may be designed for a live load equal to the basic live load multiplied by a factor $0.3 + 10/\sqrt{A}$, where A is the accumulated area in square feet.
3. An even more sophisticated formula-type method may define the maximum reduction in terms of the dead-to-live load ratio. For example, it may be specified that the maximum percentage reduction shall not exceed $[100 \times (D + L)]/4.33L$, in which D and L are the intensities of dead and live loading, respectively. This particular limit is intended to ensure that if the full live load should occur over the full tributary area, the element would not be stressed to the yield point.

3.1.2 Impact Gravity Loading

Impact loading occurs as a gravity live load in the case of an elevator being accelerated upward or brought to a rest on its way down. An increase of 100% of the static elevator load has usually been used to give a satisfactory performance of the supporting structure [3.1].

3.1.3 Construction Loads

Construction loads are often claimed to be the most severe loads that a building has to withstand. Certainly, many more failures occur in buildings under construction than in those that are complete, but it is rare for special provision to be made for construction loads in tall building design. If, however, in a building with an unusual structure, a lack of consideration for construction loading could increase the total cost of the project, an early liaison between the designer and contractor on making some provision would obviously be desirable.

Typically, the construction load that has to be supported is the weight of the floor forms and a newly placed slab, which, in total, may equal twice the floor dead load. This load is supported by props that transfer it to the three or four previously constructed floors below. Now, with the possibility of as little as 3-day cycle, or even 2-day cycle, story construction, and especially with concrete pumping, which requires a more liquid mix, the problem is more severe; this is because the newly released slab, rather than contributing to supporting the construction loads, is still in need of support itself.

The climbing crane is another common construction load. This is usually supported by connecting it to a number of floors below with, possibly, additional shoring in stories further below.

3.2 WIND LOADING

The lateral loading due to wind or earthquake is the major factor that causes the design of high-rise buildings to differ from those of low- to medium-rise buildings. For buildings of up to about 10 stories and of typical proportions, the design is rarely affected by wind loads. Above this height, however, the increase in size of the structural members, and the possible rearrangement of the structure to account for wind loading, incurs a cost premium that increases progressively with height. With innovations in architectural treatment, increases in the strengths of materials, and advances in methods of analysis, tall building structures have become more efficient and lighter and, consequently, more prone to deflect and even to sway under wind loading. This served as a spur to research, which has produced significant advances in understanding the nature of wind loading and in developing methods for its estimation. These developments have been mainly in experimental and theoretical techniques for determining the increase in wind loading due to gusting and the dynamic interaction of structures with gust forces.

The following review of some representative Code methods, which includes ones that are relatively advanced in their consideration of gust loading, summarizes the state of the art. The first method described is a static approach, in that it assumes the building to be a fixed rigid body in the wind. Static methods are appropriate for tall buildings of unexceptional height, slenderness, or susceptibility to vibration in the wind. The subsequently described dynamic methods are for exceptionally tall, slender, or vibration-prone buildings. These may be defined, for example, as in the *Uniform Building Code* [3.2], as those of height greater than 400 ft (123 m), or of a height greater than five times their width, or those with structures that are sensitive to wind-excited oscillations. Alternatively, such exceptional buildings may be defined in a more rigorous way according to the natural frequency and damping of the structure, as well as to its proportions and height [3.3].

The methods are now explained with a level of detail intended to convey for each its philosophy of approach. For more detailed information, sufficient to allow the use of the methods, the reader is referred to the particular Codes of Practice.

3.2.1 Simple Static Approach

***Uniform Building Code (1988) Method* [3.2]**. The method is representative of modern static methods of estimating wind loading in that it accounts for the effects of gusting and for local extreme pressures over the faces of the building. It also accounts for local differences in exposure between the open countryside and a city center, as well as allowing for vital facilities such as hospitals, and fire and police stations, whose safety must be ensured for use after an extreme windstorm.

The design wind pressure is obtained from the formula

$$p = C_e C_q q_s I \tag{3.1}$$

in which C_e is a coefficient to account for the combined effects of height, exposure, and gusting, as defined in Table 3.2.

C_q is a coefficient that allows for locally higher pressures for wall and roof elements as compared with average overall pressures used in the design of the primary structure. For example, C_q has a value of 1.4 when using the projected area method of calculating wind loading for structures over 40 ft in height, whereas it has a local value of 2.0 at wall corners.

The pressure q_s is a wind stagnation pressure for a minimum basic 50-year wind speed at a height of 30 ft above ground, as given for different regions of the United States in a wind speed contour map. Where local records indicate a greater than basic value of the wind speed, this value should be used instead in determining q_s.

The importance factor I is taken as 1.15 for postdisaster buildings and 1.00 for all other buildings.

TABLE 3.2 Combined Height, Exposure, and Gust Factor Coefficient (C_e)

Height above Average Level of Adjoining Ground (ft)	Exposure C[a]	Exposure B[a]
0–20	1.2	0.7
20–40	1.3	0.8
40–60	1.5	1.0
60–100	1.6	1.1
100–150	1.8	1.3
150–200	1.9	1.4
200–300	2.1	1.6
300–400	2.2	1.8

Reproduced from the 1988 edition of the Uniform Building Code, copyright © 1988, with the permission of the publishers, the International Conference of Building Officials.

[a] Exposure C represents the most severe exposure with a flat and open terrain. Exposure B has terrain with buildings, forest, or surface irregularities 20 ft or more in height.

3.2.2 Dynamic Methods

If the building is exceptionally slender or tall, or if it is located in extremely severe exposure conditions, the effective wind loading on the building may be increased by dynamic interaction between the motion of the building and the gusting of the wind. If it is possible to allow for it in the budget of the building, the best method of assessing such dynamic effects is by wind tunnel tests in which the relevant properties of the building and the surrounding countryside are modeled. For buildings that are not so extreme as to demand a wind tunnel test, but for which the simple design procedure is inadequate, alternative dynamic methods of estimating the wind loading by calculation have been developed. The wind tunnel experimental method and one of the dynamic calculation methods will be reviewed briefly.

Wind Tunnel Experimental Method. Wind tunnel tests to determine loading may be quasisteady for determining the static pressure distribution or force on a building. The pressure or force coefficients so developed are then used in calculating the full-scale loading through one of the described simple methods. This approach is satisfactory for buildings whose motion is negligible and therefore has little effect on the wind loading.

If the building slenderness or flexibility is such that its response to excitation by the energy of the gusts may significantly influence the effective wind loading, the wind tunnel test should be a fully dynamic one. In this case, the elastic structural properties and the mass distribution of the building as well as the relevant characteristics of the wind should be modeled.

Building models for wind tunnel tests are constructed to scales which vary from $\frac{1}{100}$ to $\frac{1}{1000}$, depending on the size of the building and the size of the wind tunnel, with a scale of $\frac{1}{400}$ being common. Tall buildings typically exhibit a combination of shear and bending behavior that has a fundamental sway mode comprising a flexurally shaped lower region and a relatively linear upper region. This can be represented approximately in wind tunnel tests by a rigid model with a flexurally sprung base. It is not necessary in such a model to represent the distribution of mass in the building, but only its moment of inertia about the base.

More complex models are used when additional modes of oscillation are expected including, possibly, torsion. These models consist of lumped masses, springs, and flexible rods, designed to simulate the stiffnesses and mass properties of the prototype. Wind pressure measurements are made by flush surface pressure taps on the faces of the models, and pressure transducers are used to obtain the mean, root mean square (RMS), and peak pressures.

The wind characteristics that have to be generated in the wind tunnel are the vertical profile of the horizontal velocity, the turbulence intensity, and the power spectral density of the longitudinal component. Special ''boundary layer'' wind tunnels have been designed to generate these characteristics. Some use long working sections in which the boundary layer develops naturally over a rough floor;

24 LOADING

other shorter ones include grids, fences, or spires at the test section entrance together with a rough floor, while some activate the boundary layer by jets or driven flaps. The working sections of the tunnel are up to a maximum of about 6 ft^2 and they operate at atmospheric pressure [3.4].

Detailed Analytical Method. Wind tunnel testing is a highly specialized, complex, and expensive procedure, and can be justified only for very high cost projects. To bridge the gap between those buildings that require only a simple approach to wind loading and those that clearly demand a wind tunnel dynamic test, more detailed analytical methods have been developed that allow the dynamic wind loading to be calculated [3.5, 3.6]. The method described here is based on the pioneering work of Davenport and is now included in the *National Building Code of Canada*, NBCC [3.7, 3.8].

The external pressure or suction p on the surface of the building is obtained using the basic equation

$$p = qC_e C_g C_p \tag{3.2}$$

in which the exposure factor C_e is based on a mean wind speed vertical profile, which varies according to the roughness of the surrounding terrain. Three types of exposure are considered: generally open terrain with minimal obstruction; semi-obstructed terrrain such as suburban, urban, and wooded areas, and heavily obstructed areas with heavy concentrations of tall buildings and at least 50% of all the buildings exceeding four stories. A formula expressing the value of C_e as a power of the height is given in the Code for each of the three exposure conditions.

The gust effect factor C_g is the ratio of the expected peak loading effect to the mean loading effect. It allows for the variable effectiveness of different sizes of gusts and for the load magnification effect caused by gusts in resonance with the vibrating structure. C_g is given in the Code by a series of formulas and graphs that, although not difficult to use, are too complex to describe here. They can be summarized briefly, however, as expressing the loading effect in terms of the interaction between the wind speed spectrum and the fundamental mode dynamic response of the structure, which involves the natural frequency and damping of the structure, using a transfer or admittance function.

Coefficient C_p is the external pressure coefficient averaged over the area of the surface considered. Its value is influenced by the shape of the building, the wind direction, and the profile of the wind velocity, and is usually determined from the wind tunnel experiments on small-scale models.

Details of the method are given in the *National Building Code of Canada* and in its *Supplement* [3.7, 3.8]. A similar method by Simiu [3.6] is claimed to give conservative wind loads, but of a significantly lower magnitude than those from the NBCC method. Obviously scope exists for further verification and, possibly, simplification of the dynamic load calculation methods.

3.3 EARTHQUAKE LOADING

Earthquake loading consists of the inertial forces of the building mass that result from the shaking of its foundation by a seismic disturbance. Earthquake resistant design concentrates particularly on the translational inertia forces, whose effects on a building are more significant than the vertical or rotational shaking components.

Other severe earthquake forces may exist, such as those due to landsliding, subsidence, active faulting below the foundation, or liquefaction of the local subgrade as a result of vibration. These disturbances, however, which are local effects, can be so massive as to defy any economic earthquake-resistant design, and their possibility may suggest instead the selection of an alternative site.

Where earthquakes occur, their intensity is related inversely to their frequency of occurrence; severe earthquakes are rare, moderate ones occur more often, and minor ones are relatively frequent. Although it might be possible to design a building to resist the most severe earthquake without significant damage, the unlikely need for such strength in the lifetime of the building would not justify the high additional cost. Consequently, the general philosophy of earthquake-resistant design for buildings is based on the principles that they should

1. resist minor earthquakes without damage;
2. resist moderate earthquakes without structural damage but accepting the probability of nonstructural damage;
3. resist average earthquakes with the probability of structural as well as nonstructural damage, but without collapse.

Some adjustments are made to the above principles to recognize that certain buildings with a vital function to perform in the event of an earthquake should be stronger.

The magnitude of earthquake loading is a result of the dynamic response of the building to the shaking of the ground. To estimate the seismic loading two general approaches are used, which take into account the properties of the structure and the past record of earthquakes in the region.

The first approach, termed the equivalent lateral force procedure, uses a simple estimate of the structure's fundamental period and the anticipated maximum ground acceleration, or velocity, together with other relevant factors, to determine a maximum base shear. Horizontal loading equivalent to this shear is then distributed in some prescribed manner throughout the height of the building to allow a static analysis of the structure. The design forces used in this equivalent static analysis are less than the actual forces imposed on the building by the corresponding earthquake. The justification for using lower design forces includes the potential for greater strength of the structure provided by the working stress levels, the damping provided by the building components, and the reduction in force due to the effective ductility of the structure as members yield beyond their elastic limits. The method is simple and rapid and is recommended for unexceptionally high buildings with unexceptional structural arrangements. It is also useful for the preliminary design of higher buildings and for those of a more unusual structural arrangement, which may subsequently be analyzed for seismic loading by a more appropriate method.

The second, more refined, procedure is a modal analysis in which the modal frequencies of the structure are analyzed and then used in conjunction with earthquake design spectra to estimate the maximum modal responses. These are then combined to find the maximum values of the responses. The procedure is more complex and longer than the equivalent lateral force procedure, but it is more accurate as well as being able to account approximately for the nonlinear behavior of the structure.

The two procedures are now discussed in more detail.

3.3.1 Equivalent Lateral Force Procedure

In the United States there are various code methods with similarities in some respects but having fundamental philosophical differences in the ways they express the seismicity of a region and the effect of the type of structural system; for example, the *BOCA Basic Building Code* [3.9], the *National Building Code* [3.10], the *Standard Building Code* [3.11], and the *Uniform Building Code* [3.2]. The equivalent lateral force method in the *Uniform Building Code* (UBC), which is used in the western United States and in many other locations will be discussed here. It is based on the 1988 earthquake code of the Structural Engineers Association of California [3.12].

Determination of the Minimum Base Shear Force. The UBC states that the structure shall be designed to resist a minimum total lateral seismic load V, which shall be assumed to act nonconcurrently in orthogonal directions parallel to the main axes of the structure, where V is calculated from the formula

$$V = \frac{ZIC}{R_W} W \tag{3.3}$$

in which

$$C = \frac{1.25 S}{T^{2/3}} \tag{3.4}$$

The design base shear equation (3.3) provides the level of the seismic design loading for a given structural system, assuming that the structure will undergo inelastic deformation during a major earthquake. The coefficients in Eq. (3.3) take into account the effects of the seismicity of the area, the dead load, the structural type and its ability to dissipate energy without collapse, the response of the structure, the interaction of the structure with the ground, and the importance of the structure.

The zone coefficient Z corresponds numerically to the effective peak ground acceleration (EPA) of a region, and is defined for the United States by a map that is divided into regions representing five levels of ground motion [3.2]. As an EPA value, it is used to scale the spectral shape given by the coefficient C, Eq. (3.4), so that the product of the coefficients Z and C represents an acceleration response spectrum envelope having a 10% probability of being exceeded in 50 years.

The importance factor I is concerned with the numbers of people in the building whose safety is directly at risk, and whether the building has an immediate post-earthquake role in the safety and recovery of the community.

The coefficient C represents the response of the particular structure to the earthquake acceleration spectrum. The curve given by Eq. (3.4) is a simplified multi-mode acceleration response spectrum normalized to an effective peak ground acceleration of 1 basis. It is a function of the fundamental period of the structure T, and a site coefficient S, which is included to adjust the shape of the appropriate frequency response content of the site soil conditions. The UBC has categorized the broad range of soil characteristics into four types, and a site coefficient has been assigned to each of these depending on the soil type and depth. A maximum limit on $C = 2.75$ for any structure and soil site condition is given to provide a simple seismic load evaluation for design projects where it is not practical to evaluate the site soil conditions and the structure period. In addition, to assure that a minimum base shear of 3% of the building weight is used in Seismic Zone 4, with proportional values in the lower zones, a lower limit of $C/R_W = 0.075$ is prescribed.

The structural system factor R_W is a measure of the ability of the structural system to sustain cyclic inelastic deformations without collapse. It is in the denominator of the design base shear equation (3.3) so that design loads decrease for systems with large inelastic deformation capabilities. The magnitude of R_W depends on the ductility of the type and material of the structure, the possibility of failure of the vertical load system, the degree of redundancy of the system that would allow some localized failures without overall failure, and the ability of the secondary system, in the case of dual systems, to stabilize the building when the primary system suffers significant damage.

The factor W is normally the total dead load of the building.

The value of V from Eq. (3.3) gives the magnitude of the total base shear that must be distributed over the height of the structure for the equivalent static analysis.

Distribution of Total Base Shear. Having determined a value for the total base shear it is necessary, in order to proceed with the analysis, to allocate the base shear as effective horizontal loads at the various floor levels. In deciding on an appropriate distribution for the horizontal load the following factors are considered.:

1. The effective load at a floor level is equal to the product of the mass assigned to that floor and the horizontal acceleration at that level.
2. The maximum acceleration at any level of the structure in the fundamental mode is proportional to its horizontal displacement in that mode.
3. The fundamental mode for a regular structure, consisting of shear walls and frames, is approximately linear from the base.

A reasonable distribution of the total base shear V throughout the height would be in accordance with a linear acceleration distribution, as given by

$$F_x = V \frac{w_x h_x}{\sum_{i=1}^{n} w_i h_i} \tag{3.5}$$

where w_i and w_x are those portions of W assigned to levels i and x, respectively; that is, the weight at or adjacent to levels i and x, and assigned to those levels for the purpose of the analysis.

For structures whose weight is distributed uniformly over their height, the horizontal load distribution resulting from Eq. (3.5) forms a triangle, with a maximum value at the top. Such a distribution has been found to be appropriate for buildings of relatively stocky proportions where only the fundamental mode is significant. In more slender, longer period buildings, however, higher modes become significant, causing a greater proportion of the total horizontal inertia forces to act near the top; the intensity of this effect is related to the period of the building. Consequently, this is reflected in the UBC [3.2], and in other Codes, by applying a part of the total loading as a concentrated horizontal force F_t at the top of the building. The remainder of the total base shear is then distributed over the height of the building as an inverted triangle.

Torsion in any story of the building is prescribed in the UBC [3.2], to be taken as the product of the story shear and an eccentricity resulting from the addition of a calculated eccentricity of the mass above, from the center of rigidity of the story, and an accidental eccentricity of 5% of the plan dimension of the building perpendicular to the direction of the force being considered. If torsional irregularities exist, the accidental eccentricity is to be increased by an amplification factor relating the maximum story drift at one end of the structure to the average of the story drifts of the two ends of the structure.

Explanatory material and related technical information useful to the designer in the application of the design procedure for this equivalent static approach is provided in the tentative commentary of the 1988 earthquake code of the Structural Engineers Association of California [3.12].

The Applied Technology Council (ATC) produced a report in 1978 with a second printing in 1984, *Tentative Provisions for the Development of Seismic Regulations for Buildings, ATC 3-06 Amended* [3.13], for the consideration of building authorities across the United States. Its recommendations indicate the likely developments in the equivalent lateral force procedures of the major Codes. Details in the approach of the ATC 3-06 have been reviewed by Berg [3.14]. Many of the ATC's provisions have been used by SEAOC, and consequently the International Conference of Building Officials, as well as the National Research Council of Canada, as key resource documents to develop their new code editions, *Recommended Lateral Force Requirements and Tentative Commentary* [3.12], *Uniform Building Code* [3.2], and *National Building Code of Canada* [3.7, 3.8], respectively.

3.3.2 Modal Analysis Procedure

The equivalent static load type of analysis is suitable for the majority of high-rise structures. If, however, either the lateral load resisting elements or the vertical distribution of mass are significantly irregular over the height of the building, as in buildings with large floor-to-floor variations of internal configuration, or with setbacks, an analysis that takes greater consideration of the dynamic characteristics of the building must be made. Usually, in such cases, a modal analysis would be appropriate.

A detailed explanation of the theory and procedure of modal analysis is given in Chapter 17 and in other texts [3.15, 3.16]. Reviewing the method briefly, however, in a modal analysis a lumped mass model of the building with horizontal degrees of freedom at each floor is analyzed to determine the modal shapes and modal frequencies of vibration. The results are then used in conjunction with an earthquake design response spectrum, and estimates of the modal damping, to determine the probable maximum response of the structure from the combined effect of its various modes of oscillation.

Buildings in which the mass at the floor levels is highly eccentric from the corresponding centers of resistance will be subjected to torque, causing the possibility of significant torsional vibrations and of coupling between the lateral and torsional mode. The modal method can also be applied to the analysis of such a building, by adding to the structural model a third, rotational, degree of freedom at each floor level.

The modal method is applicable, in the strictest sense, only to linear elastic systems. Consequently, the results for a building structure's response are, at best, only an approximate estimate, because of its typically being designed to suffer significant inelastic deformations in only moderate earthquakes. More accurate values of response may be obtained for some buildings by the modal analysis method, using modified design response spectra for inelastic systems [3.16].

3.4 COMBINATIONS OF LOADING

Methods of accounting for load combinations and their effects on the design of members vary according to the Code used and to the design philosophy. The combination of dead and live loading with reductions in the live loading to allow for the improbability of fully loaded tributary areas, and considering patterned live loading for the worst effects, have already been discussed.

The approaches to combinations of loading by two north American Codes, the *Uniform Building Code* [3.2] and the *National Building Code of Canada* [3.7], will be referred to as representative of many of the major building Codes.

3.4.1 Working Stress Design

The UBC and NBCC both assume that wind and earthquake loading need not be taken to act simultaneously. The UBC considers the improbability of extreme gravity and wind, or earthquake, loadings acting simultaneously by allowing for the combination a one-third increase in the permissible working stresses, which is equivalent to a 25% reduction in the sum of the gravity and wind, or earthquake, loading.

The NBCC approach to allowing for the improbability of the loads acting simultaneously is to apply a reduction factor to the combined loads rather than to allow an increase in the permissible stresses, with greater reductions for the greater number of load types combined.

3.4.2 Limit States Design

In limit states design, the adequacy of the building and its members is checked against factored loads in order to satisfy the various safety and serviceability limit states.

The UBC requires that the strength must be able to resist the actions resulting from the combination of the individually factored dead and live loads, where the load factors take into account the variability of the load and load patterns.

If a wind load or earthquake load is to be included, a reduction factor is applied to the combination of the individually factored loads to allow for the improbability of the maximum values of the wind or earthquake, and other live loads occurring simultaneously.

In the NBCC, three factors are required to account for combinations of loading in limit states design: a load factor, which accounts for the variability of the loads as before; a load combination factor, which is applied to loads other than dead loads and accounts for the improbability of their extreme values acting simultaneously; and an importance factor, which allows a reduction where collapse is not likely to have serious consequences.

In both the UBC and the NBCC the strength requirement is satisfied by ensuring that the factored resistance of the members is not less than the corresponding actions caused by the factored loads.

3.4.3 Plastic Design

In buildings in which plastic design is used for parts or the whole of the steel framed structure, available methods of analysis are based on proportional systems of loading, that is load combinations in which increasing loads maintain their relative magnitudes. Consequently, all the loads within a combination are given the same load factor.

SUMMARY

Loading on high-rise buildings differs from loading on low-rise buildings mainly in its accumulation over the height to cause very large gravity and lateral load forces within the structure. In buildings that are exceptionally slender or flexible, the building dynamics can also become important in influencing the effective loading.

Gravity loading consists of dead loading, which can be predicted reasonably accurately, and live loading, whose magnitudes are estimates based on experience and field surveys, and which are predictable with much less accuracy. The probability of not all parts of a floor supported by a beam, and of not all floors supported by a column, being subjected to the full live loading simultaneously, is provided for by reductions in the beam loading and in the column loading, respectively, in accordance with various formulas. It is sometimes necessary to consider also the effects of construction loads.

Wind loading becomes significant for buildings over about 10 stories high, and progressively more so with increasing height. For buildings that are not very tall or slender, the wind loading may be estimated by a static method. Modern static methods of determining a design wind loading account for the region of the country where the building is to be located, the exposure of the particular location, the effects of gusting, and the importance of the building in a postwindstorm situation.

For exceptionally tall, slender, or flexible buildings, it is recommended that a wind tunnel test on a model is made to estimate the wind loading. Boundary layer wind tunnels, which simulate the variation of wind speed with height, and the gusting are used for this purpose.

For buildings that do not quite fall into the category that demands a wind tunnel test, or for those that are in that category but whose budget does not allow such a test, dynamic methods of calculating the wind load have been developed.

Earthquake loading is a result of the dynamic response of the building to the shaking of the ground. Estimates of the loading account for the properties of the structure and the record of earthquakes in the region. For unexceptionally high buildings with unexceptional structural arrangements an equivalent lateral force method is recommended. In this, the loading is estimated on the basis of a simple approximation for the structure's fundamental period, its dead load, the anticipated ground acceleration or velocity, and other factors relating to the soil site conditions, structure type and the importance of its use. The method gives the value of the maximum horizontal base shear, which is then distributed as an equivalent lateral load over the height of the building so that a static analysis can be performed.

If the building is exceptionally tall, or irregular in its structure or its mass distribution, a modal analysis procedure is recommended for estimating the earthquake loading. The modal shapes and frequencies of vibration are analyzed; these are used in conjunction with an earthquake design response spectrum and estimates of the modal damping to determine the probable maximum responses. The modal method can also allow for the simultaneous torsional oscillation of the building.

Methods of combining types of loading vary according to the design method and the Code of Practice concerned. Although dead load is considered to act in full all the time, live loads do not necessarily do so. The probability of the full gravity live loading acting with either the full wind, earthquake, or temperature loading is low, and of all of them acting together is even lower. This is reflected in the Codes by applying a greater reduction factor to those combinations incorporating more different types of loading. Wind and earthquakes are assumed never to act simultaneously.

REFERENCES

3.1 *Tall Building Criteria and Loading, Monograph on Planning and Design of Tall Buildings,* Vol. CL, ASCE, New York, 1980.

3.2 *Uniform Building Code, 1988,* International Conference of Building Officials, Whittier, California.

3.3 Cook, N. J. *The Designer's Guide to Wind Loading of Building Structures, Part 1,* Building Research Establishment Report, Butterworths, London, 1985.

3.4 Simiu, E. and Scanlan, R. H. *Wind Effects on Structures,* 2nd ed., Wiley Interscience, New York, 1986.

3.5 Davenport, A. G. "Gust loading factors," *J. Struct. Div., Proc. A.S.C.E.* **93,** June 1967, 12–34.

3.6 Simiu, E. "Equivalent static wind loads for tall building design," *J. Struct. Div., Proc. A.S.C.E.* **102,** April 1976, 719–737.

3.7 *National Building Code of Canada, 1990,* National Research Council of Canada, Ottawa, Canada.

3.8 *Supplement to the National Building Code of Canada, 1990,* National Research Council of Canada, Ottawa, Canada.

3.9 *The BOCA Basic Building Code—1990,* Building Officials and Code Administrators International, Homewood, Illinois.

3.10 *The National Building Code—1976,* American Insurance Association, New York.

3.11 *Standard Building Code, 1988 Edition,* Southern Building Code Congress International, Birmingham, Alabama.

3.12 *Recommended Lateral Force Requirements and Commentary,* Seismology Committee, Structural Engineers Association of California, 1988.

3.13 *Tentative Provisions for the Development of Seismic Regulations for Buildings, ATC 3-06 Amended,* Applied Technology Council, National Bureau of Standards, Washington, D.C., 1984.

3.14 Berg, G. V. *Seismic Design Codes and Procedures*, Earthquake Engineering Research Institute, Berkeley, California, 1982.

3.15 Clough, R. W. and Penzien, J. *Dynamics of Structures*, McGraw-Hill, New York, 1975.

3.16 Newmark, N. M. and Hall, W. J. *Earthquake Spectra and Design*, Earthquake Engineering Research Institute, Berkeley, California, 1982.

CHAPTER 4

Structural Form

From the structural engineer's point of view, the determination of the structural form of a high-rise building would ideally involve only the selection and arrangement of the major structural elements to resist most efficiently the various combinations of gravity and horizontal loading. In reality, however, the choice of structural form is usually strongly influenced by other than structural considerations. The range of factors that has to be taken into account in deciding the structural form includes the internal planning, the material and method of construction, the external architectural treatment, the planned location and routing of the service systems, the nature and magnitude of the horizontal loading, and the height and proportions of the building. The taller and more slender a building, the more important the structural factors become, and the more necessary it is to choose an appropriate structural form.

In high-rise buildings designed for a similar purpose and of the same material and height, the efficiency of the structures can be compared roughly by their weight per unit floor area. In these terms, the weight of the floor framing is influenced mainly by the floor span and is virtually independent of the building height, while the weight of the columns, considering gravity load only, is approximately proportional to the height (Fig. 4.1). Buildings of up to 10 stories designed for gravity loading can usually accommodate wind loading without any increase in member sizes, because of the typically allowed increase in permissible stresses in Design Codes for the combined loading. For buildings of more than 10 stories, however, the additional material required for wind resistance increases nonlinearly with height so that for buildings of 50 stories and more the selection of an appropriate structural form may be critical for the economy, and indeed the viability, of the building.

A major consideration affecting the structural form is the function of the building. Modern office buildings call for large open floor spaces that can be subdivided with lightweight partitioning to suit the individual tenant's needs. Consequently, the structure's main vertical components are generally arranged, as far as possible, around the perimeter of the plan and, internally, in groups around the elevator, stair, and service shafts (Fig. 4.2). The floors span the areas between the exterior and interior components, leaving large column-free areas available for office planning. The services are distributed horizontally in each story above the partitioning and are usually concealed in a ceiling space. The extra depth required by this space

STRUCTURAL FORM 35

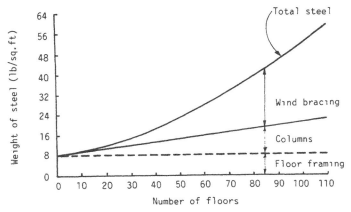

Fig. 4.1 Weight of steel in tall buildings.

causes the typical story height in an office building to be 11 ft-6 in. (3.5 m) or more.

In a residential building or hotel, accommodation is subdivided permanently and usually repetitively from floor to floor. Therefore, continuously vertical columns and walls can be distributed over the plan to form, or fit within, the partitioning (Fig. 4.3). The services can then be run vertically, adjacent to the walls and columns or in separate shafts, to emerge in each story either very close to where required, or to be distributed horizontally from there to where required, along the corridor ceiling spaces. With the exception of the corridors, therefore, a ceiling space is not required, and the soffit of the slab can serve as the ceiling. This allows the story heights in a typical residential building or hotel to be kept down to approximately 8 ft-8 in. (2.7 m). A 40-story residential building is, therefore, generally of significantly less height than a 40-story office building.

In addition to satisfying the previously mentioned nonstructural requirements, the principal objectives in choosing a building's structural form are to arrange to support the gravity, dead and live, loading, and to resist at all levels the external

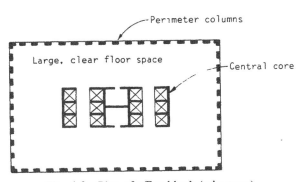

Fig. 4.2 Plan of office block (tube-type).

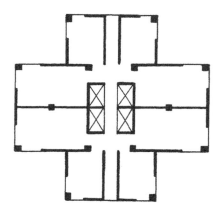

Fig. 4.3 Plan of residential block.

horizontal load shear, moment, and torque with adequate strength and stiffness. These requirements should be achieved, of course, as economically as possible.

With regard to horizontal loading, a high-rise building is essentially a vertical cantilever. This may comprise one or more individually acting vertical cantilevers, such as shear walls or cores, each bending about its own axis and acting in unison only through the horizontal in-plane rigidity of the floor slabs. Alternatively, the cantilever may comprise a number of columns or walls that are mobilized to act compositely, to some degree, as the chords of a single massive cantilever, by vertically shear-resistant connections such as bracing or beams. The lateral stiffness and strength of both of these basic cantilever systems may be further enhanced if the major vertical elements have different free deflection characteristics, in which case they will interact horizontally through the connecting slabs and beams.

Within the constraints of the selected structural form, advantage may be taken of locating the main vertical members on plan so that the dead load compressive stresses suppress the lateral load tensile stresses, thereby avoiding the possibility of net tension occurring in the vertical members and uplift on the foundations. Particular emphasis is placed in some types of structural form on routing the gravity load to the outer vertical members to achieve this purpose.

Steel framing has played a pioneering role in the history of tall buildings. It is appropriate for all heights of structure and, because of its high strength-to-weight ratio, it has always been the material of construction for the tallest buildings. It allows the possibility of longer floor spans, and of partial prefabrication, leading to reduced site work and more rapid erection. Its disadvantages, however, include needing fire and rust protection, being expensive to clad, and requiring costly diagonal bracing or rigid-frame connections.

After the earlier use of steel through the first half of the century, in the form of braced construction, it has evolved in its structural forms somewhat in parallel with reinforced concrete to include rigid-frame, shear wall, wall-frame, tube and braced-tube, and outrigger types of arrangements, as well as in forms more particular to steel such as the suspended structure and the highly efficient massive space frame.

Reinforced concrete tall buildings were introduced approximately two decades after the first steel tall buildings. Understandably, the earlier concrete building structures were influenced in form by the skeletal, column and girder arrangements of their steel counterparts, but they differed in depending on the inherent rigid-frame action of concrete construction to resist horizontal loading. Subsequently, the flat plate and flat slab forms were introduced and these, with the moment-resistant frame, continued as the main repertoire of reinforced concrete high-rise structural form until the late 1940s.

A major step forward in reinforced concrete high-rise structural form came with the introduction of shear walls for resisting horizontal loading. This was the first in a series of significant developments in the structural forms of concrete high-rise buildings, freeing them from the previous 20- to 25-story height limitations of the rigid-frame and flat plate systems. The innovation and refinement of these new forms, together with the development of higher strength concretes, has allowed the height of concrete buildings to reach within striking distance of 100 stories.

Of the following structural forms, some are more appropriate to steel and others to reinforced concrete; many are suitable for either material, while a few allow or demand a combination of materials in the same structure. They are described in a roughly historical sequence.

The structural form of tall buildings, as discussed so far, has concerned mainly the arrangement of the primary vertical components and their interconnections. This topic would not be complete, however, without including consideration of floor systems, because some of them play an integral part with the vertical components in resisting the lateral, as well as the gravity, loading. The last part of the chapter is devoted, therefore, to a brief review of the floor systems used in tall buildings. Many of these are commonly used also in low-rise buildings but are included here for completeness.

4.1 STRUCTURAL FORM

4.1.1 Braced-Frame Structures

In braced frames the lateral resistance of the structure is provided by diagonal members that, together with the girders, form the "web" of the vertical truss, with the columns acting as the "chords" (Fig. 4.4). Because the horizontal shear on the building is resisted by the horizontal components of the axial tensile or compressive actions in the web members, bracing systems are highly efficient in resisting lateral loads.

Bracing is generally regarded as an exclusively steel system because the diagonals are inevitably subjected to tension for one or the other directions of lateral loading. Concrete bracing of the double diagonal form is sometimes used, however, with each diagonal designed as a compression member to carry the full external shear.

The efficiency of bracing, in being able to produce a laterally very stiff structure for a minimum of additional material, makes it an economical structural form for

38 STRUCTURAL FORM

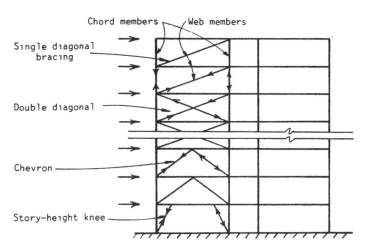

Fig. 4.4 Braced frame—showing different types of bracing.

any height of building, up to the very tallest. An additional advantage of fully triangulated bracing is that the girders usually participate only minimally in the lateral bracing action; consequently, the floor framing design is independent of its level in the structure and, therefore, can be repetitive up the height of the building with obvious economy in design and fabrication. A major disadvantage of diagonal bracing is that it obstructs the internal planning and the location of windows and doors. For this reason, braced bents are usually incorporated internally along wall and partition lines, and especially around elevator, stair, and service shafts. Another drawback is that the diagonal connections are expensive to fabricate and erect.

The traditional use of bracing has been in story-height, bay-width modules (Fig. 4.4) that are fully concealed in the finished building. More recently, however, external larger scale bracing, extending over many stories and bays (Fig. 4.5), has been used to produce not only highly efficient structures, but aesthetically attractive buildings.

Bracing and its modes of behavior are described in more detail in Chapter 6.

4.1.2 Rigid-Frame Structures

Rigid-frame structures consist of columns and girders joined by moment-resistant connections. The lateral stiffness of a rigid-frame bent depends on the bending stiffness of the columns, girders, and connections in the plane of the bent (Fig. 4.6). The rigid frame's principal advantage is its open rectangular arrangement, which allows freedom of planning and easy fitting of doors and windows. If used as the only source of lateral resistance in a building, in its typical 20 ft (6 m)-30 ft (9 m) bay size, rigid framing is economic only for buildings up to about 25 stories. Above 25 stories the relatively high lateral flexibility of the frame calls for uneconomically large members in order to control the drift.

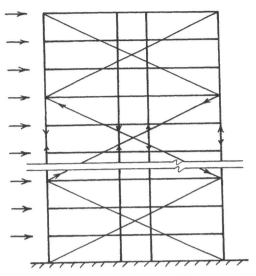

Fig. 4.5 Large-scale braced frame.

Rigid-frame construction is ideally suited for reinforced concrete buildings because of the inherent rigidity of reinforced concrete joints. The rigid-frame form is also used for steel frame buildings, but moment-resistant connections in steel tend to be costly. The sizes of the columns and girders at any level of a rigid frame are directly influenced by the magnitude of the external shear at that level, and they therefore increase toward the base. Consequently, the design of the floor framing cannot be repetitive as it is in some braced frames. A further result is that sometimes it is not possible in the lowest stories to accommodate the required depth of girder within the normal ceiling space.

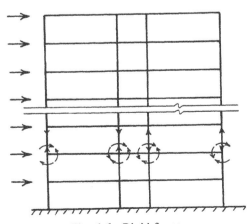

Fig. 4.6 Rigid frame.

Gravity loading also is resisted by the rigid-frame action. Negative moments are induced in the girders adjacent to the columns causing the mid-span positive moments to be significantly less than in a simply supported span. In structures in which gravity loads dictate the design, economies in member sizes that arise from this effect tend to be offset by the higher cost of the rigid joints.

While rigid frames of a typical scale that serve alone to resist lateral loading have an economic height limit of about 25 stories, smaller scale rigid frames in the form of a perimeter tube, or typically scaled rigid frames in combination with shear walls or braced bents, can be economic up to much greater heights. These structural forms are described later in this chapter. The detailed behavior of rigid frames is discussed in Chapter 7.

4.1.3 Infilled-Frame Structures

In many countries infilled frames are the most usual form of construction for tall buildings of up to 30 stories in height. Column and girder framing of reinforced concrete, or sometimes steel, is infilled by panels of brickwork, blockwork, or cast-in-place concrete.

When an infilled frame is subjected to lateral loading, the infill behaves effectively as a strut along its compression diagonal to brace the frame (Fig. 4.7). Because the infills serve also as external walls or internal partitions, the system is an economical way of stiffening and strengthening the structure.

The complex interactive behavior of the infill in the frame, and the rather random quality of masonry, has made it difficult to predict with accuracy the stiffness and strength of an infilled frame. Indeed, at the time of writing, no method of analyzing infilled frames for their design has gained general acceptance. For these reasons, and because of the fear of the unwitting removal of bracing infills at some time in the life of the building, the use of the infills for bracing tall buildings has mainly been supplementary to the rigid-frame action of concrete frames. An outline of a method for designing infilled frames is given in Chapter 8.

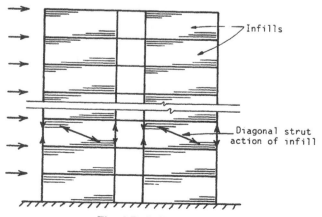

Fig. 4.7 Infilled frame.

4.1.4 Flat-Plate and Flat-Slab Structures

The flat-plate structure is the simplest and most logical of all structural forms in that it consists of uniform slabs, of 5-8 in. (12-20 cm) thickness, connected rigidly to supporting columns (Fig. 4.27). The system, which is essentially of reinforced concrete, is very economical in having a flat soffit requiring the most uncomplicated formwork and, because the soffit can be used as the ceiling, in creating a minimum possible floor depth.

Under lateral loading the behavior of a flat-plate structure is similar to that of a rigid frame, that is, its lateral resistance depends on the flexural stiffness of the components and their connections, with the slabs corresponding to the girders of the rigid frame. It is particularly appropriate for apartment and hotel construction where ceiling spaces are not required and where the slab may serve directly as the ceiling. The flat-plate structure is economical for spans of up to about 25 ft (8 m), above which drop panels can be added to create a flat-slab structure (Fig. 4.28) for spans of up to 38 ft (12 m).

Buildings that depend entirely for their lateral resistance on flat-plate or flat-slab action are economical up to about 25 stories. Previously, however, when Code requirements for wind design were less stringent, many flat-plate buildings were constructed in excess of 40 stories, and are still performing satisfactorily.

4.1.5 Shear Wall Structures

Concrete or masonry continuous vertical walls may serve both architecturally as partitions and structurally to carry gravity and lateral loading. Their very high in-plane stiffness and strength makes them ideally suited for bracing tall buildings. In a shear wall structure, such walls are entirely responsible for the lateral load resistance of the building. They act as vertical cantilevers in the form of separate planar walls, and as nonplanar assemblies of connected walls around elevator, stair, and service shafts (Fig. 4.8). Because they are much stiffer horizontally than rigid frames, shear wall structures can be economical up to about 35 stories.

In contrast to rigid frames, the shear walls' solid form tends to restrict planning where open internal spaces are required. They are well suited, however, to hotels and residential buildings where the floor-by-floor repetitive planning allows the walls to be vertically continuous and where they serve simultaneously as excellent acoustic and fire insulators between rooms and apartments.

If, in low- to medium-rise buildings, shear walls are combined with frames, it is reasonable to assume that the shear walls attract all the lateral loading so that the frame may be designed for only gravity loading. It is especially important in shear wall structures to try to plan the wall layout so that the lateral load tensile stresses are suppressed by the gravity load stresses. This allows them to be designed to have only the minimum reinforcement. Shear wall structures have been shown to perform well in earthquakes, for which case ductility becomes an important consideration in their design. The behavior and methods of analysis of shear wall structures are discussed in detail in Chapter 9.

42 STRUCTURAL FORM

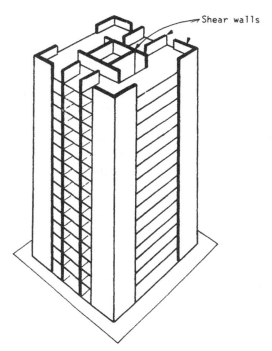

Fig. 4.8 Shear wall structure.

Coupled Wall Structures. A coupled wall structure is a particular, but very common, form of shear wall structure with its own special problems of analysis and design. It consists of two or more shear walls in the same plane, or almost the same plane, connected at the floor levels by beams or stiff slabs (Fig. 4.9). The effect of the shear-resistant connecting members is to cause the set of walls to behave in their plane partly as a composite cantilever, bending about the common centroidal axis of the walls. This results in a horizontal stiffness very much greater than if the walls acted as a set of separate uncoupled cantilevers.

Coupled walls occur often in residential construction where lateral-load resistant cross walls, which separate the apartments, consist of in-plane coupled pairs, or trios, of shear walls between which there are corridor or window openings. Coupled shear walls are considered in detail in Chapter 10.

Although shear walls are obviously more appropriate for concrete construction, they have occasionally been constructed of heavy steel plate, in the style of massive vertical plate or box girders, as parts of steel frame structures. These have been designed for locations of extremely heavy shear, such as at the base of elevator shafts.

4.1.6 Wall-Frame Structures

When shear walls are combined with rigid frames (Fig. 4.10) the walls, which tend to deflect in a flexural configuration, and the frames, which tend to deflect in

4.1 STRUCTURAL FORM 43

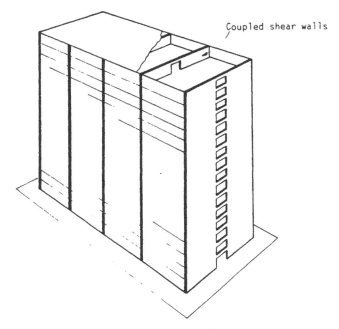

Fig. 4.9 Coupled shear wall structure.

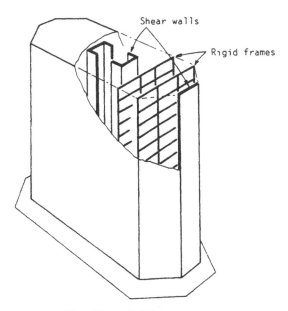

Fig. 4.10 Wall-frame structure.

44 STRUCTURAL FORM

a shear mode, are constrained to adopt a common deflected shape by the horizontal rigidity of the girders and slabs. As a consequence, the walls and frames interact horizontally, especially at the top, to produce a stiffer and stronger structure. The interacting wall-frame combination is appropriate for buildings in the 40- to 60-story range, well beyond that of rigid frames or shear walls alone.

An additional, less well known feature of the wall-frame structure is that, in a carefully "tuned" structure, the shear in the frame can be made approximately uniform over the height, allowing the floor framing to be repetitive.

Although the wall-frame structure is usually perceived as a concrete structural form, with shear walls and concrete frames, a steel counterpart using braced frames and steel rigid frames offers similar benefits of horizontal interaction. The braced frames behave with an overall flexural tendency to interact with the shear mode of the rigid frames.

Detailed descriptions of the behavior and methods of analysis for wall-frame structures are given in Chapter 11.

4.1.7 Framed-Tube Structures

The lateral resistance of framed-tube structures is provided by very stiff moment-resisting frames that form a "tube" around the perimeter of the building. The frames consist of closely spaced columns, 6-12 ft (2-4 m) between centers, joined by deep spandrel girders (Fig. 4.11). Although the tube carries all the lateral loading, the gravity loading is shared between the tube and interior columns or walls. When lateral loading acts, the perimeter frames aligned in the direction of loading act as the "webs" of the massive tube cantilever, and those normal to the direction of the loading act as the "flanges."

The close spacing of the columns throughout the height of the structure is usually unacceptable at the entrance level. The columns are therefore merged, or terminated on a transfer beam, a few stories above the base so that only a few, larger, more widely spaced columns continue to the base. The tube form was developed originally for buildings of rectangular plan, and probably its most efficient use is in that shape. It is appropriate, however, for other plan shapes, and has occasionally been used in circular and triangular configurations.

The tube is suitable for both steel and reinforced concrete construction and has been used for buildings ranging from 40 to more than 100 stories. The highly repetitive pattern of the frames lends itself to prefabrication in steel, and to the use of rapidly movable gang forms in concrete, which make for rapid construction.

The framed tube has been one of the most significant modern developments in high-rise structural form. It offers a relatively efficient, easily constructed structure, appropriate for use up to the greatest of heights. Aesthetically, the tube's externally evident form is regarded with mixed enthusiasm; some praise the logic of the clearly expressed structure while others criticize the grid-like facade as small-windowed and uninterestingly repetitious.

The tube structure's structural efficiency, although high, still leaves scope for improvement because the "flange" frames tend to suffer from "shear lag"; this

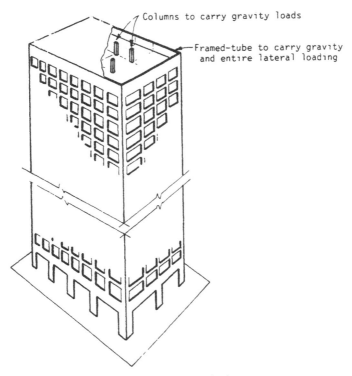

Fig. 4.11 Framed-tube.

results in the mid-face "flange" columns being less stressed than the corner columns and, therefore, not contributing as fully as they could to the flange action.

Tube-in-Tube or Hull-Core Structures. This variation of the framed tube consists of an outer framed tube, the "hull," together with an internal elevator and service core (Fig. 4.12). The hull and core act jointly in resisting both gravity and lateral loading. In a steel structure the core may consist of braced frames, whereas in a concrete structure it would consist of an assembly of shear walls.

To some extent, the outer framed tube and the inner core interact horizontally as the shear and flexural components of a wall–frame structure, with the benefit of increased lateral stiffness. However, the structural tube usually adopts a highly dominant role because of its much greater structural depth.

Bundled-Tube Structures. This structural form is notable in its having been used for the Sears Tower in Chicago—the world's tallest building. The Sears Tower consists of four parallel rigid steel frames in each orthogonal direction, interconnected to form nine "bundled" tubes (Fig. 4.13a). As in the single-tube structure, the frames in the direction of lateral loading serve as "webs" of the vertical cantilever, with the normal frames acting as "flanges."

46 STRUCTURAL FORM

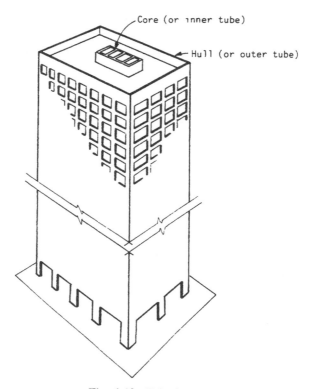

Fig. 4.12 Tube-in-tube.

The introduction of the internal webs greatly reduces the shear lag in the flanges; consequently their columns are more evenly stressed than in the single-tube structure, and their contribution to the lateral stiffness is greater. This allows columns of the frames to be spaced further apart and to be less obtrusive. In the Sears Tower, advantage was taken of the bundled form to discontinue some of the tubes, and so reduce the plan of the building at stages up the height (Fig. 4.13b, c, and d).

Braced-Tube Structures. Another way of improving the efficiency of the framed tube, thereby increasing its potential for use to even greater heights as well as allowing greater spacing between the columns, is to add diagonal bracing to the faces of the tube. This arrangement was first used in a steel structure in 1969, in Chicago's John Hancock Building (Fig. 4.14), and in a reinforced concrete structure in 1985, in New York's 780 Third Avenue Building (Fig. 4.15). In the steel tube the bracing traverses the faces of the rigid frames, whereas in the concrete structure the bracing is formed by a diagonal pattern of concrete window-size panels, poured integrally with the frame.

Because the diagonals of a braced tube are connected to the columns at each intersection, they virtually eliminate the effects of shear lag in both the flange and

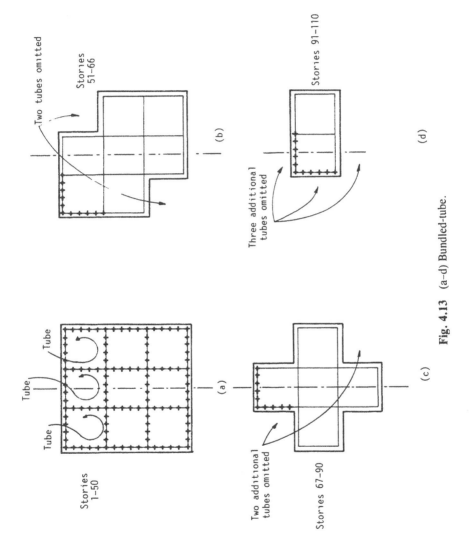

Fig. 4.13 (a–d) Bundled-tube.

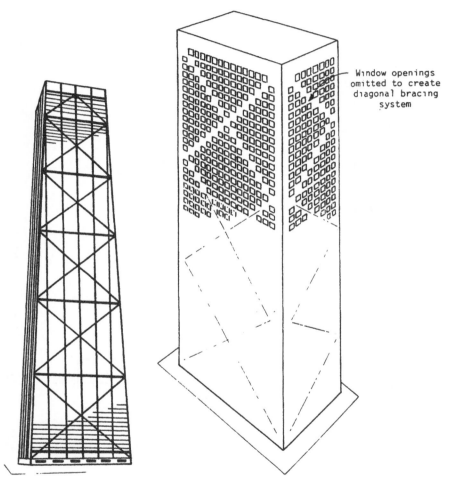

Fig. 4.14 Steel-braced tube.   **Fig. 4.15** Concrete-braced tube.

web frames. As a result, the structure behaves under lateral loading more like a braced frame, with greatly diminished bending in the members of the frames. Consequently, the spacing of the columns can be larger and the depth of the spandrels less, thereby allowing larger size windows than in the conventional tube structure.

In the braced-tube structure the bracing contributes also to the improved performance of the tube in carrying gravity loading: differences between gravity load stresses in the columns are evened out by the braces transferring axial loading from the more highly to the less highly stressed columns.

4.1.8 Outrigger-Braced Structures

This efficient structural form consists of a central core, comprising either braced frames or shear walls, with horizontal cantilever "outrigger" trusses or girders connecting the core to the outer columns (Fig. 4.16a). When the structure is loaded horizontally, vertical plane rotations of the core are restrained by the outriggers through tension in the windward columns and compression in the leeward columns (Fig. 4.16b). The effective structural depth of the building is greatly increased, thus augmenting the lateral stiffness of the building and reducing the lateral deflections and moments in the core. In effect, the outriggers join the columns to the core to make the structure behave as a partly composite cantilever.

Perimeter columns, other than those connected directly to the ends of the outriggers, can also be made to participate in the outrigger action by joining all the perimeter columns with a horizontal truss or girder around the face of the building

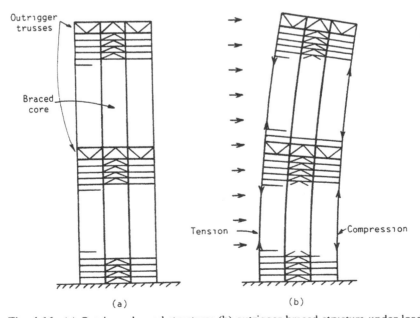

Fig. 4.16 (a) Outrigger-braced structure; (b) outrigger-braced structure under load.

at the outrigger level. The large, often two-story, depths of the outrigger and perimeter trusses make it desirable to locate them within the plant levels in the building.

The degree to which the perimeter columns of an outrigger structure behave compositely with the core depends on the number of levels of outriggers and their stiffnesses. Multilevel outrigger structures show a considerable increase in their effective moment of resistance over single outrigger structures. This increase diminishes, however, with each additional level of outriggers, so that four or five levels appears to be the economic limit. Outrigger-braced structures have been used for buildings from 40 to 70 stories high, but the system should be effective and efficient for much greater heights.

4.1.9 Suspended Structures

The suspended structure consists of a central core, or cores, with horizontal cantilevers at roof level, to which vertical hangers of steel cable, rod, or plate are attached. The floor slabs are suspended from the hangers (Fig. 4.17a).

The advantages of this structural form are primarily architectural in that, except for the presence of the central core, the ground story can be entirely free of major vertical members, thereby allowing an open concourse; also, the hangers, because they are in tension and consequently can be of high strength steel, have a minimum sized section and are therefore less obtrusive. The potential of this latter benefit tends to be offset, however, by the need to proof the hangers against fire and rust, thereby significantly increasing their bulk. The suspended structure has some construction advantages in allowing the core, cantilevers, and hangers to be constructed while the slabs are being poured on top of each other at ground level; the slabs are then lifted in sets and fixed in position (Fig. 4.17b).

The structural disadvantages of the suspended structure are that it is inefficient in first transmitting the gravity loads upward to the roof-level cantilevers before returning them through the core to the ground, and that the structural width of the building at the base is limited to the relatively narrow depth of the core, which restricts the system to buildings of lesser height. A further problem is caused by the vertical extension of the slender hangers that, over the range from zero to full live loading, can result in significant changes in the levels of the edges of the slabs. This effect increases at each level down the length of the hanger and, consequently, is worst at the lowest hung floor. The problem can be limited by restricting the maximum number of floors supported by a single length of hanger to about 10, and by having multilevel cantilever systems (Fig. 4.18). Similarly to outrigger structures, and for the same reasons, the cantilevers are normally incorporated within the plant levels.

Variations from the single-core hanging structure include two- and four-core structures, in which vertical hangers are suspended from massive girders that span between the cores, or in which hangers are draped, catenary fashion, between the cores. The benefits of such multicore hanging structures include large open floor spaces at all levels, and the possiblity of a column-free ground story.

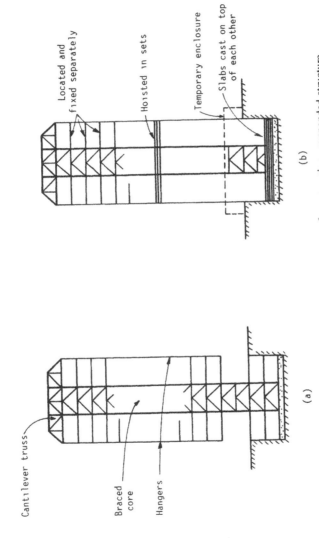

Fig. 4.17 (a) Suspended structure; (b) sequence of construction—suspended structure.

52 STRUCTURAL FORM

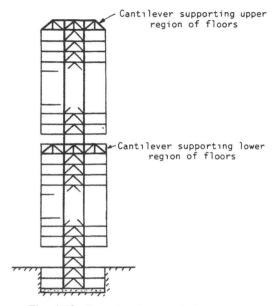

Fig. 4.18 Two-tiered suspended structure.

4.1.10 Core Structures

In these structures a single core serves to carry the entire gravity and horizontal loading (Fig. 4.19). In some, the slabs are supported at each level by cantilevers

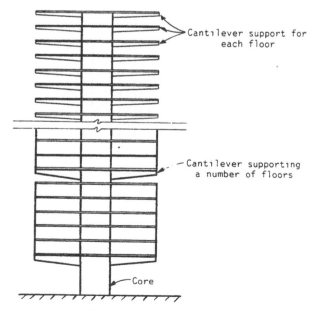

Fig. 4.19 Core structure.

from the core. In others, the slabs are supported between the core and perimeter columns, which terminate either on major cantilevers at intervals down the height, or on a single massive cantilever a few stories above the ground.

Similarly to the suspended building, the merits of the system are mainly architectural, in providing a column-free perimeter at the ground level and at other levels just below the cantilevers. The structural penalties are considerable, however, in having only the small effective structural depth of the core and, therefore, being inefficient in resisting lateral loading, as well as in supporting the floor loading by cantilevers—a highly inefficient structural component.

4.1.11 Space Structures

The primary load-resisting system of a space structure consists essentially of a three-dimensional triangulated frame—as distinct from an assembly of planar bents—whose members serve dually in resisting both gravity and horizontal loading. The result is a highly efficient, relatively lightweight structure with a potential for achieving the greatest heights. The 76-story Hong Kong Bank of China Building (Fig. 4.20) is a classic example of this structural form.

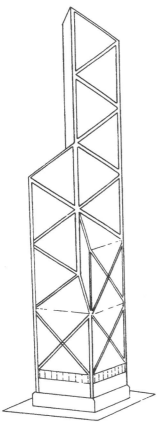

Fig. 4.20 Space structure.

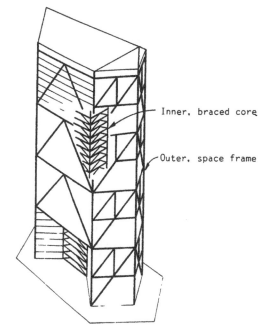

Fig. 4.21 Space structure.

Although simple in their overall concept, space structures are usually geometrically complex, which calls for considerable structural ingenuity in transferring both the gravity loading and the lateral loading from the floors to the main structure. One solution is to have an inner braced core, which serves to collect the lateral loading, and the inner region gravity loading, from the slabs over a number of multistory regions. At the bottom of each region, the lateral and gravity loads are transferred out to the main joints of the space frame (Fig. 4.21).

Although the multidirectional inclined members of the space frame are structurally awkward and costly to connect, as well as making the fenestration difficult, the structural form is visually interesting and aesthetically very pleasing in its apparent simplicity.

4.1.12 Hybrid Structures

Many of the previously described structural arrangements are particularly suitable for prismatically shaped, tower or block, so-called "modern" buildings, which can be completely structured by a single identifiable system, for example, a tube or a wall–frame.

Partly as a reaction to an increasingly monotonous urban environment consisting of regularly shaped and repetitive "modern" buildings, and partly because the analysis and design of much more complex structures have become feasible, architects have responded with a new generation of "postmodern" buildings that are emphatically nonregular in shape, with large-scale cut-outs, flutings, facets, and crowns that defy classification in their intricacy and variety.

Buildings of a nonprismatic shape are less amenable to a single form of structure and, therefore, the engineer has to improvise in developing a satisfactory structural solution. In such situations combinations of two or even more of the basic structural forms have often been used in the same building, either by direct combination as, for example, in a superimposed tube and outrigger system (Fig. 4.22), or by adopting different forms in different parts of the structure as, for example, in a tube system on three faces of the building and a space frame on a faceted fourth face (Fig. 4.23).

During the earlier period of the rapid development of structural form, that is from the 1950s until the mid-1970s, the single form high-rise structure had the advantage that it was usually possible to make an approximate but acceptable structural analysis either by hand or by the use of a small computer. Now, with the ready availability of powerful computers and highly efficient structural analysis programs, an engineer possessing a sound knowledge of structural form and behavior should be able to devise and analyze a structure to suit a building of almost any conceivable irregularity.

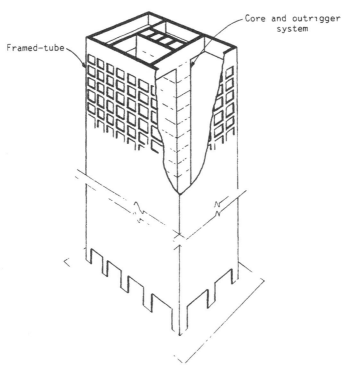

Fig. 4.22 Hybrid structure.

4.2 FLOOR SYSTEMS—REINFORCED CONCRETE

An appropriate floor system is an important factor in the overall economy of the building. Some of the factors that influence the choice of floor system are architectural. For example, in residential buildings, where smaller permanent divisions of the floor space are required, shorter floor spans are possible; whereas, in modern office buildings, that require more open, temporarily subdivisible floor spaces, longer span systems are necessary. Other factors affecting the choice of floor system are related to its intended structural performance, such as whether it is to participate in the lateral load-resisting system, and to its construction, for example, whether there is urgency in the speed of erection.

Reinforced concrete floor systems are grouped into two categories: one-way, in which the slab spans in one direction between supporting beams or walls, and two-way, in which the slab spans in orthogonal directions. In both systems, advantage is taken of continuity over interior supports by providing negative moment reinforcement in the slab.

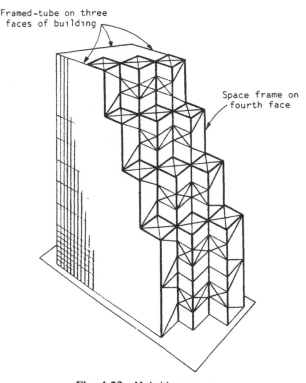

Fig. 4.23 Hybrid structure.

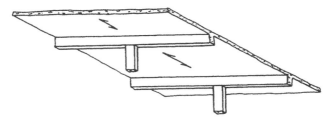

Fig. 4.24 One-way slab.

4.2.1 One-Way Slabs on Beams or Walls

A solid slab of up to 8 in. (0.2 m) thick, spanning continuously over walls or beams up to 24 ft (7.4 m) apart (Fig. 4.24), provides a floor system requiring simple formwork, possibly flying formwork, with simple reinforcement. The system is heavy and inefficient in its use of both concrete and reinforcement. It is appropriate for use in cross-wall and cross-frame residential high-rise construction and, when constructed in a number of uninterrupted continuous spans, lends itself to prestressing.

4.2.2 One-Way Pan Joists and Beams

A thin, mesh-reinforced slab sits on closely spaced cast-in-place joists spanning between major beams which transfer the load to the columns (Fig. 4.25). The slab may be as thin as 2.5 in. (6 cm) while the joists are from 6 in. (15 cm) to 20 in. (51 cm) in depth and spaced from 20 to 30 in. (76 cm) centers. The compositely acting slab and joists form in effect a set of closely spaced T-beams, capable of large, up to 40 ft (12.3 m), spans. The joists are formed between reusable pans that are positioned to set the regular width of the joist, as well as any special widths.

4.2.3 One-Way Slab on Beams and Girders

A one-way slab spans between beams at a relatively close spacing while the beams are supported by girders that transfer the load to the columns (Fig. 4.26). The

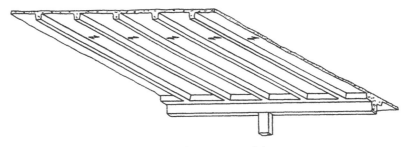

Fig. 4.25 One-way pan joists.

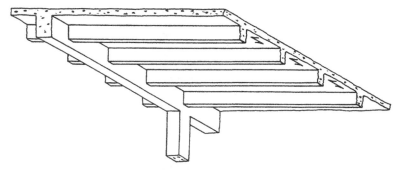

Fig. 4.26 One-way slab on beams and girders.

short spanning slab may be thin, from 3 to 6 in. (7.6–15 cm) thick, while the system is capable of providing long spans of up to 46 ft (14 m). The principal merits of the system are its long span capability and its compatibility with a two-way lateral load resisting rigid-frame structure.

4.2.4 Two-Way Flat Plate

A uniformly thick, two-way reinforced slab is supported directly by columns or individual short walls (Fig. 4.27). It can span up to 26 ft (8 m) in the ordinary reinforced form and up to 36 ft (11 m) when posttensioned. Because of its simplicity, it is the most economical floor system in terms of formwork and reinforcement. Its uniform thickness allows considerable freedom in the location of the supporting columns and walls and, with the possibility of using the clear soffit as a ceiling, it results in minimum story height.

4.2.5 Two-Way Flat Slab

The flat slab differs from the flat plate in having capitals and/or drop panels at the tops of the columns (Fig. 4.28). The capitals increase the shear capacity, while

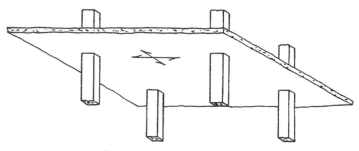

Fig. 4.27 Two-way flat plate.

4.2 FLOOR SYSTEMS—REINFORCED CONCRETE

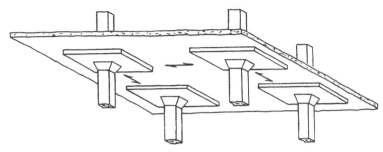

Fig. 4.28 Two-way flat slab.

the drop panels increase both the shear and negative moment capacities at the supports, where the maximum values occur. The flat slab is therefore more appropriate than the flat plate for heavier loading and longer spans and, in similar situations, would require less concrete and reinforcement. It is most suitably used in square, or near-to-square, arrangements.

4.2.6 Waffle Flat Slabs

A slab is supported by a square grid of closely spaced joists with filler panels over the columns (Fig. 4.29). The slab and joists are poured integrally over square, domed forms that are omitted around the columns to create the filler panels. The forms, which are of sizes up to 30 in. (76 m) square and up to 20 in. (50 cm) deep, provide a geometrically interesting soffit, which is often left without further finish as the ceiling.

4.2.7 Two-Way Slab and Beam

The slab spans two ways between orthogonal sets of beams that transfer the load to the columns or walls (Fig. 4.30). The two-way system allows a thinner slab and is economical in concrete and reinforcement. It is also compatible with a lateral load-resisting rigid-frame structure. The maximum length-to-width ratio for a slab to be effective in two directions is approximately 2.

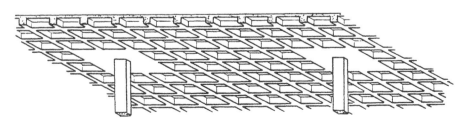

Fig. 4.29 Waffle flat slab.

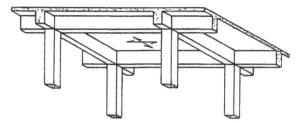

Fig. 4.30 Two-way slab and beam.

4.3 FLOOR SYSTEMS—STEEL FRAMING

The steel-framed floor system is characterized by a reinforced concrete slab supported on a steel framework consisting variously of joists, beams, and girders that transfer the gravity loading to the columns. The slab component is usually one-way with either a cast-in-place solid reinforced concrete slab from 4 in. (10 cm) to 7 in. (18 cm) thick, or a concrete on metal deck slab with a variety of possible section shapes and a minimum slab thickness from 2.5 in. (6 cm) (Fig. 4.31), or a slab of precast units laid on steel beams and covered by a thin concrete topping (Fig. 4.32).

A major consideration in the weight and cost of a steel frame building is the weight of the slab. A floor arrangement with shorter spanning, thinner slabs is desirable. Longer span, closer spaced beams supporting a short-spanning slab is a typical arrangement meeting these requirements. The following types of steel floor framing are categorized according to the spanning arrangement of the supporting steel framework.

4.3.1 One-Way Beam System

A rectangular grid of columns supports sets of parallel longer span beams at a relatively close spacing, with the slab spanning the shorter spans transversely to the beams (Figs. 4.33). In cross-frame structures, the beams at partition lines may be deepened to participate in lateral load resisting rigid frames or braced bents.

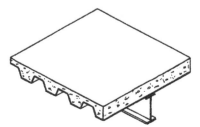

Fig. 4.31 Concrete on metal deck.

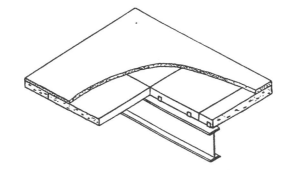

Fig. 4.32 Precast units with topping.

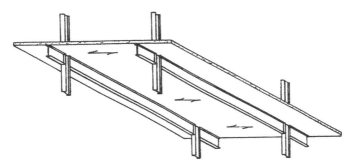

Fig. 4.33 One-way beam system in steel.

4.3.2 Two-Way Beam System

In buildings in which columns are required to be farther apart in both directions, a two-way frame system of girders and beams is often used, with the slab spanning between the beams (Fig. 4.34). To minimize the total structural depth of the floor frame, the heavily loaded girders are aligned with the shorter span and the relatively lightly loaded secondary beams with the longer span.

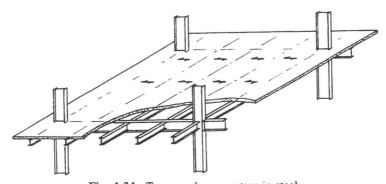

Fig. 4.34 Two-way beam system in steel.

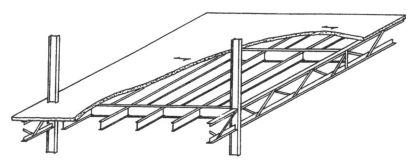

Fig. 4.35 Three-way beam system in steel.

4.3.3 Three-Way Beam System

In buildings in which the columns have to be very widely spaced to allow large internal column-free areas, a three-way beam system may be necessary (Fig. 4.35). A deep lattice girder may form the primary component with beams or open web joists forming the secondary and tertiary systems. In each case the system is arranged to provide relatively short spans for the supported concrete slab.

4.3.4 Composite Steel–Concrete Floor Systems

The use of steel members to support a concrete floor slab offers the possibility of composite construction in which the steel members are joined to the slab by shear connectors so that the slab serves as a compression flange.

In one simple and constructionally convenient slab system, steel decking, which is often used to act merely as rapidly erected permanent formwork for a bar-reinforced slab, serves also as the reinforcement for the concrete slab in a composite role, using thicker wall sections with indentations or protrusions for shear connectors (Fig. 4.36).

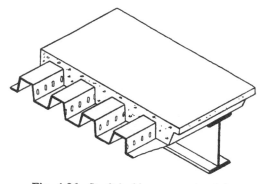

Fig. 4.36 Steel decking composite slab.

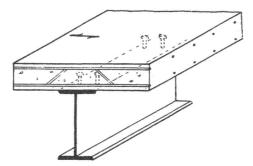

Fig. 4.37 Composite frame system.

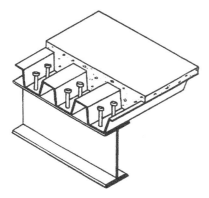

Fig. 4.38 Composite frame and steel decking.

Slabs may also be designed to act compositely with the supporting beams by the more usual forms of stud, angle, or channel shear connectors, so that the slab alone spans the short distance between the beams while the compositely acting slab and beam provide the supporting system (Fig. 4.37). The further combination of a concrete slab on metal decking with shear connectors welded through to the supporting beam or truss is an efficient floor system (Fig. 4.38).

SUMMARY

The structural form of a high-rise building is influenced strongly by its function, while having to satisfy the requirements of strength and serviceability under all probable conditions of gravity and lateral loading. Other influential factors include the building's material of construction, its accommodation of services and, of course, its overall economy. The taller a building, the more important it is economically to select an appropriate structural form.

The basic structural forms of the first half of the twentieth century were the braced frame, which is unrestricted in height but limited to steel structures, and the rigid frame or the flat plate, which are economical to only about 25 stories in height and appropriate particularly to concrete structures. These forms have now been augmented by a variety of other forms that allow structures of greater efficiency and height to be achieved in both steel and concrete. Advances have occurred mainly in the use of shear walls, framed tubes, large-scale braced systems, and space frames, and in better recognizing and accounting for the various types of vertical and horizontal interaction between the major vertical components.

The single structural forms used in the vertically prismatic "modern" high-rise buildings of the 1950s, 1960s, and early 1970s have given way to some extent to hybrid, or mixed, forms in the less regularly shaped "postmodern" buildings of the later 1970s and 1980s. In these mixed forms, combinations of two or more of the single forms are used to fit the "postmodern" buildings' irregular shapes or cut-outs.

Floors slabs are invariably of reinforced concrete. The most appropriate type of floor framing system may depend on the material of construction of the building, whether the building is for office use—requiring larger spans, or residential use—allowing shorter spans, and whether the floor system is expected to participate in the lateral load resistance of the building.

Reinforced concrete systems include one- or two-way spanning slabs on a system of beams or beams and girders. Alternatively, two-way spanning slabs or waffle slabs with or without drop panels, and supported directly by columns, allow the possibility of lesser floor depths and a nonuniform column grid.

Steel-framed floor systems consist of a slab, which may be of solid one-way reinforced concrete, or of concrete on metal decking, or of precast concrete units, supported by a one-, two- or three-way steel beam system.

Composite steel–concrete floor systems consist of a steel frame supporting either a solid reinforced-concrete slab joined to the frame by shear connectors, or a concrete on steel decking slab with or without shear connectors joining it to the frame.

CHAPTER 5

Modeling for Analysis

A building's response to loading is governed by the components that are stressed as the building deflects. Ideally, for ease and accuracy of the structural analysis, the participating components would include only the main structural elements: the slabs, beams, girders, columns, walls, and cores. In reality, however, other, nonstructural, elements are stressed and contribute to the building's behavior; these include, for example, the staircases, partitions, and cladding. To simplify the problem it is usual, in modeling a building for analysis, to include only the main structural members and to assume that the effects of the nonstructural components are small and conservative.

To identify the main structural elements, it is necessary to recognize the dominant modes of action of the proposed building structure and to assess the extent of the various members' contribution to them. Then, by neglecting consideration of the nonstructural components, and the less essential structural components, the problem of analyzing a tall building structure can be reduced to a more viable size.

For extremely large or complex building structures, it may be essential to reduce even further the size of the analysis problem by representing some of the structure's assemblies by simpler analogous components. This chapter reviews the more usual approaches to analysis, the most commonly made assumptions, and the principles and techniques employed in forming a model for structural analysis.

5.1 APPROACHES TO ANALYSIS

The modeling of a tall building structure for analysis is dependent to some extent on the approach to analysis, which is in turn related to the type and size of structure and the stage of design for which the analysis is made. The usual approach is to conduct approximate rapid analyses in the preliminary stages of design, and more detailed and accurate analyses for the final design stages. A hybrid approach is also possible in which a simplified model of the total structure is analyzed first, after which the results are used to allow part by part detailed analyses of the structure.

5.1.1 Preliminary Analyses

The purpose of preliminary analyses, that is analyses for the early stages of design, may be to compare the performance of alternative proposals for the structure, or to determine the deflections and major member forces in a chosen structure so as to allow it to be properly proportioned. The formation of the model and the procedure for a preliminary analysis should be rapid and should produce results that are dependable approximations. The model and its analysis should therefore represent fairly well, if not absolutely accurately, the principal modes of action and interaction of the major structural elements.

The simplifications adopted in making a preliminary analysis are often in the formation of the structural model. Sometimes the approximation is large, as, for example, when numerous hinges are inserted at assumed points of contraflexure in the beams and columns of a rigid frame to convert it from a highly statically indeterminate into a statically determinate system, thus allowing a simple solution using the equilibrium equations. Or the approximation may be to assume a simple cantilever to represent a complex bent, or that a bent is uniform throughout its height and that its beams are "smeared" to allow a continuum solution. These are just a few of the gross approximations that may be made in a structural model to allow a relatively simple preliminary analysis to be achieved.

Alternative and sometimes additional simplifications adopted for the preliminary analysis concern the loading. For example, it is common to make an assumption for the distribution, between the individual bents, of the total external loading on the building, and then to analyze each bent in turn for its assumed loading. In structures with different types of bents, this is a highly uncertain approach. Or, if a continuum analysis is to be made, it will be assumed that the load is applied in some continuous distribution over the height of the structure rather than as it really occurs, at discrete cladding connection levels.

Even with the gross approximations made in simplifying the structure and the loading, it is generally expected that a preliminary analysis should give results for deflections and main member forces that are dependably within about 15% of the values from an accurate analysis.

5.1.2 Intermediate and Final Analysis

The requirement of intermediate and final analyses is that they should give, as accurately as possible, results for deflections and member forces. The model should, therefore, be as detailed as the analysis program and computer capacity will allow for its analysis. All the major modes of action and interaction, and as many as possible of the lesser modes, should be incorporated. Except where a structure is symmetrical in plan and loading, the effects of the structure's twisting should be included.

The most complete approach to satisfying the above requirements would be a three-dimensional stiffness matrix analysis of a fully detailed finite element model of the structure. The columns, beams, and bracing members would be represented by beam elements, while shear wall and core components would be represented by assemblies of membrane elements.

Certain reductions in the size or complexity of the model might be acceptable while allowing it to still qualify in accuracy as a final analysis; for example, if the structure and loading are symmetrical, a three-dimensional analysis of a half-structure model, or even a two-dimensional analysis of a fully interactive two-dimensional model, would be acceptable. Or, if repetitive regions up the height of a structure can be simplified by a lumping technique, this also would be acceptable.

In contrast to the reductions above, however, certain final analyses may require separate, more detailed analyses of particular parts, using the forces or applied displacements from the main analysis, for example, in deep beams at transition levels of the structure, or around irregularities or holes in shear walls.

5.1.3 Hybrid Approach to Preliminary and Final Analyses

If the three-dimensional analysis of a fully detailed model of a structure presents too formidable a task of bookkeeping or computation, an alternative might be to use a hybrid, two-stage approach that would serve dually for the preliminary and final analyses.

When a structure consists of bents that are representable by simple equivalent cantilevers, a three-dimensional model of the structure can be formed by an assembly of the cantilevers that can be analyzed to find the approximate deflections and bent loadings. Detailed, two-dimensional models of the individual bents subjected to the determined loadings are then analyzed individually to find the member forces and to allow the member sizes to be adjusted. This first cycle of the overall three-dimensional and individual two-dimensional analyses would be considered preliminary.

In the second cycle, the three-dimensional model, with the cantilevers modified to represent the adjusted bents, would then be analyzed to obtain the corrected bent loadings. These bent loadings would then be used to reanalyze the two-dimensional bents to obtain the final member forces and structure deflections.

Using this two-stage procedure, a single large three-dimensional analysis of a detailed model can be avoided, and replaced by a number of simpler analyses.

All approaches to analysis call for a sound understanding of high-rise structural behavior and a knowledge of modeling techniques. The hybrid approach in particular requires special care in forming the three-dimensional cantilever model to obtain reliable results. An understanding of high-rise behavior and modeling is valuable not only for analysis, but also for deciding on and developing the structural forms of proposed tall buildings.

5.2 ASSUMPTIONS

An attempt to analyze a high-rise building and account accurately for all aspects of behavior of all the components and materials, even if their sizes and properties were known, would be virtually impossible. Simplifying assumptions are necessary to reduce the problem to a viable size.

Although a wide variety of assumptions is available, some more valid than others, the ones adopted in forming a particular model will depend on the arrangement of the structure, its anticipated mode of behavior, and the type of analysis. The most common assumptions are as follows.

5.2.1 Materials

The material of the structure and the structural components are linearly elastic. This assumption allows the superposition of actions and deflections and, hence, the use of linear methods of analysis. The development of linear methods and their solution by computer have made it possible to analyze large complex statically indeterminate structures.

Although nonlinear methods of analysis have been and are still being developed, their use at present for high-rise buildings is more for research than for the design office.

5.2.2 Participating Components

Only the primary structural components participate in the overall behavior. The effects of secondary structural components and nonstructural components are assumed to be negligible and conservative. Although this assumption is generally valid, exceptions occur. For example, the effects of heavy cladding may be not negligible and may significantly stiffen a structure; similarly, masonry infills may significantly change the behavior and increase the forces unconservatively in a surrounding frame.

5.2.3 Floor Slabs

Floor slabs are assumed to be rigid in plane. This assumption causes the horizontal plane displacements of all vertical elements at a floor level to be definable in terms of the horizontal plane rigid-body rotation and translations of the floor slab. Thus the number of unknown displacements to be determined in the analysis is greatly reduced.

Although valid for practical purposes in most building structures, this assumption may not be applicable in certain cases in which the slab plan is very long and narrow, or it has a necked region, or it consists of precast units without a topping.

5.2.4 Negligible Stiffnesses

Component stiffnesses of relatively small magnitude are assumed negligible. These often include, for example, the transverse bending stiffness of slabs, the minor-axis stiffness of shear walls, and the torsional stiffness of columns, beams, and walls. The use of this assumption should be dependent on the role of the component in the structure's behavior. For example, the contribution of a slab's bending resistance to the lateral load resistance of a column-and-beam rigid-frame structure is negligible, whereas its contribution to the lateral resistance of a flat plate structure is vital and must not be neglected.

5.2.5 Negligible Deformations

Deformations that are relatively small, and of little influence, are neglected. These include the shear and axial deformations of beams, the previously discussed in-plane bending and shear deformations of floor slabs, and, in low- to medium-rise structures, the axial deformations of columns.

5.2.6 Cracking

The effects of cracking in reinforced concrete members due to flexural tensile stresses are assumed representable by a reduced moment of inertia. The gross inertias of beams are usually reduced to 50% of their uncracked values, while the gross inertias of columns are reduced to 80%.

5.3 HIGH-RISE BEHAVIOR

A reasonably accurate assessment of a proposed high-rise structure's behavior is necessary to form a properly representative model for analysis. A high-rise structure is essentially a vertical cantilever that is subjected to axial loading by gravity and to transverse loading by wind or earthquake.

Gravity live loading acts on the slabs, which transfer it horizontally to the vertical walls and columns through which it passes to the foundation. The magnitude of axial loading in the vertical components is estimated from the slab tributary areas, and its calculation is not usually considered to be a difficult problem. Horizontal loading exerts at each level of a building a shear, a moment, and sometimes, a torque, which have maximum values at the base of the structure that increase rapidly with the building's height. The response of a structure to horizontal loading, in having to carry the external shear, moment, and torque, is more complex than its first-order response to gravity loading. The recognition of the structure's behavior under horizontal loading and the formation of the corresponding model are usually the dominant problems of analysis. The principal criterion of a satisfactory model is that under horizontal loading it should deflect similarly to the prototype structure.

The resistance of the structure to the external moment is provided by flexure of the vertical components, and by their axial action acting as the chords of a vertical truss. The allocation of the external moment between the flexural and axial actions of the vertical components depends on the vertical shearing stiffness of the "web" system connecting the vertical components, that is, the girders, slabs, vertical diaphragms, and bracing. The stiffer the shear connection, the larger the proportion of the external moment that is carried by axial forces in the vertical members, and the stiffer and more efficiently the structure behaves.

The described flexural and axial actions of the vertical components and the shear action of the connecting members are interrelated, and their relative contributions define the fundamental characteristics of the structure. It is necessary in forming a model to assess the nature and degree of the vertical shear stiffness between the vertical components so that the resulting flexural and axially generated resisting moments will be apportioned properly.

The horizontal shear at any level in a high-rise structure is resisted by shear in the vertical members and by the horizontal component of the axial force in any diagonal bracing at that level. If the model has been properly formed with respect to its moment resistance, the external shear will automatically be properly apportioned between the components.

Torsion on a building is resisted mainly by shear in the vertical components, by the horizontal components of axial force in any diagonal bracing members, and by the shear and warping torque resistance of elevator, stair, and service shafts. If the individual bents, and vertical components with assigned torque constants, are correctly simulated and located in the model, and their horizontal shear connections are correctly modeled, their contribution to the torsional resistance of the structure will be correctly represented also.

A structure's resistance to bending and torsion can be significantly influenced also by the vertical shearing action between connected orthogonal bents or walls. It is important therefore that this is properly included in the model by ensuring the vertical connections between orthogonal components.

The preceding discussion of a high-rise structure's behavior has emphasized the importance of the role of the vertical shear interaction between the main vertical components in developing the structure's lateral load resistance. An additional mode of interaction between the vertical components, a horizontal force interaction, can also play a significant role in stiffening the structure, and this also should be recognized when forming the model. Horizontal force interaction occurs when a horizontally deflected system of vertical components with dissimilar lateral deflection characteristics, for example, a wall and a frame, is connected horizontally. In constraining the different vertical components to deflect similarly, the connecting links or slabs are subjected to horizontal interactive forces that redistribute the horizontal loading between the vertical components. For this reason, in a tall wall–frame structure the wall tends to restrain the frame near the base while the frame restrains the wall near the top. Similarly, horizontal force interaction occurs when a structure consisting of dissimilar vertical components twists. In constraining the different vertical components to displace about a center of rotation and to twist identically at each level, the connecting slabs are subjected to horizontal forces that redistribute the torque between the vertical components and increase the torque resistance of the structure.

Having assessed a proposed structure's dominant modes of behavior, the formation of an appropriate model requires next a knowledge of the available modeling elements and their methods of connection.

5.4 MODELING FOR APPROXIMATE ANALYSES

Approximate analyses are often made at the preliminary design stage to estimate quickly a proposed structure's stiffness and hence its feasibility. They are also used to estimate the allocation of external loading between the bents to allow for more detailed individual bent analyses.

The requirements of simplicity and rapidity for a preliminary analysis usually call for large approximations in forming the model. An approximate analysis may be a numerical analysis of a very simplified, discrete member model or, for certain types of structure, the analysis may consist of a closed solution to the characteristic differential equation of an equivalent continuum structure. Some approximations used in these two types of model are now described. The accuracy of an approximate solution depends on how closely the approximations made in forming the model represent the real structure.

5.4.1 Approximate Representation of Bents

Bents consisting of shear walls or of moment-resisting frames can be modeled approximately provided that the flexural and shear characteristics of the original assembly are reproduced in the model.

An axially concentric tall shear wall (Fig. 5.1a), consisting of relatively uniform regions, can be modeled by a column located on the centroidal axis of the wall (Fig. 5.1b). The column segments are assigned to have the inertias and shear areas of the corresponding regions of the wall. If the centroidal axis of the wall is not concentric, as in Fig. 5.2a, the analogous columns on the respective wall axes should be connected by horizontal rigid arms (Fig. 5.2b). When using a column to model a wall, the wall stresses are evaluated by applying the resulting column moment and shear to the appropriate sectional properties of the wall.

A multibay rigid frame (Fig. 5.3a) can be modeled very closely with regard to its lateral behavior by a single-bay rigid frame (Fig. 5.3b). The criteria for equivalence are that the racking shear rigidity (GA) as defined by the column and beam

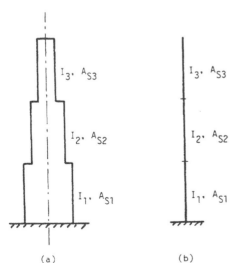

Fig. 5.1 (a) Axially concentric shear wall; (b) equivalent column.

72 MODELING FOR ANALYSIS

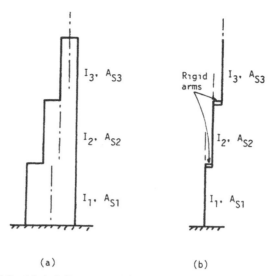

Fig. 5.2 (a) Axially eccentric shear wall; (b) equivalent column.

flexural inertias, the sum of the column inertias, I_i and the overall flexural inertia, I_g, as defined by the column sectional areas, are at each level the same in the equivalent single-bay frame as they are in the multibay frame. These properties and their equivalence are discussed in Chapter 7. Rigid-frame and braced-frame bents (Figs. 5.3a and 5.4a) whether single or multibay, can be represented in a very approximate way by single-column models (Fig. 5.4b). In these, the shear area of the analogous column is assigned to provide the same shear rigidity GA as the racking shear rigidity (GA) of the bent. Formulas for evaluating the racking shear rigidities (GA) for braced frames are given in Chapter 6. The flexural inertia of the equivalent column is assigned to have the same value as the inertia of the

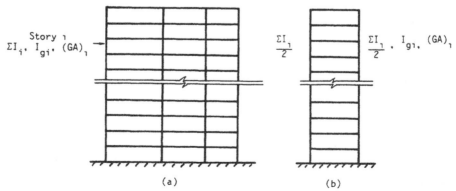

Fig. 5.3 (a) Multibay rigid frame; (b) equivalent single-bay frame.

5.4 MODELING FOR APPROXIMATE ANALYSES

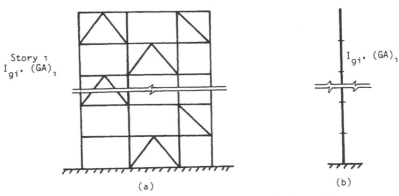

Fig. 5.4 (a) Multibay braced frame; (b) equivalent column.

column areas about their common centroid in the braced or the rigid frame. In this approximation, the single curvature flexure of the columns in the braced and rigid frames, which usually has only a minor influence on the frames' overall behavior, is neglected in the column model.

If a shear wall has beams connecting to it in-plane, causing it to interact vertically, as well as horizontally, with another shear wall or with other parts of the structure (Fig. 5.5a) the wall can be represented by an analogous "wide column." This is a column placed at the wall's centroidal axis and assigned to have the wall's inertia and axial area, and having rigid arms that join the column to the connecting beams at each framing level (Fig. 5.5b). In this way the rotations and vertical displacements at the edges of the wall are transferred to the connecting beams.

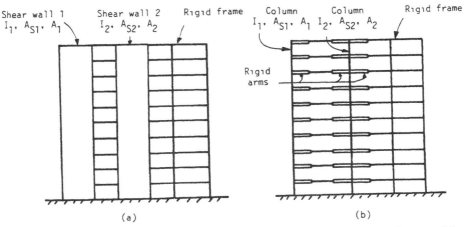

Fig. 5.5 (a) Shear walls and frame joined by beams; (b) equivalent wide-column model.

Nonplanar assemblies of shear walls that form elevator cores (Fig. 5.6a and b) in structures that translate but do not twist under lateral loading, can be simulated by a single column located at the shear center of the section and assigned to have the principal second moments of area of the core section (Fig. 5.7a). If the structure twists as well as translates, and the core has an effectively closed, box-like section, as in Fig. 5.6b, the single column should be additionally assigned the torsion constant J of the core (Fig. 5.7b).

If the structure twists and translates, and the core walls form an I, U, as in Fig. 5.6a, or more complex open-section shape, warping torsional effects may be important, in which case it is possible to use a two-column model (Fig. 5.7c) to give an approximate representation of all the bending and torsional properties. Details of such a model are given in Chapter 13.

5.4.2 Approximate Modeling of Slabs

In-Plane Effects. In structures that do not depend on the transverse bending resistance of slabs as part of their lateral load resisting system, the slabs are taken to serve only as rigid diaphragms that distribute the horizontal loading to the vertical

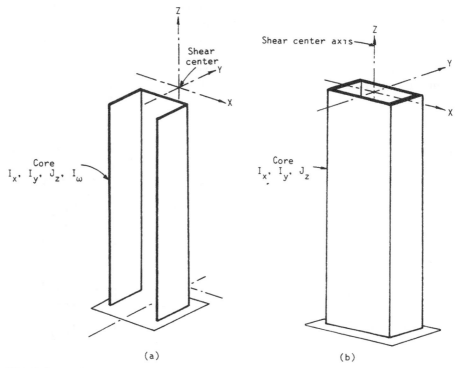

Fig. 5.6 (a) Open section nonplanar shear wall assembly; (b) closed section shear wall assembly.

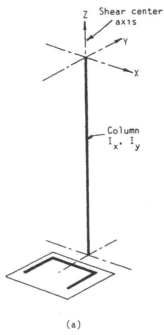

(a)

Fig. 5.7 (a) Equivalent flexural column; (b) equivalent flexural-torsional column; (c) equivalent two-column flexural-torsional-warping model.

elements and that hold the building plan in shape as the structure translates and twists. The slab then serves to constrain the horizontal displacements of the vertical components at each floor to be related to the horizontal two displacements and rotation of the slab. In a three-dimensional analysis of a structure (Fig. 5.8a) the in-plane rigidity of the slab can be represented at each floor by a horizontal frame of rigid beams joining the vertical elements (Fig. 5.8b) or, if the computer program includes a "rigid-floor" option for simulating a rigid in-plane slab, its use is simpler and more accurate.

Transverse Bending Effects. Flat plate structures, and structures with shear walls coupled by slabs, employ the transverse bending stiffness of slabs as part of the lateral load-resisting system, similar to the girders of a rigid frame, as well as using the in-plane rigidity of the slabs to hold the plan shape of the building. In modeling the structure, the bending action of a slab between in-line columns or walls can be represented by a connecting beam of equivalent flexural stiffness (Fig. 5.9). This model will result in the correct horizontal deflections, and forces in the vertical members, but it gives only the concentrated moments and shears applied to the slabs. The inertia of equivalent connecting beams to represent the slab bending action is discussed in Chapter 7 and Appendix 1.

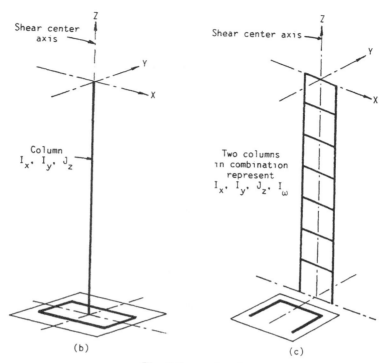

Fig. 5.7 continued.

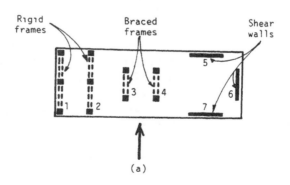

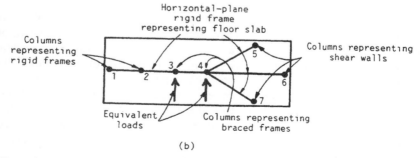

Fig. 5.8 (a) Plan asymmetric structure; (b) representation of slab diaphragm action.

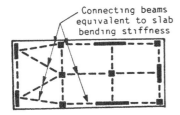

Fig. 5.9 Equivalent beam representation of slab bending action.

5.4.3 Modeling for Continuum Analyses

So far, all the considered approximations have been for discrete member models, that is incorporating individual vertical and horizontal members, for solution by a stiffness matrix analysis. For certain structures with relatively uniform properties over the height, alternative continuum analogy models may be formed that can be analyzed by a closed solution of the characteristic differential equation. In a continuum model, the horizontal slabs and beams connecting the vertical elements are assumed to be smeared as a continuous connecting medium—a continuum—having equivalent distributed stiffness properties. Although continuum methods are limited in their facility to represent variations of a structure over its height, they can give very rapid approximate solutions and are valuable in providing a general understanding of a structure's behavior. Two examples of the types of structure that can be solved using continuum techniques are a coupled wall and a wall–frame structure (Figs. 5.10a and 5.11a). In the coupled wall, the connecting beams are represented by a continuum with equivalent bending and shear properties (Fig. 5.10b). In a wall–frame structure, the connecting links between the wall and the frame are represented by a horizontally incompressible medium, while the beams in the frame are smeared into the general shear property of the equivalent shear column (Fig. 5.11b).

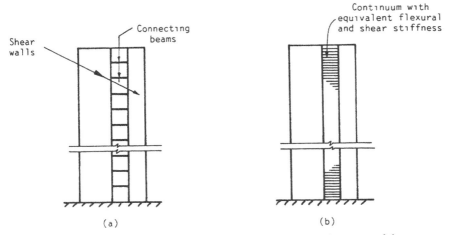

Fig. 5.10 (a) Coupled shear walls; (b) equivalent continuum model.

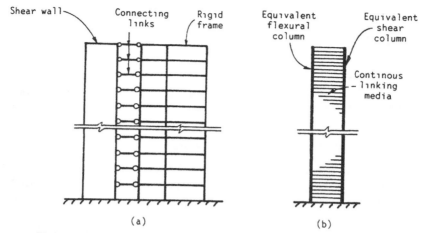

Fig. 5.11 (a) Wall–frame structure; (b) equivalent continuum model.

5.5 MODELING FOR ACCURATE ANALYSIS

It is necessary for the intermediate and final stages of design to obtain a reasonably accurate estimate of the structure deflections and member forces. With the wide availability of structural analysis programs and powerful computers it is now possible to solve very large and complex structural models. Some of the more gross approximations used for a preliminary analysis, such as representing braced frames and rigid frames by single columns, are too approximate for a detailed analysis, and they do not yield the detailed forces necessary for sizing and reinforcing the individual members. The structural model for an accurate analysis should represent in a more detailed way all the major active components of the prototype structure. The principal ones are the columns, walls, and cores, and their connecting slabs and beams.

The major structural analysis programs typically offer a variety of finite elements for structural modeling. As an absolute minimum for accurately representing high-rise structures, a three-dimensional program with beam elements and quadrilateral membrane elements (Fig. 5.12a and b) will suffice. Beam elements are used to represent beams and columns and, by making their inertias negligibly small or by releasing their end rotations, they can also be used to represent truss members. Membrane elements, which are used for shear walls and wall assemblies, should preferably include an incompatible mode option to better allow for the characteristic in-plane bending of shear walls.

If truss elements (Fig. 5.12c), quadrilateral plate elements, (Fig. 5.12d), and combined membrane-plate elements are also available, they can be used to advantage in representing, respectively, truss members, slabs in bending, and shear walls subjected to out of plane bending.

Some typical high-rise structural components and assemblies, and their representation by finite elements, will now be discussed.

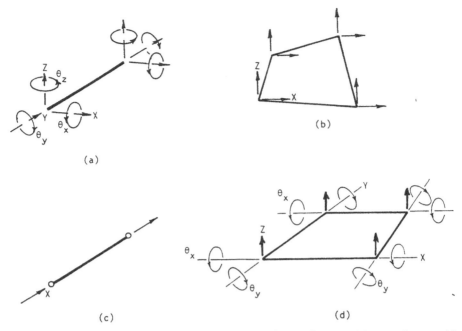

Fig. 5.12 (a) Beam element; (b) quadrilateral membrane element; (c) truss element; (d) quadrilateral plate bending element.

5.5.1 Plane Frames

A plane rigid frame, which is probably the simplest assembly to be modeled, has both its column and beam members represented by beam elements (Fig. 5.13). Shear deformations of the members are normally neglected except for beams with a span-to-depth ratio of less than about 5. The results of the analysis include the vertical and horizontal displacements, and the vertical plane rotations of the nodes,

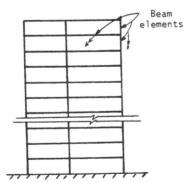

Fig. 5.13 Rigid frame using beam elements.

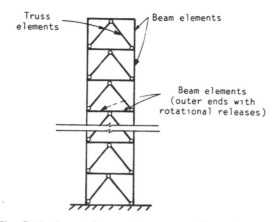

Fig. 5.14 Braced frame using truss and beam elements.

together with the members' axial force, shear force, and bending moments. In a braced frame (Fig. 5.14) the braces are represented by truss elements or small-inertia beam elements, the columns by beam elements, and the beams by beam elements with their end rotations released. The results for the truss elements give axial forces only.

5.5.2 Plane Shear Walls

Similar to the modeling of walls for an approximate analysis, a tall slender shear wall that is not connected by beams to other parts of the structure (Figs. 5.1a and 5.2a) can be modeled for an accurate analysis by a stack of beam elements (Figs. 5.1b and 5.2b) located on the centroidal axis of the wall, and assigned to have the principal inertia and corresponding shear areas of the wall. Shear walls connected by beams to other parts of the structure (Fig. 5.5a) can be similarly represented by vertical stacks of beam elements located on the centroidal axes of the walls with rigid horizontal beam elements attached at the framing levels to represent the effect of the walls' width (Fig. 5.5b). In the case of a beam-connected wall, axial forces will be induced in the wall, so it is necessary to assign to the analogous column an axial area as well as an inertia and a shear area.

Walls that are not slender, or that have openings, cannot be well represented by simple equivalent columns and are better represented by an assembly of plane-stress membrane elements (Fig. 5.15a). Because the segments of a shear wall and the membrane elements that are used to model it are subjected to in-plane bending, incompatible mode elements that are formed to include this deformation invariably give more accurate results, as well as allowing the use of rectangular elements of much greater height-to-width proportions with acceptably accurate results. The results for a plane-stress element typically include the horizontal and vertical dis-

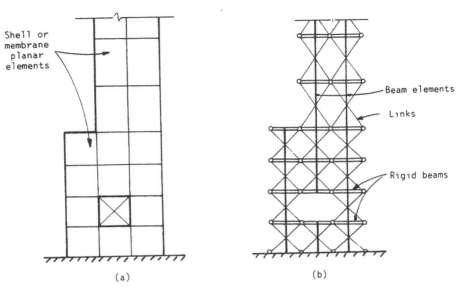

Fig. 5.15 (a) Shear wall: membrane element model; (b) shear wall: analogous frame model.

placements of the nodes, and the vertical and horizontal direct stresses and shear stresses at either the corners or the mid-sides of the element.

If the available structural analysis computer program does not include plane-stress elements, a shear wall can be modeled alternatively using an analogous frame, such as in Fig. 5.15b, which can be assembled entirely from beam elements. The stresses resulting from such a model are usually within 1 or 2% of those from a membrane element model analysis. Details of an analogous frame are given in Chapter 9.

Nonrectangular walls can be modeled using quadrilateral elements, and, if more detailed stresses are required in a particular region of the wall, a finer mesh can be used in that area, with quadrilateral elements being used to make the transition (Fig. 5.16). For greater accuracy, quadrilateral elements should be proportioned to be as close as possible to equal-sided parallelograms.

When modeled by membrane elements, shear walls with in-plane connecting beams require special consideration. Membrane elements do not have a degree of freedom to represent an in-plane rotation of their corners; therefore, a beam element connected to a node of a membrane element is effectively connected only by a hinge. A remedy for this deficiency is to add a fictitious, flexurally rigid, auxiliary beam to the edge wall element, in one of the ways shown in Fig. 5.17. The adjacent ends of the auxiliary beam and the external beam are both constrained to rotate with the wall–edge node. Consequently, the rotation of the wall, as defined by the relative transverse displacements of the ends of the auxiliary beam, and a moment, are transferred to the external beam.

82 MODELING FOR ANALYSIS

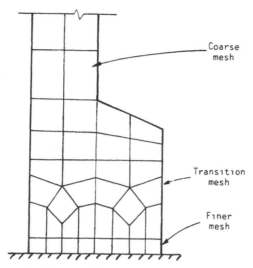

Fig. 5.16 Nonrectangular shear wall with transition, represented by quadrilateral elements.

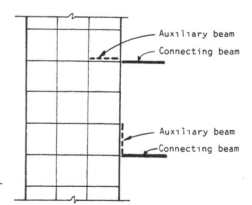

Fig. 5.17 Connection of beams to membrane element shear wall.

5.5.3 Three-Dimensional Frame and Wall Structures

The high-rise rigid frame structure has moment-resisting joints, and its columns and beams are modeled by three-dimensional beam elements (Fig. 5.18). These elements deform axially, in shear and bending in two transverse directions, and in twist. Generally, therefore, they have to be assigned an axial area, two shear areas, two flexural inertias, and a torsion constant. Often, however, shear deformations of the columns and beams, and axial deformations of the beams, are assumed

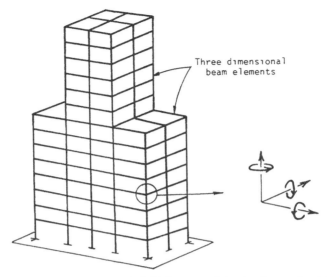

Fig. 5.18 Three-dimensional rigid-frame model using beam elements.

negligible. These are usually allowed for by omitting the assignment of a shear area and by assigning either a fictitiously large axial area, or constraints between the axial displacements of the member ends. In addition, the torsional stiffness of practically proportioned beams and columns is usually negligible, which is allowed for by omitting the assignment of a torsion constant. The usual results of significance are, therefore, the translations and rotations of the nodes, the shear forces, bending moments and axial force in the columns, and the shear forces and moments in the beams.

Three-dimensional shear wall assemblies often form the most important major lateral load-resisting components in a high-rise building. They occur variously in multibranch open sectional shapes (Fig. 5.19a), in effectively closed sections (Fig. 5.19b), and in beam-connected sections (Fig. 5.19c). Whether of closed or open-section form, the principal actions of the individual walls in an assembly are in-plane shear and flexure, and the principal interaction between the walls of an assembly is vertical shear along the joints. Consequently, plane stress membrane elements are highly suitable for modeling three-dimensional shear wall components (Fig. 5.20a and b). Story-height wall-width elements give an acceptably accurate representation for most purposes.

Plane stress elements alone are not adequate for modeling three-dimensional wall systems because they lack the transverse stiffness necessary at orthogonal wall connections to allow a stiffness matrix analysis of the problem. Nor, when used alone, can plane stress membrane elements provide the out-of-plane rigidity required to maintain the sectional shape of the core, as it is held in reality by the in-

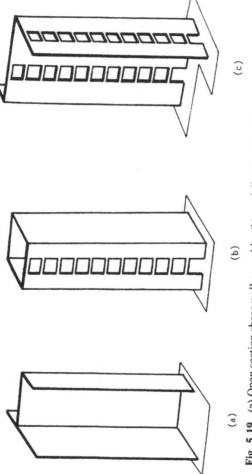

Fig. 5.19 (a) Open section shear wall assembly; (b) partially closed section shear wall assembly; (c) nonplanar walls connected by beams.

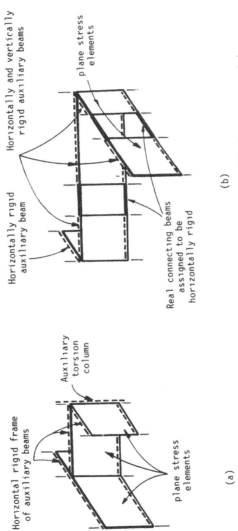

Fig. 5.20 (a) Membrane element and auxiliary beam model; (b) model for beam-connected shear walls.

86 MODELING FOR ANALYSIS

plane rigidity of the floor slabs. The remedy for these deficiencies is to add at each nodal level a horizontal frame of fictitious, rigid auxiliary beams (Fig. 5.20a). If any of the walls are connected in-plane to each other, or to other parts of the structure, by beams, the auxiliary beams adjacent to the wall edges can be made vertically rigid also, to cause the transfer of moment (Fig. 5.20b) as described in Section 5.5.2.

Another action, which would automatically be accounted for if shell elements were used for the model, but not in the case of plane stress elements, is the torsional stiffness corresponding to twisting of the walls. Although this is usually relatively insignificant, in open-section wall assemblies it can be important and should be incorporated. It is introduced by adding to the model a fictitious column located on any one of the vertical sets of nodes (Fig. 5.20a) and assigning it a torsion constant with a value equal to the sum of the individual walls' torsion constants, as discussed in Chapter 13. The axial area and inertia of the column are assigned to be zero.

An alternative way of representing beams connecting shear walls in the same plane is to represent them by story-height membrane elements with a vertical shearing stiffness equal to the vertical-displacement stiffness of the represented beam, as shown in Fig. 5.21. In such a model, auxiliary beams are still required to form a horizontally rigid frame around each level of the wall assembly, but the beams adjacent to the openings do not have to be vertically rigid.

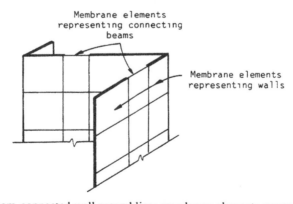

Fig. 5.21 Beam-connected wall assemblies: membrane elements representing beams.

5.5.4 P-Delta Effects

Second order P-Delta effects of gravity loading can be included in a single first-order computer analysis of the structure by adding to the first-order model a fictitious column with a negative stiffness.

The translational P-Delta effects in a nontwisting structure can be incorporated in the two-dimensional model by adding a shear column, connected to the model by rigid links at the framing levels (Fig. 5.22a). The column is assigned a negative

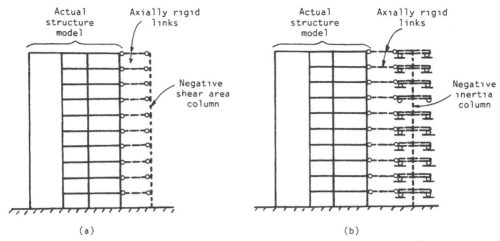

Fig. 5.22 (a) P-Delta negative shear column model; (b) P-Delta negative inertia column model.

shear area to simulate the lateral softening of the structure due to gravity loading. The column is assigned to be rigid in flexure. Alternatively, the same result can be achieved by using a flexural column with its rotation restrained at the framing levels (Fig. 5.22b) and its inertia assigned a negative value. The column is specified to be rigid in shear. The resulting deflections and member forces in the model then include the P-Delta effects of gravity loading. Details of the technique are given Chapter 16.

When making a full three-dimensional analysis of an asymmetric structure, the P-Delta effects of twisting, as well as of translating parallel to the building's major axes, can also be represented in the model by a fictitious negative stiffness column. The column is located in each story at the centroid of the resultant gravity loading acting through the story, and is assigned to have either negative shear areas, or negative inertias, as described before, corresponding to the directions of the building's two major axes. The column is additionally assigned a negative torsion constant to allow for the twisting P-Delta effects. This technique also is discussed in detail in Chapter 16.

5.5.5 The Assembled Model

By combining the previously described techniques, a complete three-dimensional model can be formed for any high-rise structure consisting of a combination of frames, walls, and cores with beam and slab connections.

If the bending resistance of the slabs contributes to the lateral load resistance of the structure, it is usual to model the slabs by beams of equivalent flexural stiffness connecting the vertical components. Although an even more accurate model could be formed by representing each slab as an assembly of plate elements, such a detailed representation would vastly increase the size of the problem.

In the complete detailed model, therefore, beam elements are used to represent beams and columns, and story-height plane-stress membrane elements are used to represent shear walls and cores. At all floor levels an auxiliary beam is added to the top of each membrane element. The auxiliary beams, and the real beams, are assigned extremely high axial areas and horizontal bending inertias in order to simulate the rigid diaphragm effect of the slab. Auxiliary beams are also used at each floor level to interconnect frames, walls and cores, as well as any isolated columns. Where a real beam connects in plane with a wall, the auxiliary beam on the connected wall element is assigned to be rigid in the vertical, as well as the horizontal, plane so as to transfer moment between the wall and the external beam. For each open section shear wall assembly, a vertical column assigned to have the walls' torsion constant is added to the assembly.

The requirement for providing auxiliary beams, joining the columns, walls, and cores to form a rigid horizontal diaphragm at each floor, and to connect shear walls to beams in their planes, has been avoided in at least one tall building structure analysis program [5.1].

5.6 REDUCTION TECHNIQUES

When the detailed model of a high-rise structure is so large and complex that its analysis presents a formidable task of bookkeeping and computation, it may be preferable to try to simplify the model, provided the accuracy of the results is not seriously compromised. The following techniques are among those used to simplify the model. Some of the techniques do not diminish at all the accuracy of the analysis, while others, although losing a little in accuracy, are still good enough for a final design analysis. The reductions are therefore applicable to both detailed and to simplified models for anlaysis.

5.6.1 Symmetry and Antisymmetry

A structure that is symmetric in plan about the axis of horizontal loading (Fig. 5.23a) can be analyzed as a half-structure, to one side of the line of symmetry, subjected to half the loads (Fig. 5.23b). The ends of the members cut by the line of symmetry must be constrained to represent the omitted half of the structure. That is, they must be constrained against rotation and horizontal displacement in the plane perpendicular to the direction of loading, and against rotation about a vertical axis, while simultaneously being free to displace vertically and to translate in the direction of the loading. The results for the deflections and forces for the analyzed half-structure will apply symmetrically to the corresponding nodes and members in the omitted half-structure.

A structure that is symmetric in plan about a horizontal axis perpendicular to the axis of horizontal loading (Fig. 5.24a) behaves antisymmetrically about the axis of symmetry. In this case only half of the structure, to one side of the axis of symmetry, and subjected to loads of half value, needs to be analyzed (Fig. 5.24b).

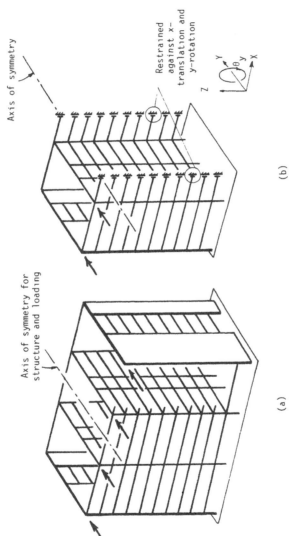

Fig. 5.23 (a) Plan symmetric structure with symmetric loading; (b) half-structure model.

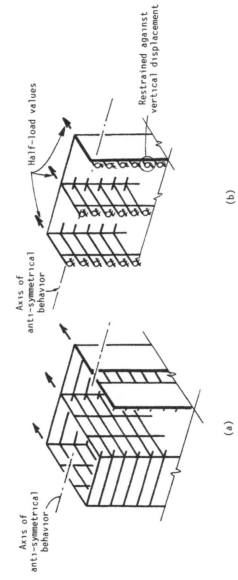

Fig. 5.24 (a) Antisymmetrically behaving structure; (b) half-structure model.

The ends of the cut members are constrained on the line of symmetry to represent their connection to the omitted antisymmetrically behaving other half of the structure. That is, they are constrained against vertical displacement, but are free to rotate in the vertical plane parallel to the direction of loading. The values of the results for the analyzed half-structure apply antisymmetrically to the omitted half-structure.

Thus, if a structure is doubly symmetric on plan and subjected to horizontal loading along one of its axes of symmetry, it can be analyzed by considering just one-quarter of the structure, with appropriate constraints applied to the ends of members cut on the lines of symmetry, to represent the symmetrical and antisymmetrical aspects of behavior.

5.6.2 Two-Dimensional Models of Nontwisting Structures

The assumption that the floor slabs are rigid in plane, which permits the horizontal displacements of all vertical elements at a floor level to be defined in terms of the slab's horizontal translation and rotation, allows the possibility of representing a three-dimensional structure by a two-dimensional model. An explanation of this can be developed by first considering techniques for the planar representation of nontwisting structures, and then extending them to twisting structures.

Symmetrical Structure Consisting of Parallel Bents. A structure that is symmetric on plan and symmetrically loaded does not twist. Adding to this the assumption of the slab's in-plane rigidity means that the horizontal displacements of all the vertical components at a floor level are identical. Now considering the symmetrical structure in Fig. 5.25a, and allowing for symmetry by analyzing only one-half of the structure, the identity of displacements at the floor levels can be established in a planar model by assembling the bents in the same plane, in any order and at an arbitrary spacing, as in Fig. 5.25b, and providing a horizontal constraint between the bents at each level. The constraint can be formed in two alternative ways. If the analysis program has a dependent node option, sets of nodes, one in each bent, at the same level, can be assigned to have the same horizontal displacements. If a dependent node option is not available, pairs of nodes at the same level in adjacent bents may be joined by axially rigid pin-ended links, as in Fig. 5.25b. The half-structure model is then subjected at the floor levels to loads of half the value of the total load per level.

As far as the validity of the assumptions allow, the resulting moments, shears, and vertical axial forces in the model will correctly represent those in the structure. The shear in the slabs between bents must be found by considering the differences between the shears in successive stories of each bent, and the relative plan location of the bents. The axial forces in the beams and links of the model are not meaningful because both the application of the loading and the horizontal connections in the planar model do not properly represent their on-plan locations in the real structure.

92 MODELING FOR ANALYSIS

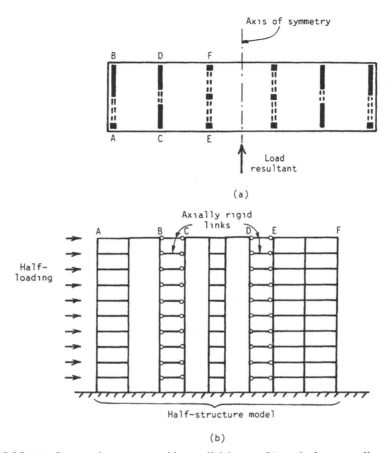

Fig. 5.25 (a) Symmetric structure with parallel bents; (b) equivalent two-dimensional model.

Symmetrical Structure with Connected Orthogonal Bents.
Structures that consist of an orthogonal system of connected bents, which are symmetrically located about the axis of horizontal loading, as in Fig. 5.26a, can be modeled for analysis by an extension of the planar modeling technique described above.

Considering half the structure, and assuming that, perpendicular to their planes, the bents have negligible stiffness, the structure's shear resistance in the direction of loading is provided by bents AB and CD, as they displace horizontally in their planes parallel to the direction of loading (Fig. 5.26a). Bents AE and BF, perpendicular to the loading, do not displace horizontally in their planes, but interact vertically with bents AB and CD along their vertical lines of connection A, B, C, and D. This vertical interaction causes the perpendicular bents to act as "flanges" to the parallel bent "webs," as part of the structure's overall flexural action.

In the equivalent half-structure planar model, half the parallel bents and the perpendicular half-bents are assembled in plane, with the parallel bents in one

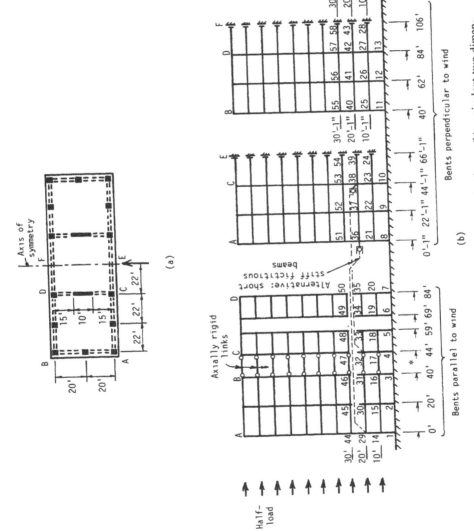

Fig. 5.26 (a) Symmetric structure with orthogonal interacting bents; (b) equivalent two-dimensional model.

94 MODELING FOR ANALYSIS

group and the perpendicular half-bents in another (Fig. 5.26b). A column at the intersection of orthogonal bents appears twice in the planar model, once in a parallel bent and once in a perpendicular bent. In each bent the column is assigned an inertia appropriate to its bending in the plane of that bent. So that the axial area of an intersection line column is not represented twice, it is arbitrarily assigned entirely to the column in the parallel bent with a zero area assigned to that in the perpendicular frame. The nodes in the model are numbered so that those on the vertical lines of intersection, which are represented twice, are assigned two different numbers, as in Fig. 5.26b.

The identical horizontal displacements of the parallel bents are established in the model as before, either by using the dependent node facility or by including fictitious axially rigid links, as between B and C in Fig. 5.26b.

The compatibility of vertical displacements between the parallel and perpendicular bents may also be achieved in alternative ways. If a dependent node option is available, vertical compatibility can be established by constraining the connection nodes that are duplicated in the parallel and horizontal bents to have the same vertical displacements. The zero horizontal in-plane displacement of the perpendicular bents is arranged by constraining horizontally at least one vertical line of nodes in each of those bents.

If a dependent node option is not available, there are two alternative ways of using fictitious members to establish the connection in the planar model. The first is to dimension the model horizontally so that the vertical intersection lines on each perpendicular frame, as for example lines A and lines C in Fig. 5.26b, are located immediately adjacent, say as close as 1/200 of the adjacent span, to the duplicate intersection lines of the connected "parallel" frames. Each pair of duplicated connection nodes is then joined by a very stiff horizontal beam with a horizontal and rotational release at one end, as, for example, nodes 32 and 38 in Fig. 5.26b. The alternative is to dimension the model horizontally so that the vertical connection lines on the perpendicular frames are, in effect, coincident with those on the parallel frames, as, for example, lines B and lines D in Fig. 5.26b, and to dimension them vertically so that the connection nodes on the perpendicular frame are displaced upward slightly, say 1/100 story height, from the corresponding nodes on the parallel frame. Each pair of duplicated connection nodes, as, for example, nodes 46 and 55 in Fig. 5.26b, is then joined by a vertical axially rigid link. In either of these ways, the fictitious links establish vertical compatibility, while avoiding horizontal interaction and vertical plane rotational interaction between the orthogonal bents.

The technique can be used for structures whose bents consist of walls, or frames, or combinations of both.

5.6.3 Two-Dimensional Models of Structures That Translate and Twist

The common assumption for analysis, that the floor slabs are rigid in their planes, implies that for an arbitrary origin and a pair of axes parallel to the orthogonally oriented bents of a laterally loaded structure (Fig. 5.27) the resulting displaced

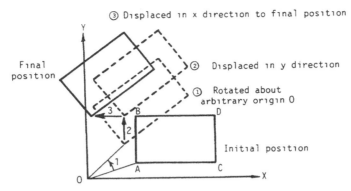

Fig. 5.27 Displacements of bending and twisting structure.

location of any floor slab can be defined in terms of the rotation of the slab about the origin, and two displacements parallel to the axes. Further, for the horizontal equilibrium of any slab, the external X- and Y-direction forces on the slab and their combined moment about the vertical axis through the origin must be in equilibrium with, respectively, the X- and Y-direction resultants of the reactions from the bents and their resultant moment about the origin.

Assuming that the structure consists of a plan-asymmetric system of orthogonal bents that are stiff in their planes but have zero transverse and torsional stiffnesses (Fig. 5.28a), a two-dimensional model can be formed to satisfy the above conditions of displacement and equilibrium, as follows.

First, select an arbitrary origin 0 (Fig. 5.28a) that is located to the left of and below the lower left-hand corner of the structural plan. Bents AB and CD are parallel to, and at distances x_1 and x_2 from the Y axis, while the orthogonal bents AC and BD are parallel to, and at y_1 and y_2 from the X axis.

Next, form the two-dimensional model by assembling all the bents in the same plane with the X-direction bents in one group and the Y-direction bents in the other,

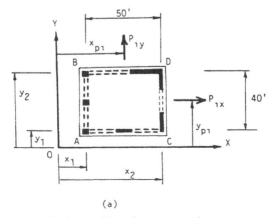

(a)

Fig. 5.28 (a) Plan of nonsymmetric structure.

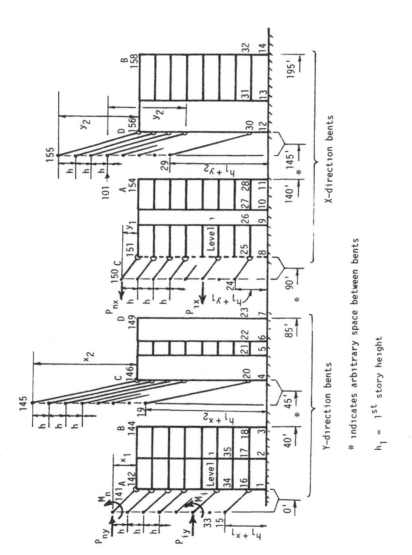

Fig. 5.28 (b) Equivalent planar model.

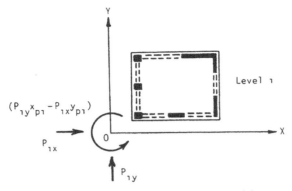

Fig. 5.28 (c) Equivalent loading on model.

as shown in Fig. 5.28b. To make the viewed faces of the bents consistent with the location of the origin as specified above, the bents are displayed in the model (Fig. 5.28b) as viewed in Fig. 5.28a looking negatively along the X and Y axes, respectively (i.e., A to the left of B in bent AB, and C to the left of A in bent CA). This is to ensure that a horizontal plane rigid body rotation of a slab about the origin 0, in Fig. 5.28a, corresponds to all the bents moving in the same direction in the planar model (Fig. 5.28b). For example, a counterclockwise rotation of the slab about 0 in Fig. 5.28a corresponds to a rightward displacement of the bents in the model of Fig. 5.28b.

Because the bents are shown separately from each other, the columns on lines of intersection of orthogonal bents will appear twice in the model, as, for example, column A appearing in each of bents AB and CA. The flexural inertias and axial areas of the duplicated columns are assigned in the way described previously for intersecting bent structures.

Establish on the model, for the left-hand edge of each bent and on the same vertical line as the edge, a set of "governing" nodes, one node for each floor level. Each governing node is located above its associated floor level by a height equal to the distance on-plan of the bent from the X or Y axis to which it is parallel. For example, governing node 141, for the top floor of bent AB, is on the vertical line A at a height x_1 above the top floor, while governing node 101, for the third-to-top floor of bent DB, is on line D at a height y_2 above the third-to-top floor (Fig. 5.28b).

In Fig. 5.28b, the governing nodes are shown, for clarity, offset to the left from the left-hand edges of their associated bents, but they are assigned horizontal coordinates to locate them in the model on the same vertical line.

Now connect each governing node to its corresponding floor-level node by an effectively rigid vertical arm, with a rotational release at the floor-level node. All the nodes of the structure are then numbered in sequence, starting from left to right across the base of the model, and then level by level upward, including the governing nodes, as shown in Fig. 5.28b.

Consider now, for example, the top levels of bents AB and CD in the model.

98 MODELING FOR ANALYSIS

By constraining governing nodes 141 and 145 to displace horizontally together, using the dependent node option, a horizontal translation without rotation of governing node 141, and hence of 145, will cause through their connecting arms a translation in the Y direction of the whole top floor. Similarly, in bents CA and DB, constraining governing nodes 150 and 155 to translate identically means that a horizontal translation without rotation of governing node 150, and hence of 155, will cause a translation in the X direction of the top floor. Further, by constraining nodes 141, 145, 150, and 155 to rotate together means that a rotation without translation of node 141, and hence of 145, 150, and 155, will cause, through their connecting arms, the top floors of the bents in the planar model to translate with the same relative displacements as they would in rotating in plan as a rigid body about the origin 0 in Fig. 5.27. Specifying similar translational and rotational constraints between the governing nodes for each of the other floor levels will cause the horizontal displacements of the planar model to properly represent in-plane displacements of each level of each bent, due to the translations and rotations of the structure.

Considering again the top levels of bents AB and CD in the planar model, their horizontal reactions will be transmitted to governing nodes 141 and 145 as horizontal forces and vertical plane moments, having the same magnitudes, respectively, as the Y-direction reactions, and the horizontal plane moments of those reactions about the origin, of the top levels of bents AB and CD in the plan view (Fig. 5.28a). Similarly, the horizontal reactions of bents CA and DB in the planar model will be transmitted to nodes 150 and 155 as horizontal forces and vertical plane moments having the same magnitudes as the X-direction reactions, and the horizontal plane moments of those reactions about the origin, as the top levels of bents CA and DB in the plan view (Fig. 5.28a).

Because nodes 141 and 145 are constrained to displace identically in the Y direction, the sum of the Y-direction reactions P_{ny} can be assumed to act at node 141. Similarly, the sum of the X-direction reactions P_{nx} can be assumed to act at node 150. Since nodes 141, 145, 150, and 155 are constrained to rotate together, the sum of the moments from all the X- and Y-direction bents can be assumed to occur at node 141. By this reasoning, the resulting horizontal reactions at governing nodes 141 and 150, and the resulting vertical plane moment of the reactions at node 141, are the same as the resultant Y- and X-direction reactions, and their resultant horizontal plane moment about the origin, in Fig. 5.28a.

When orthogonal bents in a structure intersect, as they do in Fig. 5.28a, the vertical interaction along their vertical lines of intersection is an important factor in the structure's behavior and must be represented in the model. For this, additional constraints have to be applied in the planar model to establish vertical compatibility between the orthogonal bents. In the model (Fig. 5.28b) these constraints are on the vertical lines of intersection A, B, C, and D. The constraints are applied using the dependent node option by assigning each pair of duplicated connection nodes, for example, 16 and 28 at A, 18 and 32 at B, 20 and 25 at C, and 23 and 30 at D, to have the same vertical displacements.

Having formed the planar model, it remains only to transform the loading on the structure into equivalent loads for application to the model. Referring to Fig. 5.28a, the load in the Y direction at level i, P_{iv}, acts at a distance x_{pi} from the origin 0. This may be transformed into a force P_{iv} acting at the origin and a torque $P_{iv}x_{pi}$ (Fig. 5.28c). Similarly, the load P_{ix} in the X direction, at a distance y_{pi} from the origin, can be transformed into a force P_{ix} at the origin and a torque $-P_{ix}y_{pi}$. These equivalent actions may be applied to the model (Fig. 5.28b) as P_{iv} to one of the governing nodes for level i of the Y-direction bents, P_{ix} to one of the governing nodes for level i of the X-direction bents, and a torque $P_{iv}x_{pi} - P_{ix}y_{pi}$ to one of the governing nodes for level i of all the bents. For example, at the top, nth level, P_{nv} is applied to node 141, P_{nx} is applied to node 150, and a counterclockwise torque M_n equal to $P_{nv}x_{pn} - P_{nx}y_{pn}$ is applied to node 141. A similar transformation of the loads at each other level to equivalent loads and a torque about the origin, and their application to the corresponding governing nodes in the planar model, will make the model ready for analysis. Note that loads that act in the A-to-B and A-to-C directions of the bents on the plan of the structure are applied in the A-to-B and A-to-C directions to the governing nodes of the planar model. A horizontal plane counterclockwise torque on the plan of the structure is applied as a vertical plane counterclockwise torque on the planar model.

A two-dimensional stiffness method analysis of the planar model subjected to the transformed loads will yield results for deflections and member forces identical to those from a full three-dimensional analysis of the structure, provided that in the latter analysis the assumptions of the slabs' in-plane rigidity and the bents' zero transverse and torsional stiffness are also adopted. If a structure includes a core consisting of an assembly of shear walls, this can also be included in the planar model by treating each individual wall of the core as a bent, representing it by a stack of plane-stress finite elements, and assigning to it a set of governing nodes, rigid arms, and constraints, as though it were just another bent.

A final necessary comment concerns the flexural stiffnesses to be assigned to the connecting arms to cause them to behave as rigid. It is recommended that each should be assigned an inertia such that, if the arm were considered as a vertical cantilever fixed at its governing node, its lateral stiffness at the lower end would be of an order 1000 times greater than the estimated lateral stiffness of the bent at the level where the arm connects. If the arm stiffnesses were assigned to be not stiff enough, they would bend and not enforce proper translations on the bent, whereas, if they were excessively stiff they could cause numerical instability in the analysis.

An explanation of this modeling technique is given in Ref. [5.2].

5.6.4 Lumping

"Lumping" means the combination of several of a structure's similar, and similarly behaving, components or assemblies of components into an equivalent single component or assembly in order to reduce the size of the model for analysis. The resulting forces in the equivalent component or assembly are subsequently distributed to give the forces in the original units.

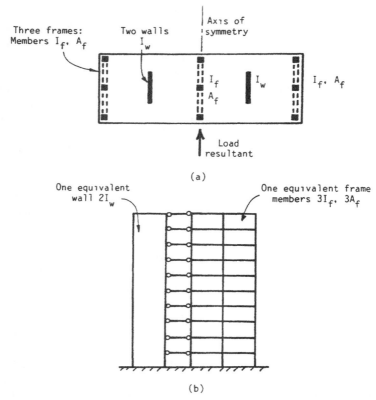

Fig. 5.29 (a) Symmetric structure with repetitive bents; (b) equivalent lumped model.

Lateral Lumping. Consider as an example the symmetrical and symmetrically loaded (and therefore nontwisting) structure in Fig. 5.29a, which consists of two identical shear walls and three identical rigid frames. The walls can be lumped laterally into a single wall, with twice the inertia of an individual wall, and the frames lumped into a single frame with member properties three times those of an individual frame. The lumped wall and frame can then be assembled as a planar model (Fig. 5.29b) and analyzed relatively simply. The resulting forces in the wall and frame of the lumped structure are divided by two and three, respectively, to give the forces in the individual walls and frames. This simple lateral lumping technique may be applied only to structures that do not twist, because the forces in the bents of a nontwisting structure are independent of their lateral location.

Vertical Lumping. More usual examples of lumping occur in tall multistory coupled-wall or rigid-frame structures in which the story heights and beam sizes are repetitive, as in Figs. 5.30a and 5.31a. The detailed models can be simplified by vertically combining groups of three or five beams into single beams, at the middle beam location, and assigning to them the lumped properties of inertia and

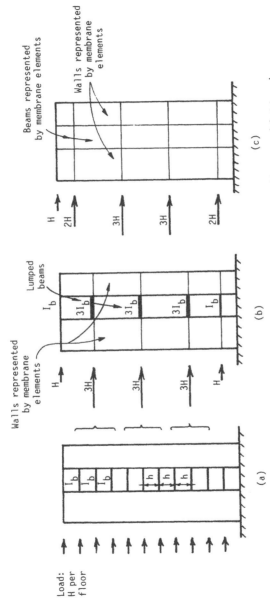

Fig. 5.30 (a) Coupled walls with repetitive beams; (b) equivalent lumped beam model; (c) equivalent membrane element reduced model.

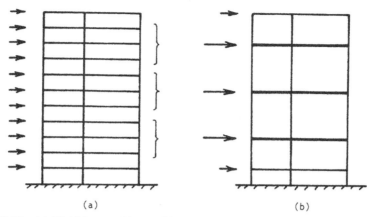

Fig. 5.31 (a) Rigid frame with repetitive beams; (b) equivalent lumped beam model.

shear area (Figs. 5.30b and 5.31b). It is advisable to leave the bottom one or two beams, and the top one or two beams, of the structure in their original locations to better represent the localized effects at the base and the top.

In the case of lumping beams that connect shear walls, as in Fig. 5.30a, the sectional properties of the membrane elements, or the analogous wide columns, representing the walls would be the same in the lumped model (Fig. 5.30b) as in the nonlumped model, because of the predominantly single-curvature behavior of the walls. In a rigid frame, however (Fig. 5.31a), the predominantly story-height double-curvature bending of the columns would require their inertias to be increased in the lumped model with its increased story heights, to make the lateral racking stiffnesses of the two models identical. The axial areas of the columns in the two models would, however, be the same. The lateral loads are also lumped and applied at the lumped beam levels. Details of this technique are given in Chapter 7.

When coupled walls are being represented by membrane finite elements, a variation of the lumping technique is to represent sets of n successive connecting beams, as well as the shear walls by n-story-height membrane elements (Fig. 5.30c). The wall elements are assigned the same sectional dimensions as the walls, while the elements representing the beams are assigned a thickness to represent the distributed vertical flexural and shear stiffnesses of the connecting beams. Details of this technique are also given in Chapter 7.

The results for the member forces of a lumped model analysis must be interpreted to obtain the forces in the members of the original structure. The resulting moment and shear in the original middle beam of a lumped set of n beams are one-nth of the resulting values for the lumped beam. The forces in the other beams of the original structure must be estimated by interpolation between the values obtained for the middle beams above and below. The distribution of horizontal shear between the vertical members at any level in the original structure will be in the same ratio as between the corresponding members in the lumped model structure, while the sum of shears will be equal to the external shear at that level. The moment at any level in a shear wall of the original structure will be given by the moment at that level of the wall in the lumped structure, while moments in the columns of a rigid frame will be given approximately by the product of the column shear, determined as above, and the original half-story height.

5.6.5 Wide-Column Deep-Beam Analogies

It has been explained earlier how horizontally loaded shear walls connected by beams, as in Fig. 5.32a, can be modeled by equivalent wide columns that consist of a column on the centroidal axis of the wall, with rigid arms at the beam levels to represent the effects of the walls' width (Fig. 5.32b). Some frame analysis programs include a rigid-end member option that includes the wide-column effects and therefore allows the beam to be considered as a single member between the column axes. If the available program does not have such an option, the rigid-end beam may be simplified in the model to a full-span uniform beam with an increased inertia to allow for the wide-column effects (Fig. 5.32c). An expression for the increase in effective inertia, which is given in Chapter 10, is dependent on the assumption of the wall cross sections rotating in-plane identically at the same level. This is generally valid for coupled shear walls and for rigid frames with a pattern of regularly spaced equally sized columns such as occur in framed-tube structures.

In rigid-frame systems with deep beams (Fig. 5.33a), the stiffening effect of the beam depth on the columns can be represented by rigid vertical arms (Fig. 5.33b). This also can be accommodated in an analysis by a rigid-end member option.

If the analysis program does not have a rigid-end member facility, however, the rigid-end column can be replaced in the model by a uniform full-height column between the beam axes (Fig. 5.33c) with modified stiffness properties to allow for the deep beam effect. The inertia of the full-height column will be increased to allow for the rigid-end effect by a factor that depends on whether the vertical

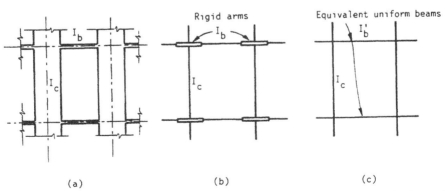

Fig. 5.32 (a) Coupled shear walls; (b) equivalent wide-column model; (c) equivalent uniform beam model.

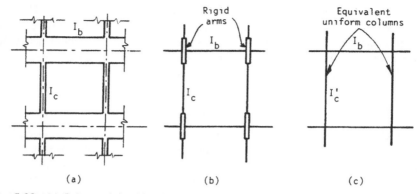

Fig. 5.33 (a) Columns joined by deep beams; (b) equivalent deep beam model; (c) equivalent uniform column model.

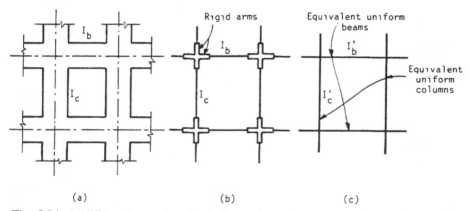

Fig. 5.34 (a) Wide-column, deep beam frame; (b) wide-column, deep beam model; (c) equivalent uniform member model.

members are deflecting primarily in single curvature, as would shear walls, or in story-height double curvature, as would slender columns.

A frame combining wide columns and deep beams, such as a reinforced concrete frame tube (Fig. 5.34a), can be represented either by an analogous wide-column deep-beam frame (Fig. 5.34b) or more simply by a frame of equivalent full length beams and columns with appropriately increased stiffnesses (Fig. 5.34c).

SUMMARY

In modeling a structure for analysis it is usual to represent only the main structural members and to assume that the effects of nonstructural members are small and conservative. Additional assumptions are made with regard to the linear behavior of the material, the in-plane rigidity of the floor slabs, and the neglect of certain member stiffnesses and deformations, in order to further simplify the model for analysis.

The extent to which a model will be simplified is related to the stage of analysis: a simple model will be used for an approximate preliminary analysis, and a relatively detailed one for a more accurate final analysis. In approximate modeling, whole bents, which may be rigid frames, braced frames, shear walls, or cores, may be reduced to equivalent single-column members, for a computer stiffness matrix analysis. Or sets of connecting beams or links between major vertical components may be represented by an equivalent continuous medium to allow a closed solution of the governing differential equation.

In more accurate modeling, the columns and beams of frames will be represented individually by beam finite elements, while shear walls and cores will be represented by assemblies of membrane finite elements. In cases where the transverse bending of slabs is important, they will be represented by equivalent beams. For an accurate solution, a computer analysis using a general structural analysis program is usually accepted as the best method.

Certain reductions of a detailed model are possible while still producing an acceptably accurate solution. These reductions include halving the model to allow for symmetrical or antisymmetrical behavior, or representing the structures by a planar model and conducting a two-dimensional analysis, or lumping similar frames together in a nontwisting structure, or lumping vertically adjacent beams in a frame or connected wall structure.

The ability to model high-rise structures successfully for analysis requires an understanding of their behavior under load, while a good grasp of the techniques of modeling serves in return as an aid in generally assessing a tall building's behavior, as well as assisting in the selection and development of structural forms for tall buildings.

REFERENCES

5.1 ETABS, Three Dimensional Analysis of Building Systems. Computers and Structures Inc., Berkeley, California, 1989.

5.2 Stafford Smith, B. and Cruvellier, M. "Planar Modelling Techniques for Asymmetric Building Structures." *Proc. Inst. Civil Engineers* Part 2, **89**, March 1990, 1-14.

CHAPTER 6

Braced Frames

Bracing is a highly efficient and economical method of resisting horizontal forces in a frame structure. A braced bent consists of the usual columns and girders, whose primary purpose is to support the gravity loading, and diagonal bracing members that are connected so that the total set of members forms a vertical cantilever truss to resist the horizontal loading. The braces and girders act as the web members of the truss, while the columns act as the chords. Bracing is efficient because the diagonals work in axial stress and therefore call for minimum member sizes in providing stiffness and strength against horizontal shear.

Historically, bracing has been used to stabilize laterally the majority of the world's tallest building structures, from the earliest examples at the end of the nineteenth century to the present time. The Statue of Liberty, constructed in New York in 1883, was one of the first major braced structures. In the following three decades large numbers of braced steel-frame tall buildings were erected in Chicago and New York. The 57-story, 792-ft-high, braced steel Woolworth Tower, completed in 1913, established a height record, which it held until the 77-story, 1046-ft-high Chrysler Building and the 102-story, 1250-ft-high Empire State Building (Fig. 6.1) were completed in 1930 and 1931, respectively.

One- or two-story-height bracing, as used generally in the earlier high-rise steel structures, is an effective and still widely used arrangement. Recently, however, a much larger scale form of bracing, traversing many stories and bays, has also been used to considerable structural and architectural advantage in medium- and high-rise buildings, thereby extending significantly the repertoire of bracing concepts.

6.1 TYPES OF BRACING

Diagonal bracing is inherently obstructive to the architectural plan and can pose problems in the organization of internal space and traffic as well as in locating window and door openings. For this reason it is usually concentrated in vertical panels or bents that are located to cause a minimum of obstruction while satisfying the structural requirements of resisting the shear and torque on the building. In many locations the type of bracing has to be selected primarily on the basis of allowing the necessary openings through the bay, often at the expense of efficiency in resisting the lateral forces.

6.1 TYPES OF BRACING   107

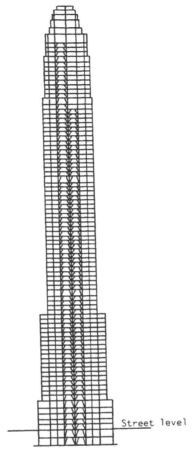

Fig. 6.1 Empire State Building: typical braced bent.

In low- or moderate-rise buildings that are not particularly slender, it is usually possible for the engineer to arrange the bracing in the structure without the architect having to consider it in planning the building. In a slender, moderate-rise building or a truly high-rise building, the location of the lateral load-resisting bents is more important and, indeed, might be all important to the viability of the structure. In such cases the architect and the structural engineer should liaise in the early stages of design.

The most efficient, but also the most obstructive, types of bracing are those that form a fully triangulated vertical truss. These include the single-diagonal, double-diagonal, and K-braced types (Fig. 6.2a–e). The full-diagonal types of braced bent are usually located where passage is not required, such as beside and between elevator, service, and stair shafts, which are unlikely to be relocated in the lifetime of the building.

Other types of braced bent that allow window and door openings, but whose

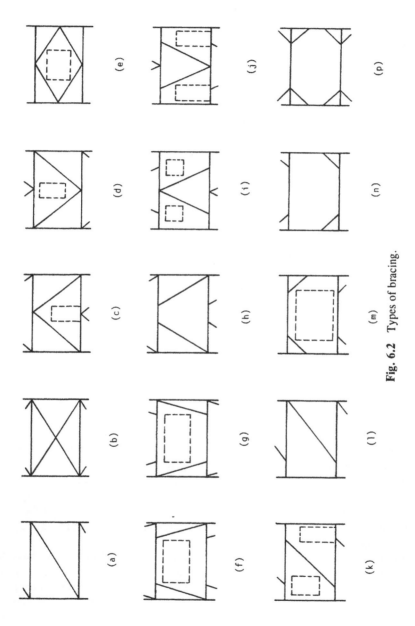

Fig. 6.2 Types of bracing.

arrangements cause bending in the girder, are shown in Fig. 6.2f-l. Some other types, which introduce bending in both the columns and the girders, are shown in Fig. 6.2m, n, and p. Generally, the types of braced bent that respond to lateral loading by bending of the girders, or of the girders and columns, are laterally less stiff and, therefore, less efficient, weight for weight, than the fully triangulated trusses, which respond with axial member forces only.

6.2 BEHAVIOR OF BRACING

Because lateral loading on a building is reversible, braces will be subjected in turn to both tension and compression; consequently, they are usually designed for the more stringent case of compression. For this reason, bracing systems with shorter braces, for example the K-types, may be preferred to the full-diagonal types. As an exception to designing braces for compression, the braces in the double-diagonal system are sometimes assumed to buckle in compression, and each diagonal is designed to carry in tension the full shear in the panel.

A significant advantage of the fully triangulated bracing types (Fig. 6.2a-e) is that the girder moments and shears are independent of the lateral loading on the structure. Consequently, the floor framing, which, in this case, is designed for gravity loading only, can be repetitive throughout the height of the structure with obvious economy in the design and construction.

In bracing systems in which the diagonals connect to the girder at a significant distance from the girder ends, for example, those in Fig. 6.2c, d, e, and h, the girder can be designed more economically as continuous over the connection, thus helping to offset the cost of the bracing. A further advantage of this type of bracing system is that the braces, in having one or both ends connected to the beam, which is relatively flexible vertically, do not attract a significant load as the columns shorten under gravity loading.

Eccentric bracing systems (i.e., systems in which the braces are not concentric with the main joints) may be used to design a ductile structure for an earthquake-resistant steel-framed building. The bracing acts in its usual elastic manner when controlling drift against wind or minor earthquakes. In the event of an overload during a major earthquake, the short link in the beam between the brace connection and the column in Fig. 6.2f, g, k, and l, and the link in the beam between brace connections in Fig. 6.2h, serves as a "fuse" by deforming plastically in shear to give a ductile response of the structure. Such braced systems combine high elastic stiffness and a large inelastic energy dissipation capacity that can be sustained over many cycles.

The roles of the "web" members in resisting shear on a bent can be understood by following the path of the horizontal shear down the bent from story to story. Referring to Fig. 6.3 and considering four typical types of bracing subjected to the total external shear, that is, neglecting the lesser effects of the horizontal forces applied locally at the floor levels, the vertical transmission of horizontal shear can be traced. In Fig. 6.3a the diagonal in each story is in compression, causing the beams to be in axial tension; therefore, the shortening of the diagonals and exten-

110 BRACED FRAMES

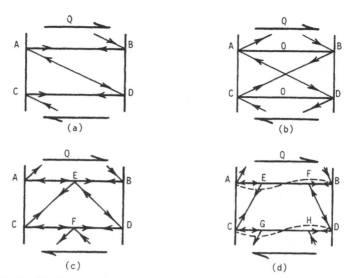

Fig. 6.3 Path of horizontal shear through web members. (a) Single-diagonal bracing; (b) double-diagonal bracing; (c) K-bracing; (d) story-height knee bracing.

sion of the beams give rise to the shear deformation of the bent. In Fig. 6.3b, the forces in the braces connecting to each beam end are in equilibrium horizontally, with the beam carrying an insignificant axial load. In Fig. 6.3c half of each beam is in compression and the other half in tension, whereas in Fig. 6.3d the end parts of the beam are in compression and tension with the whole beam subjected to double curvature bending. With a reverse in the direction of the horizontal load on the structure the actions and deformations in each member of the bracing will also be reversed.

The roles, if any, of the web members in picking up compressive force as the structure shortens vertically under gravity loading can be traced similarly. As the columns in Fig. 6.4a and b shorten, the diagonals are subjected to compression, which can be developed because of the tying action of the beams. In Fig. 6.4c the ends of the beams where diagonals are not connected are not stiffly restrained by the columns' bending rigidity; therefore the beams cannot provide the horizontal restraint that the diagonals need to develop a force. Consequently, the diagonals will not attract significant gravity load forces. Similarly, in Fig. 6.4d the vertical restraint from the flexural stiffness of the beam is not large; therefore, as in the previous case, the diagonals experience only negligible gravity load forces. If the type of bracing system allows the diagonals to attract compressive loading due to gravity loading on the structure, the diagonals should be either designed to carry the compressive forces or, to avoid backlash in the lateral load behavior of the structure due to the braces having buckled, they must be detailed short and prestressed in tension during erection.

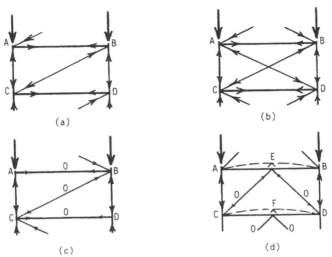

Fig. 6.4 Path of gravity loading down bent. (a) Single-diagonal, single-direction bracing; (b) double-diagonal bracing; (c) single-diagonal, alternate-direction bracing; (d) K-bracing.

6.3 BEHAVIOR OF BRACED BENTS

A braced bent behaves under horizontal loading as a vertical cantilever truss. The columns act as the chords in carrying the external load moment, with tension in the windward column and compression in the leeward column. The diagonals and girders serve as the web members in carrying the horizontal shear, with the diagonals in axial tension or compression depending on their direction of inclination. The girders act axially and, in some cases, in bending also.

The effect of the chords' axial deformations on the lateral deflection of the frame is to tend to cause a "flexural" configuration of the structure, that is, with concavity downwind and a maximum slope at the top (Fig. 6.5a). The effect of the web member deformations, however, is to tend to cause a "shear" configuration of the structure (i.e., with concavity upwind, a maximum slope at the base, and a zero slope at the top; Fig. 6.5b). The resulting deflected shape (Fig. 6.5c) is a combination of the effects of the flexural and shear curves with a resultant configuration depending on their relative magnitudes, as determined mainly by the type of bracing.

In bents that are braced in a single bay, horizontal loading causes a maximum tension at the base of the windward column of the braced bay. The more slender the bay, the larger the tensile force. Depending on the tributary area of slab supported by the column, the tension will be partly or wholly suppressed by the dead load of the structure. For height-to-width ratios of braced bays greater than about 10, however, the probability arises of uplift forces that are too large to handle. In

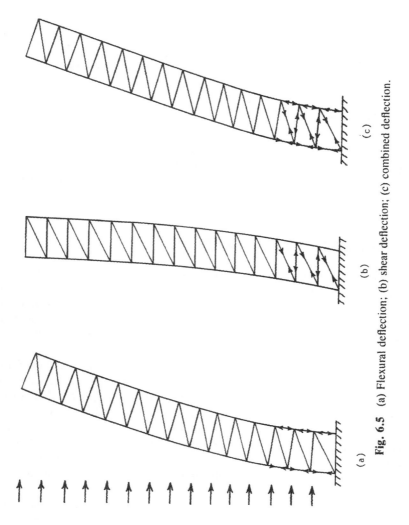

Fig. 6.5 (a) Flexural deflection; (b) shear deflection; (c) combined deflection.

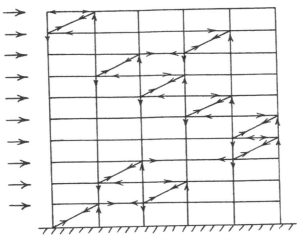

Fig. 6.6 Bracing in different bays of a bent.

multibay bents this problem can be avoided by placing successive story bracing in different bays of the bent, as in Fig. 6.6. In this arrangement the column axial forces caused by horizontal loading will be significantly smaller.

In providing for architectural requirements it is sometime necessary to use different types of bracing in different bays of the same bent, or in bays of different parallel bents. This does not present a particular problem, except that care should be taken to ensure that the lateral stiffnesses of the individual braced bays are comparable. Combinations of full-diagonal or K-type braced panels, both of which are usually very stiff in shear, with knuckle-type braced panels, which are usually much less stiff, may prove unsatisfactory, because the stiff panels will attract an unacceptably large proportion of the lateral load. In determining the individual panel stiffness, the total height behavior of the braced panel should be considered. This means that the lateral flexural flexibility due to axial deformations of the columns, as well as the lateral shear flexibility due to deformations of the braces and girders, should be taken into account.

In some situations, because of setbacks or transition levels, it is not possible to locate the braces in a single vertical plane throughout the entire height of the structure. In these cases the shear can be transferred from the braced bents above the setback or transition to those below by the horizontal-plane rigidity of the floor slab or by horizontal bracing in the plane of the floor.

6.4 METHODS OF ANALYSIS

6.4.1 Member Force Analysis

In the majority of modern design offices all but the simplest of braced high-rise structures are now analyzed by computer using a frame analysis program. To remind the reader of other possibilities, however, simple hand methods of analysis that may be used for statically determinate, or certain low-redundancy, braced structures will be reviewed.

An analysis of the forces in a statically determinate triangulated braced bent can be made using the method of sections. For example, in the single-diagonal braced panel of Fig. 6.7, subjected to an external shear Q_i in story i and external moments M_i and M_{i-1} at floor levels i and $i-1$, respectively, and assuming the frame to be pin-jointed so that the members carry only axial forces, the force in the brace can be found by considering the horizontal equilibrium of the free body above section XX, thus,

$$F_{BC} \cos \theta = Q_i \tag{6.1}$$

hence,

$$F_{BC} = \frac{Q_i}{\cos \theta} \tag{6.2}$$

The force F_{BD} in the column BD is found by considering moment equilibrium of the upper free body about C, thus

$$F_{BD} L = M_{i-1} \tag{6.3}$$

hence

$$F_{BD} = \frac{M_{i-1}}{L} \tag{6.4}$$

while the force F_{AC} in column AC is obtained similarly from the moment equilibrium of the upper free body about B, to give

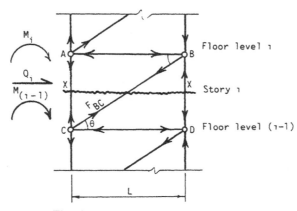

Fig. 6.7 Single diagonal braced panel.

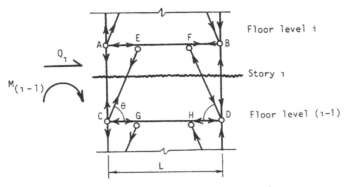

Fig. 6.8 Story-height knee-braced panel.

$$F_{AC} = \frac{M_i}{L} \tag{6.5}$$

This procedure can be repeated for the members in each story of the bent.

The member forces in more complex types of braced bents can also be obtained by taking horizontal sections. For example, in the story-height knee-braced bent of Fig. 6.8, it could be assumed that the shear in story i is shared equally between the braces. Then, from horizontal equilibrium of the upper free body,

$$F_{EC} = \frac{Q_i}{2\cos\theta} \tag{6.6}$$

and, from moment equilibrium of the upper free body about D,

$$(F_{AC} + F_{EC}\sin\theta)L = M_{i-1} \tag{6.7}$$

from which

$$F_{AC} = \frac{M_{i-1}}{L} - F_{EC}\sin\theta \tag{6.8}$$

As the frame responds to horizontal shear, the girder in this type of bracing system is subjected to bending throughout its length and to axial forces $Q_i/2$ in the lengths AE and FB. The bending is caused by the vertical components of the forces in the braces, while the axial forces are caused by the horizontal components.

6.4.2 Drift Analysis

In considering the deflected shape of a braced frame it is important to appreciate the relative influence of the flexural and shear mode contributions, due to the column axial deformations and to the diagonal and girder deformations, respectively. In typically proportioned low-rise braced structures, the shear mode displacements are the most significant and, incidentally, will largely determine the lateral stiffness of the structure. In medium- to high-rise structures, however, the higher axial forces and deformations in the columns, and the accumulation of their effects over a greater height, cause the flexural component of displacement to become dominant. In a panel with single diagonal bracing and a height-to-width ratio of 8, the total drift may be typically 60–70% attributable to the flexural component, with the remainder due to the shear component. In knee-braced bents, in which lateral loading subjects the girders—and in some arrangements the girders and columns—to bending, as well as the braces to axial deformation, the proportion of the total drift attributable to the shear component would be significantly greater.

The story drift, that is, the increment of lateral deflection in a story height, which is often the limiting drift criterion and which in a braced bent is a maximum at or close to the top of the structure, is more strongly influenced by the flexural component of deflection. This is because the inclination of the structure caused by the flexural component accumulates up the structure, while the story shear component diminishes toward the top. Consequently, in a single-diagonal braced frame, such as the one previously cited, the flexural component may contribute as much as 95% of the top-story drift.

One virtue of a hand analysis for drift is that it easily allows the drift contributions of the individual frame members to be seen, thereby providing guidance as to which members should be increased in size to most effectively reduce an excessive total drift or story drift.

Virtual Work Drift Analysis. In this method a force analysis of the structure subjected to the design horizontal loading is first made in order to determine the axial force P_j in each member j, as well as the bending moment M_{xj} at sections X along those members subjected to bending (Fig. 6.9a). A second force analysis is then made with the structure subjected to only a unit imaginary or "dummy" horizontal load at the level N whose drift is required (Fig. 6.9b) to give the axial force $\bar{p}_{jN}$, and moment $\bar{m}_{xjN}$ at section X in the bending members. The resulting horizontal deflection at N is then given by

$$\Delta_N = \Sigma \bar{p}_{jN} \left(\frac{PL}{EA}\right)_j + \Sigma \int_0^{L_j} \bar{m}_{xjN} \left(\frac{M_x}{EI}\right)_j dx \qquad (6.9)$$

in which L_j, A_j, and I_j are the length, sectional area, and moment of inertia for each member j, and E is the elastic modulus. The first summation in (6.9) refers to all members subjected to axial loading, while the second refers to only those members subjected to bending.

If the drift is required at another level, n, of the structure, another dummy unit load analysis will have to be made, but with the unit load applied only at level n. The resulting values $\bar{p}_{jn}$ and $\bar{m}_{xjn}$ will be substituted in Eq. (6.9) to give the drift.

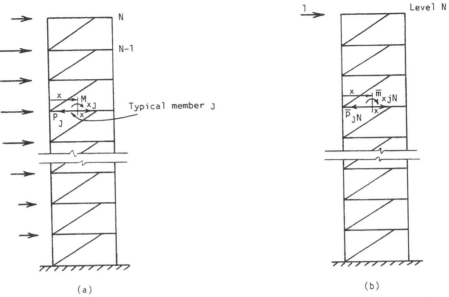

Fig. 6.9 (a) Member forces due to design horizontal loading; (b) member forces due to unit dummy loading.

The virtual work method is exact and can easily be systematized by tabulation. An adequate assessment of the deflected configuration, the total drift, and the story drifts can be obtained by plotting the deflection diagram from the deflections at just three or four equally spaced points up the height of the structure, requiring one design load force analysis plus three or four "dummy" unit load analyses.

Combined Moment-Area and Shear Formula Approximate Drift Analysis.

An approximate calculation of the drift can be made by using the moment-area method to obtain the flexural component (i.e., the component resulting from column axial deformations) and by applying a shear deflection formula to calculate the shear component. The method is appropriate for braced bents in which the flexural mode stiffness is entirely attributable to the axial areas of the columns; these include the majority of bracing types. It has the advantage that a detailed member force analysis of the frame is not necessary; only the external moment and the total shear force at each level are required.

Flexural Component. The procedure for obtaining the flexural component of drift is to first calculate for the structure (Fig. 6.10a) the external moment diagram (Fig. 6.10b). Then, to compute for the different vertical regions of the bent, the second moments of area I of the column sectional areas about their common centroid. For example, the value for the lower region of the braced bent in Fig. 6.10a is

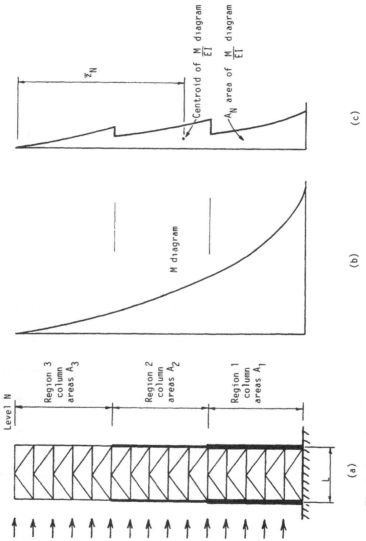

Fig. 6.10 (a) Braced frame: approximate deflection analysis; (b) external load moment diagram; (c) M/EI diagram.

$$I_1 = 2A_1\left(\frac{L}{2}\right)^2 = \frac{A_1 L^2}{2} \tag{6.10}$$

The moment diagram and the values of I are used to construct an M/EI diagram, as in Fig. 6.10c.

The story drift in story i, δ_{if}, due to the flexure of the structure, is then obtained from

$$\delta_{if} = h_i \theta_{if} \tag{6.11}$$

in which h_i is the height of story i, and θ_{if} is the inclination of story i, which is equal to the area under the M/EI curve between the base of the structure and the mid-height of story i.

The total drift at floor n, due to flexure, is then given by the sum of the story drifts from the first to the nth stories.

$$\Delta_{nf} = \sum_1^n \delta_{if} \tag{6.12}$$

Shear Component. The shear component of the story drift in story i, δ_{is}, is a function of the external shear and the properties of the braces and girder in that story. The shear component of the total drift at floor level n, Δ_{ns}, is equal to the sum of the story shear components of drift from the first to the nth stories, that is

$$\Delta_{ns} = \sum_1^n \delta_{is} \tag{6.13}$$

Formulas for the shear component of the story drift, δ_{is}, are given for various types of braced bent in Table 6.1.

Having obtained the flexural and shear components of drift, the total drift at level n is given by

$$\Delta_n = \Delta_{nf} + \Delta_{ns} \tag{6.14}$$

6.4.3 Worked Example for Calculating Drift by Approximate Methods

A 15-story single-diagonally braced frame (Fig. 6.11) consists of three 5-story regions. It is required to determine the drift at floors 5, 10, and 15 (i.e., where floor n is at the *top* of story n) for a uniform wind load of 10 kips per story. Assume the elastic modulus $E = 4.2 \times 10^6$ kip/ft^2.

The flexural and shear components of drift will be determined separately, as follows.

TABLE 6.1 Braced Bents: Shear Deflection per Story

TYPE OF BRACING	DIMENSIONS	SHEAR DEFLECTION PER STORY
SINGLE DIAGONAL		$\delta^S = \dfrac{Q}{E}\left(\dfrac{d^3}{L^2 A_d} + \dfrac{L}{A_g}\right)$
DOUBLE DIAGONAL		$\delta^S = \dfrac{Q}{2E}\left(\dfrac{d^3}{L^2 A_d}\right)$
K-BRACE		$\delta^S = \dfrac{Q}{E}\left(\dfrac{2d^3}{L^2 A_d} + \dfrac{L}{4A_g}\right)$
STORY HEIGHT KNEE-BRACE		$\delta^S = \dfrac{Q}{E}\left(\dfrac{d^3}{2m^2 A_d} + \dfrac{m}{2A_g} + \dfrac{h^2(L-2m)^2}{12 I_g L}\right)$
OFFSET DIAGONAL		$\delta^S = \dfrac{Q}{E}\left(\dfrac{d^3}{(L-2m)^2 A_d} + \dfrac{(L-2m)}{A_g} + \dfrac{h^2 m^2}{3 I_g L}\right)$

Q is the story shear
A_d is the sectional area of a diagonal
A_g and I_g are, respectively, the sectional area and inertia of the upper girder
E is the elastic modulus

Flexural Component. Using Table 6.2 to record the steps of the computation:

1. Compute the moment of inertia of the column sectional areas about their common centroid for each of the three height regions and record the values in column 3.

In the frame under consideration the column areas are equal, therefore their common centroid is mid-way between the columns

$$I = 2 \times A_c \left(\frac{L}{2}\right)^2 = \frac{A_c L^2}{2}$$

As an example, for the lowest region, stories 1–5, where $A_c = 35$ in.^{2}

$$I = \frac{A_c L^2}{2} = \frac{35 \times 20^2}{144 \times 2} = 48.6 \text{ ft}^4$$

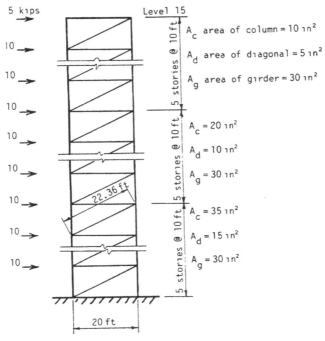

Fig. 6.11 Example frame.

2. Compute the value of the external moment M at each mid-story level and enter the values in column 4. For example, in story 12

$$M = 10(5 + 15 + 25) + 5 \times 35 = 625 \text{ kip ft}$$

3. Determine for each story the value of hM/EI, retaining E as a symbol, and enter the result in column 5. These are the changes in inclination in each story i due to flexure, $\delta\theta_{if}$.

For example, in story 5,

$$hM/EI = \delta\theta_{5f} = 10 \times 5525/48.6E = 1136.8/E$$

4. Determine for each story i the accumulation of $\delta\theta_{it}$, from story 1 up to and including story i, θ_{it} and record it in column 6.

For example, the accumulation of $\delta\theta_{if}$ up to story 5 is

$$\sum_{1}^{5} h \frac{M}{EI} = (2165.6 + 1877.6 + 1610.1 + 1363.2 + 1136.8)/E = 8153.3/E$$

Such accumulated values give the inclination of each story i due to flexure, θ_{if}.

TABLE 6.2 Evaluation of Flexural Components of Drift

Story	Story Height h_i (ft)	Frame Inertia I_i (ft^4)	External Moment M_i (k · ft)	$\delta\theta_i$ (rads/E)	Story Inclination θ_{it} (rads/E)	Story Drift δ_{it} (ft/E)	Drift $\Sigma\delta_{it}$ (ft/E)	Drift Δ_{nl} (ft)
15	10	13.9	25	18.0	14942.8	149428	159735	0.380
14	10	13.9	125	89.9	14924.8	149248		
13	10	13.9	325	233.8	14834.9	148349		
12	10	13.9	625	449.6	14601.1	146011		
11	10	13.9	1025	734.4	14151.5	141515		
10	10	27.8	1525	548.6	13414.1	134141	862791	0.205
9	10	27.8	2125	764.4	12865.5	128655		
8	10	27.8	2825	1016.1	12101.1	121011		
7	10	27.8	3625	1304.0	11085.5	110855		
6	10	27.8	4525	1627.7	9781.0	97810		
5	10	48.6	5525	1136.8	8153.3	81533	270319	0.064
4	10	48.6	6625	1363.2	7016.5	70165		
3	10	48.6	7825	1610.1	5653.3	56533		
2	10	48.6	9125	1877.6	4043.2	40432		
1	10	48.6	10525	2165.6	2165.6	21656		

5. Record the product of h_i and θ_{if} in column 7. $h_i\theta_{if}$ is the drift in story i, δ_{if}, due to flexure.

For example, the drift in story 5 due to flexure is

$$\delta_{5f} = 10 \times 8153.3/E = 81533/E \text{ ft}$$

6. At each level where the value of the lateral drift is required, evaluate the accumulation of the story drifts, δ_{if}, from story 1 up to the considered nth floor, to give the drift Δ_{nf} due to flexure. Enter these in column 8.

For example, at floor 5:

$$\Delta_{5f} = (21656 + 40432 + 56533 + 70165 + 81533)/E = 270319/4.2$$
$$\times 10^6 = 0.064 \text{ ft}$$

Shear Components. Using Table 6.3 to record the steps of the computation:

1. Compute the value of the external shear Q_i acting in each story i and enter in column 2.

2. Compute for each story i the story drift due to shear, δ_{is}, by substituting the value of the story shear and member properties into the appropriate formula from Table 6.1. Record the resulting values of δ_{is} in column 3.

TABLE 6.3 Evaluation of Shear Components of Drift

Story	Shear Q (kips)	Story Drift δ_{is} (ft)	Drift Δ_{ns} (ft)
15	5	0.0011	0.126
14	15	0.0032	
13	25	0.0054	
12	35	0.0075	
11	45	0.0097	
10	55	0.0065	0.099
9	65	0.0077	
8	75	0.0089	
7	85	0.0101	
6	95	0.0113	
5	105	0.0091	0.054
4	115	0.0100	
3	125	0.0109	
2	135	0.0117	
1	145	0.0125	

For example, the shear deflection formula for the single-diagonally braced example frame is

$$\delta_{is} = \frac{Q_i}{E}\left[\frac{d^3}{L^2 A_d} + \frac{L}{A_g}\right]_i$$

and using this to compute the drift in story 8 due to shear

$$\delta_{8s} = \frac{75}{4.2 \times 10^6}\left(\frac{22.36^3 \times 144}{20^2 \times 10} + \frac{20 \times 144}{30}\right)$$

$$= 0.0089 \text{ ft}$$

3. Sum the story drifts due to shear up to and including stories 5, 10, and 15 to obtain the total shear drift at floor levels 5, 10, and 15, and record the values in column 4.

For example, the drift due to shear at floor 5

$$\Delta_{5s} = 0.0125 + 0.0117 + 0.0109 + 0.0100 + 0.0091 = 0.054 \text{ ft}$$

Total Drift. The total drift at any floor level is the sum of the flexural and shear drifts at that level; for example, the total drift at the top of the 15-story frame in question is

$$\Delta_{15} = \Delta_{15f} + \Delta_{15s} = 0.380 + 0.126 = 0.506 \text{ ft}$$

A computer stiffness matrix analysis of the same structure gave the result $\Delta_{15} = 0.477$ ft; hence, in this case the approximate hand method was $+6.1\%$ in error. This error was probably due to the assumption implicit in the method of calculating the flexural component of drift, that the axial forces in the two columns of any particular story of a single-bay frame are equal in value. In the single-diagonally braced frame considered, the column axial forces in each story do not have exactly the same value. One is always smaller than assumed in the calculation, because of the vertical component of the force in the bracing member; hence the deflection calculated by the approximate method is larger.

Figure 6.12a shows the relative contributions of the columns', diagonals', and girders' deformations to the drift of the example structure. It is evident that although the diagonals have the largest influence on the drift in the lowest region, the column axial deformations tend to dominate the drift further up the structure, thus causing an overall flexural mode of behavior of the structure.

Figure 6.12b shows the relative contributions of the columns, diagonals, and girders to the story drifts. In the upper part of the structure, the axial deformations of the columns dominate the story drifts even more than they do the total drift.

6.5 USE OF LARGE-SCALE BRACING

Traditionally, and indeed currently, the typical arrangement of bracing in tall building structures is in story-height, bay-width modules. In this form it is usually possible to conceal the bracing within the walls or facade of a building to leave little evidence of its being a braced structure.

Over the last two decades the high efficiency of bracing in resisting lateral loading has been further exploited by using it on a larger modular scale, both within the building and externally across the faces. In the latter form the massive diagonals have sometimes been emphasized as an architectural feature of the facade.

A simple and elegant example of the use of massive K-braced trusses in resisting wind loading is in the 35-story Mercantile Tower in St. Louis, Missouri (Fig. 6.13). Four vertical trusses, each consisting of three-story height K-braced panels, are aligned diagonally in plan across the cut-off corners of the building. Each pair of vertical trusses at the ends of the building is joined by a rigid frame. The trusses are also connected to a single-bay rigid frame on each of the wide faces to form a stiff vertical U-section assembly at each end of the building. These provide resistance to wind in both the transverse and longitudinal directions of the building.

The 27-story Alcan Building in San Francisco (Fig. 6.14) uses six-story height panels of double-diagonal bracing between the main full-height columns on each of the building's four faces. At each mid-panel crossover point the braces connect to intermediate columns that rise from the first floor, transition girder level. In this arrangement the braces serve several roles:

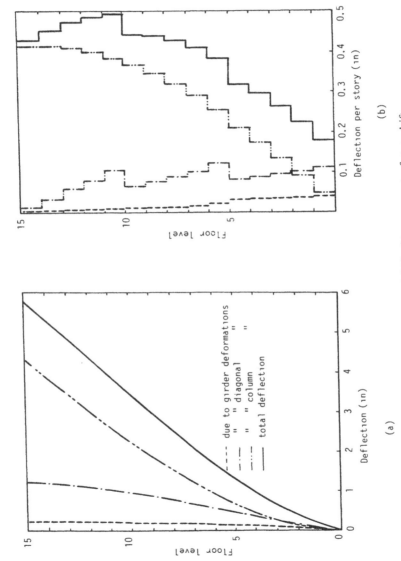

Fig. 6.12 (a) Components of total drift. (b) components of story drift.

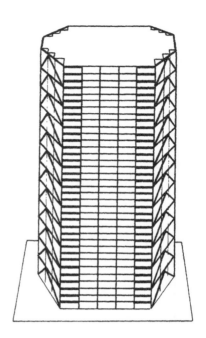

Fig. 6.13 Mercantile Tower, St. Louis, Missouri.

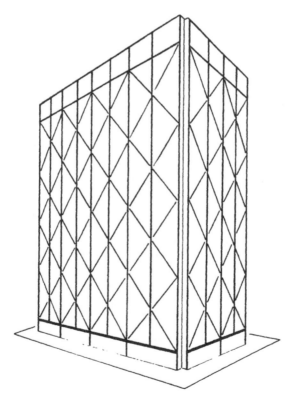

Fig. 6.14 Alcan Building, San Francisco.

1. to carry the lateral shear on the building;
2. to mobilize the intermediate columns axially so that they participate with the main columns in resisting the lateral load moment;
3. to shift gravity loading from the intermediate columns to the main columns and thus reduce the load on the transfer girder.

The 100-story John Hancock Building in Chicago is a braced tube (Fig. 4.14). In this hybrid form of structure the four rigid frame faces of the building are stiffened by overall diagonal bracing. The rigid frames form a vertical tube-type cantilever in which the frames parallel to the wind act as the webs of the cantilever, while the frames normal to the wind act as the flanges. The role of the bracing is again multi-purpose in:

1. resisting the horizontal shear;
2. reducing the shear lag in the flange column axial forces and hence making the whole cross-section of the building structure stiffer against horizontal load bending;
3. helping to equalize the gravity load stresses in the columns.

An important consequence of the reduced shear lag in the flange frames of the braced-tube structure is that the demand on the rigid-frame action is reduced so much that the columns can be spaced further apart, and the spandrel beams can be shallower than in unbraced tube structures, thereby allowing larger window openings.

The 914-ft-tall Citicorp Building in New York City has a frame structure (Fig. 6.15), which, although completely concealed by cladding, depends heavily on diagonal members. The square plan tower is supported by a full-height central core and four nine-story braced legs that are located under the middle of the tower faces. Each braced leg supports a two-story transfer truss from the top of which a "major" mast column extends in line with the leg to the top of the tower. "Minor" columns are located at the corners and quarter points of the tower faces. The column system is K-braced by eight-story-high major diagonals that form chevron-like eight-story tiers supported by the mast columns. Gravity loads are shared between the core and the outer frames. In the frames the load is transferred from the minor columns to the mast column by the diagonals at eight-story intervals. Wind shear is collected by the core over eight-story-height regions and transferred to the braced outer frames at the base of each tier. At the base of the tower the entire shear is transferred back to the core and hence to the ground. Wind moment is carried mainly by the mast columns and legs in the faces normal to the wind, and partly by the core. The unique structure of the Citicorp Building was developed to satisfy a requirement for the building to overhang an existing church on the site. Since the diagonals carry a significant part of the gravity loading, the structure may be classified as either a space truss or a braced frame.

128 BRACED FRAMES

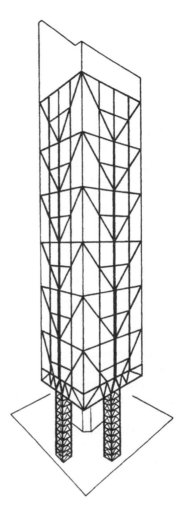

Fig. 6.15 Citicorp Building, New York City.

SUMMARY

A braced frame is an efficient structural form for resisting horizontal loading. It acts as a vertical truss, with the columns as chords and the braces and girders as web members. The most efficient type of bracing, using full diagonals, is also the most obstructive to door and window openings. Other arrangements are available that are more amenable to allowing openings but that, weight for weight, are less stiff horizontally. The bracing arrangement is usually dictated by the requirements for openings.

An advantage of some types of braced framing is that horizontal loading does not contribute significantly to the girder forces; consequently, the girders can be uniform over the height of the structure with economy in design and construction.

Some forms, in which the braces connect part way along the girder, allow the girder to be designed for gravity loading as continuous over the brace connections, again with resulting economy.

Braced bents deflect with a combination of flexural and shear components: the flexural component results from the column axial deformations, and the shear component from the brace and girder deformations. Low-rise structures deflect in a predominantly shear mode while high-rise braced bents deflect in a predominantly flexural mode.

Braced-frame member forces may usually be analyzed by the method of joints or by the method of sections. To allow a statically determinate analysis, it is usually assumed either that the shear is shared equally between the tension and compression braces, or that the compression brace has buckled and the tension brace carries all the shear.

Deflections may be analyzed by hand, either exactly, using the virtual work method, or approximately, using a combination of the moment area method and a shear deflection formula. An advantage of the virtual work method is that it indicates which members contribute most significantly to the deflection, therefore providing guidance as to which members should be adjusted to control the deflection.

Although bracing has been used typically in story-height bay-width modules, a recent development for very tall buildings has been to incorporate it in larger scale, multistory multibay arrangements. The effect of these has been to cause a more integral behavior of the column–girder system in resisting both gravity and horizontal loading, creating highly efficient structural forms for very tall buildings. In some notable examples, the large scale bracing has been exposed on the buildings' faces to give a characteristic architectural effect.

REFERENCES

6.1 Rathbun, J.C. "Wind Forces on a Tall Building." *Proc. ASCE* Paper 2056, September 1938, 1–41.

6.2 Morris, Clyde T. "Practical Design of Wind Bracing." *Proc. Am. Inst.* Steel Const., October 1927.

6.3 *Wind Bracing in Steel Buildings*. Second Progress Report of Sub-Committee No. 31, Committee on Steel of the Structural Division, Proc. A.S.C.E., February 1932, pp. 214–230.

6.4 Hart, F., Henn, W., and Sontag, H. *Multi-Story Buildings in Steel*. Crosby, Lockwood, Staples, 1978.

CHAPTER 7

Rigid-Frame Structures

A rigid-frame high-rise structure typically comprises parallel or orthogonally arranged bents consisting of columns and girders with moment resistant joints. Resistance to horizontal loading is provided by the bending resistance of the columns, girders, and joints. The continuity of the frame also contributes to resisting gravity loading, by reducing the moments in the girders.

The advantages of a rigid frame are the simplicity and convenience of its rectangular form. Its unobstructed arrangement, clear of bracing members and structural walls, allows freedom internally for the layout and externally for the fenestration. Rigid frames are considered economical for buildings of up to about 25 stories, above which their drift resistance is costly to control. If, however, a rigid frame is combined with shear walls or cores, the resulting structure is very much stiffer so that its height potential may extend up to 50 stories or more. A flat plate structure is very similar to a rigid frame, but with slabs replacing the girders. As with a rigid frame, horizontal and vertical loadings are resisted in a flat plate structure by the flexural continuity between the vertical and horizontal components.

As highly redundant structures, rigid frames are designed initially on the basis of approximate analyses, after which more rigorous analyses and checks can be made. The procedure may typically include the following stages:

1. Estimation of gravity load forces in girders and columns by approximate method.
2. Preliminary estimate of member sizes based on gravity load forces with arbitrary increase in sizes to allow for horizontal loading.
3. Approximate allocation of horizontal loading to bents and preliminary analysis of member forces in bents.
4. Check on drift and adjustment of member sizes if necessary.
5. Check on strength of members for worst combination of gravity and horizontal loading, and adjustment of member sizes if necessary.
6. Computer analysis of total structure for more accurate check on member strengths and drift, with further adjustment of sizes where required. This stage may include the second-order P-Delta effects of gravity loading on the member forces and drift.
7. Detailed design of members and connections.

This chapter considers methods of analysis for the deflections and forces for both gravity and horizontal loading. The methods are included in roughly the order of the design procedure, with approximate methods initially and computer techniques later. Stability analyses of rigid frames are discussed in Chapter 16.

7.1 RIGID FRAME BEHAVIOR

The horizontal stiffness of a rigid frame is governed mainly by the bending resistance of the girders, the columns, and their connections, and, in a tall frame, by the axial rigidity of the columns. The accumulated horizontal shear above any story of a rigid frame is resisted by shear in the columns of that story (Fig. 7.1). The shear causes the story-height columns to bend in double curvature with points of contraflexure at approximately mid-story-height levels. The moments applied to a joint from the columns above and below are resisted by the attached girders, which also bend in double curvature, with points of contraflexure at approximately midspan. These deformations of the columns and girders allow racking of the frame and horizontal deflection in each story. The overall deflected shape of a rigid frame structure due to racking has a shear configuration with concavity upwind, a maximum inclination near the base, and a minimum inclination at the top, as shown in Fig. 7.1.

The overall moment of the external horizontal load is resisted in each story level by the couple resulting from the axial tensile and compressive forces in the columns on opposite sides of the structure (Fig. 7.2). The extension and shortening of the columns cause overall bending and associated horizontal displacements of the structure. Because of the cumulative rotation up the height, the story drift due to overall bending increases with height, while that due to racking tends to decrease. Consequently the contribution to story drift from overall bending may, in the uppermost stories, exceed that from racking. The contribution of overall bending to the total drift, however, will usually not exceed 10% of that of racking,

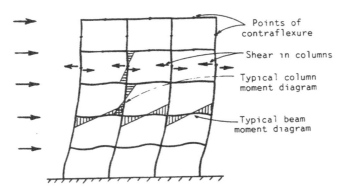

Fig. 7.1 Forces and deformations caused by external shear.

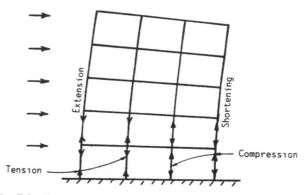

Fig. 7.2 Forces and deformations caused by external moment.

except in very tall, slender, rigid frames. Therefore the overall deflected shape of a high-rise rigid frame usually has a shear configuration.

The response of a rigid frame to gravity loading differs from a simply connected frame in the continuous behavior of the girders. Negative moments are induced adjacent to the columns, and positive moments of usually lesser magnitude occur in the mid-span regions. The continuity also causes the maximum girder moments to be sensitive to the pattern of live loading. This must be considered when estimating the worst moment conditions. For example, the gravity load maximum hogging moment adjacent to an edge column occurs when live load acts only on the edge span and alternate other spans, as for A in Fig. 7.3a. The maximum hogging moments adjacent to an interior column are caused, however, when live load acts only on the spans adjacent to the column, as for B in Fig. 7.3b. The maximum mid-span sagging moment occurs when live load acts on the span under consideration, and alternate other spans, as for spans AB and CD in Fig. 7.3a.

The dependence of a rigid frame on the moment capacity of the columns for resisting horizontal loading usually causes the columns of a rigid frame to be larger than those of the corresponding fully braced simply connected frame. On the other hand, while girders in braced frames are designed for their mid-span sagging mo-

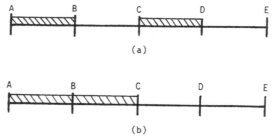

Fig. 7.3 (a) Live load pattern for maximum positive moment in AB and CD, and maximum negative moment at A; (b) live load pattern for maximum negative moment at B.

ment, girders in rigid frames are designed for the end-of-span resultant hogging moments, which may be of lesser value. Consequently, girders in a rigid frame may be smaller than in the corresponding braced frame. Such reductions in size allow economy through the lower cost of the girders and possible reductions in story heights. These benefits may be offset, however, by the higher cost of the more complex rigid connections.

7.2 APPROXIMATE DETERMINATION OF MEMBER FORCES CAUSED BY GRAVITY LOADING

A rigid frame is a highly redundant structure; consequently, an accurate analysis can be made only after the member sizes are assigned. Initially, therefore, member sizes are decided on the basis of approximate forces estimated either by conservative formulas or by simplified methods of analysis that are independent of member properties. Two approaches for estimating girder forces due to gravity loading are given here.

7.2.1 Girder Forces—Code Recommended Values

In rigid frames with two or more spans in which the longer of any two adjacent spans does not exceed the shorter by more than 20%, and where the uniformly distributed design live load does not exceed three times the dead load, the girder moment and shears may be estimated from Table 7.1. This summarizes the recommendations given in the *Uniform Building Code* [7.1]. In other cases a conventional moment distribution or two-cycle moment distribution analysis should be made for a line of girders at a floor level.

7.2.2 Two-Cycle Moment Distribution [7.2]

This is a concise form of moment distribution for estimating girder moments in a continuous multibay span. It is more accurate than the formulas in Table 7.1, especially for cases of unequal spans and unequal loading in different spans.

The following is assumed for the analysis:

1. A counterclockwise restraining moment on the end of a girder is positive and a clockwise moment is negative.
2. The ends of the columns at the floors above and below the considered girder are fixed.
3. In the absence of known member sizes, distribution factors at each joint are taken equal to $1/n$, where n is the number of members framing into the joint in the plane of the frame.

Two-Cycle Moment Distribution—Worked Example. The method is demonstrated by a worked example. In Fig. 7.4, a four-span girder AE from a rigid-

TABLE 7.1 Gravity Load Forces in Girders

	Location on Girder	Value of Moment[a]
Sagging moment	End spans: discontinuous end unrestrained	$\dfrac{wL^2}{11}$
	End spans: discontinuous end integral with support	$\dfrac{wL^2}{14}$
	Interior spans	$\dfrac{wL^2}{16}$
Hogging moment	At exterior face of first interior support: for two spans	$\dfrac{wL^2}{9}$
	At exterior face of first interior support: for more than two spans	$\dfrac{wL^2}{10}$
	At other faces of interior supports	$\dfrac{wL^2}{11}$
	At face of all supports where, at each end of each span Σ column stiffnesses/beam stiffness > 8	$\dfrac{wL^2}{12}$
	At interior face of exterior support for member built integrally with spandrel beam or girder	$\dfrac{wL^2}{24}$
	At interior face of exterior support for member built integrally with column	$\dfrac{wL^2}{16}$
Shear	In end members at face of first interior support	$1.15 \dfrac{wL}{2}$
	At face of all other supports	$\dfrac{wL}{2}$

[a] w is load per unit length of distributed load, L is the clear span for sagging moment or shear, and the average of adjacent clear spans for hogging moment.

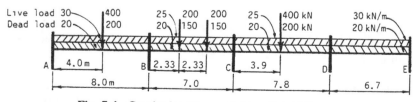

Fig. 7.4 Gravity loading on girders of rigid frame.

7.2 DETERMINATION OF MEMBER FORCES CAUSED BY GRAVITY LOADING

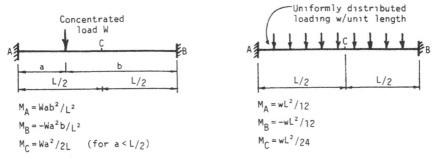

Fig. 7.5 Formulas for fixed-end moments.

frame bent is shown with its loading. The fixed-end moments in each span are calculated for dead loading and total loading using the formulas given in Fig. 7.5. The moments are summarized in Table 7.2.

The purpose of the moment distribution is to estimate for each support the maximum girder moments that can occur as a result of dead loading and pattern live loading. A different load combination must be considered for the maximum moment at each support, and a distribution made for each combination.

The five distributions are presented separately in Table 7.3, and in a combined form in Table 7.4. Distributions a in Table 7.3 are for the exterior supports A and E. For the maximum hogging moment at A, total loading is applied to span AB with dead loading only on BC. The fixed-end moments are written in rows 1 and 2. In this distribution only the resulting moment at A is of interest. For the first cycle, joint B is balanced with a correcting moment of $-(-867 + 315)/4 = -U/4$ assigned to M_{BA} where U is the unbalanced mo-

TABLE 7.2 Fixed-End Moments for Two-Cycle Moment Distribution Worked Example

Span	Loading	Dead Load Moment (kNm)	Dead + Live Load Moment (kNm)
AB	Concentrated	200	600
	Uniform distribution	107	267
	Total	307	867
BC	Concentrated	233	544
	Uniform distribution	82	184
	Total	315	728
CD	Concentrated	195	585
	Uniform distribution	101	228
	Total	296	813
DE	Concentrated	0	0
	Uniform distribution	75	187
	Total	75	187

TABLE 7.3 Two-Cycle Moment Distribution

a. Maximum Moments at A and E

	A	B			D		E
Distribution Factors	M_{AB} 1/3	M_{BA} 1/4	M_{BC} 1/4		M_{DC} 1/4	M_{DE} 1/4	M_{ED} 1/3
1. D.L. FEM[a]			315		−296		
2. T.L. FEM[a]	867	−867				187	−187
3. Carryover	69						14
4. Addition	936						−173
5. Distribution	−312						58
6. Maximum moments	624						−115

b. Maximum Moment at B

	A	B		C	
Distribution Factors	M_{AB} 1/3	M_{BA} 1/4	M_{BC} 1/4	M_{CB} 1/4	M_{CD} 1/4
1. D.L. FEM					296
2. T.L. FEM	867	−867	728	−728	
3. Carryover		−145	54		
4. Addition		−1012	782		
5. Distribution		58	58		
6. Maximum moments		−954	840		

c. Maximum Moment at C

	B		C		D	
Distribution Factors	M_{BA} 1/4	M_{BC} 1/4	M_{CB} 1/4	M_{CD} 1/4	M_{DC} 1/4	M_{DE} 1/4
1. D.L. FEM	−307					75
2. T.L. FEM		728	−728	813	−813	
3. Carryover			−53	92		
4. Addition			−781	905		
5. Distribution			−31	−31		
6. Maximum moments			−812	874		

ment. This is not recorded, but half of it, $(-U/4)/2$, is carried over to M_{AB}. This is recorded in row 3 and then added to the fixed-end moment and the result recorded in row 4.

The second cycle involves the release and balance of joint A. The unbalanced moment of 936 is balanced by adding $-U/3 = -936/3 = -312$ to M_{AB}

7.2 DETERMINATION OF MEMBER FORCES CAUSED BY GRAVITY LOADING

TABLE 7.3 (*Continued*)

d. Maximum Moment at D

	C		D		E
Distribution Factors	M_{CB} 1/4	M_{CD} 1/4	M_{DC} 1/4	M_{DE} 1/4	M_{ED} 1/3
1. D.L. FEM		−315			
2. T.L. FEM		813	−813	187	−187
3. Carryover			−62	31	
4. Addition			−875	218	
5. Distribution			164	164	
6. Maximum moments			−711	382	

D.L., dead loading; T.L., total loading; FEM, fixed-end moments.

(row 5), implicitly adding the same moment to the two column ends at A. This completes the second cycle of the distribution. The resulting maximum moment at A is then given by the addition of rows 4 and 5, 936 − 312 = 624. The distribution for the maximum moment at E follows a similar procedure.

Distribution *b* in Table 7.3 is for the maximum moment at B. The most severe loading pattern for this is with total loading on spans AB and BC and dead load only on CD. The operations are similar to those in Distribution *a*, except that the first cycle involves balancing the two adjacent joints A and C while recording only their carryover moments to B. In the second cycle, B is balanced by adding $-(-1012 + 782)/4 = 58$ to each side of B. The addition of rows 4 and 5 then gives the maximum hogging moments at B. Distributions *c* and *d*, for the moments at joints C and D, follow patterns similar to Distribution *b*.

The complete set of operations can be combined as in Table 7.4 by initially recording at each joint the fixed-end moments for both dead and total loading. Then the joint, or joints, adjacent to the one under consideration are balanced for

TABLE 7.4 Combined Two-Cycle Moment Distribution

	A	B		C		D		E
Distribution Factors	M_{AB} 1/3	M_{BA} 1/4	M_{BC} 1/4	M_{CB} 1/4	M_{CD} 1/4	M_{DC} 1/4	M_{DE} 1/4	M_{ED} 1/3
1. D.L. FEM[a]	307	−307	315	−315	296	−296	75	−75
2. T.L. FEM[a]	867	−867	728	−728	813	−813	187	−187
3. Carryover	69	−145	54	−53	92	−62	31	14
4. Addition	936	−1012	782	−781	905	−875	218	−173
5. Distribution	−312	58	58	−31	−31	164	164	58
6. Maximum moments	624	−954	840	−812	874	−711	382	−115

[a] For abbreviations, see the footnote to Table 7-3.

the appropriate combination of loading, and carryover moments assigned to the considered joint and recorded. The joint is then balanced to complete the distribution for that support.

Maximum Mid-Span Moments. The most severe loading condition for a maximum mid-span sagging moment is when the considered span and alternate other spans carry total loading. A concise method of obtaining these values may be included in the combined two-cycle distribution, as shown in Table 7.5. Adopting the convention that sagging moments at mid-span are positive, a mid-span total loading moment is calculated for the fixed-end condition of each span and entered in the mid-span column of row 2. These mid-span moments must now be corrected to allow for rotation of the joints. This is achieved by multiplying the carryover moment, row 3, at the left-hand end of the span by $(1 + 0.5 \text{ D.F.})/2$, and the carryover moment at the right-hand end by $-(1 + 0.5 \text{ D.F.})/2$, where D.F. is the appropriate distribution factor, and recording the results in the middle column. For example, the carryover to the mid-span of AB from $A = [(1 + 0.5/3)/2] \times 69 = 40$ and from $B = -[(1 + 0.5/4)/2] \times (-145) = 82$. These correction moments are then added to the fixed-end mid-span moment to give the maximum mid-span sagging moment, that is, $733 + 40 + 82 = 855$.

7.2.3 Column Forces

The gravity load axial force in a column is estimated from the accumulated tributary dead and live floor loading above that level, with reductions in live loading as permitted by the local Code of Practice. The gravity load maximum column moment is estimated by taking the maximum difference of the end moments in the connected girders and allocating it equally between the column ends just above and below the joint. To this should be added any unbalanced moment due to eccentricity of the girder connections from the centroid of the column, also allocated equally between the column ends above and below the joint.

7.3 APPROXIMATE ANALYSIS OF MEMBER FORCES CAUSED BY HORIZONTAL LOADING

7.3.1 Allocation of Loading between Bents

A first step in the approximate analysis of a rigid frame is to estimate the allocation of the external horizontal force to each bent. For this it is usual to assume that the floor slabs are rigid in plane and, therefore, constrain the horizontal displacements of all the vertical bents at a floor level to be related by the horizontal translations and rotation of the floor slab.

Symmetric Plan Structures Subjected to Symmetric Loading. A symmetric structure subjected to symmetric loading (Fig. 7.6a) translates but does not twist. From the assumption of slab rigidity, the bents translate identically. The

TABLE 7.5 Two-Cycle Moment Distribution with Maximum Mid-Span Moments

	A	B		C		D		E
Distribution Factors	M_{AB} 1/3	M_{BA} 1/4	M_{BC} 1/4	M_{CB} 1/4	M_{CD} 1/4	M_{DC} 1/4	M_{DE} 1/4	M_{ED} 1/3
1. D.L. FEM[a]	307	−307	315	−315	296	−296	75	−75
2. T.L. FEM[a]	867	−867	728	−728	813	−813	187	−187
3. Carryover	69 → 40	−145	54	−53	92 → 52	−62	31 → 17	14
4. Addition	936	−1011	782	−781	905	−875	218	−173
	82							−8
5. Distribution	−312	58	58	−31	−31	164	164	62
6. Maximum moments	624	−954	840	−812	874	−711	382	−115
	855		424		786		103	

[a] For abbreviations, see the footnote to Table 7-3.

140 RIGID-FRAME STRUCTURES

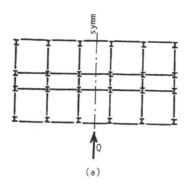

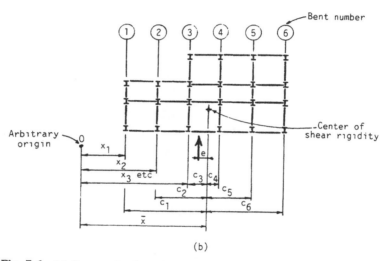

Fig. 7.6 (a) Symmetric-plan rigid frame; (b) asymmetric-plan rigid frame.

total external shear at a level will be distributed between the bents in proportion to their shear rigidities (*GA*) at that level. An explanation of the shear rigidity parameter (*GA*) is given in a later section but, for now, it may be obtained for level i in a bent simply by using

$$(GA) = \frac{12E}{h_i \left(\dfrac{1}{G} + \dfrac{1}{C} \right)_i} \tag{7.1}$$

in which h_i is the height of story i, $G = \Sigma(I_g/L)$ for all the girders of span L across floor i of the bent, and $C = \Sigma(I_c/h_i)$ for all the columns in story i of the bent. E is the modulus of elasticity, and I_c and I_g are the moments of inertia of the columns and girders, respectively.

Asymmetric Plan Structures. The effect of lateral loading on a structure having an asymmetric plan is to cause a horizontal plane torque in addition to transverse shear. Therefore, the structure will twist as well as translate.

Referring to the asymmetric structure in Fig. 7.6b, and defining the location of the center of shear rigidity of the set of parallel bents in story i, relative to an arbitrary origin 0, as given by

$$\bar{x}_i = \left[\frac{\Sigma (GA) x_j}{\Sigma (GA)} \right]_i \qquad (7.2)$$

An estimate of the shear Q_{ji} carried by bent j at level i is given by

$$Q_{ji} = \frac{Q_i (GA)_{ji}}{\Sigma (GA)_i} + \frac{Q_i e_i [(GA) c]_{ji}}{\Sigma [(GA) c^2]_i} \qquad (7.3)$$

in which for level i, Q_i is the total shear, $(GA)_{ji}$ is the shear rigidity of bent j in story i, e_i is the eccentricity of Q_i from the center of shear rigidity in story i, c_i is the distance of bent j from the center of shear rigidity, and the two summations refer to the full set of bents parallel to the direction of loading. The signs of c and e are the same when they are on the same side of the center of rigidity.

7.3.2 Member Force Analysis by Portal Method

The portal method [7.3] allows an approximate hand analysis for rigid frames without having to specify member sizes and, therefore, it is very useful for a preliminary analysis. The method is most appropriate to rigid frames that deflect predominantly by racking. It is suitable, therefore, for structures of moderate slenderness and height, and is commonly recommended as useful for structures of up to 25 stories in height with a height-to-width ratio not greater than 4:1 [7.4]. Its name is derived from the analogy between a set of single-bay portal frames and a single story of a multibay rigid frame (Fig. 7.7a and b). When each of the separate portals carries a share of the horizontal shear, tension occurs in the windward columns and compression in the leeward columns. If these are superposed to simulate the multibay frame, the axial forces of the interior columns are eliminated, leaving axial forces only in the extreme windward and leeward columns.

The reduction of the highly redundant multistory frame to allow a simple analysis is achieved by making the following assumptions:

1. Horizontal loading on the frame causes double curvature bending of all the columns and girders, with points of contraflexure at the mid-height of columns and mid-span of girders (Fig. 7.1).
2. The horizontal shear at mid-story levels is shared between the columns in proportion to the width of aisle each column supports.

142 RIGID-FRAME STRUCTURES

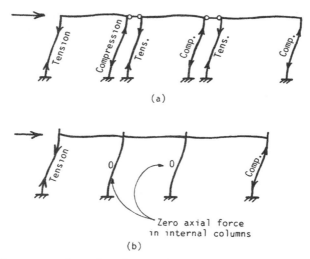

Fig. 7.7 (a) Separate portals analogy for portal method; (b) separate portals superposed.

The method may be used to analyze the whole frame, or just a portion of the frame at a selected level. The analysis of the whole frame considers in turn the equilibrium of separate frame modules, each module consisting of a joint with its column and beam segments extending to the nearest points of contraflexure. The sequence of analyzing the modules is from left to right, starting at the top and working down to the base.

The procedure for a whole frame analysis is as follows:

1. Draw a line diagram of the frame and indicate on it the horizontal shear at each mid-story level (Fig. 7.8).
2. In each story allocate the shear to the columns in proportion to the aisle widths they support, indicating the values on the diagram.
3. Starting with the top-left module (Fig. 7.9a), compute the maximum moment just below the joint from the product of the column shear and the half-story height.
4. Find the girder-end moment just to the right of the joint from the equilibrium of the column and girder moments at the joint. The moment at the other end of the girder is of the same magnitude but corresponds to the opposite curvature.
5. Evaluate the girder shear by dividing the girder end-moment by half the span.
6. Consider next the equilibrium of the second joint (Fig. 7.9b), repeating steps 3 to 5 to find the maximum moment in the second column, and the moment and shear in the second girder from the left.

This is repeated for each successive module working across to the right, and is then continued in the level below, starting again from the left. The values of shear

7.3 ANALYSIS OF MEMBER FORCES CAUSED BY HORIZONTAL LOADING

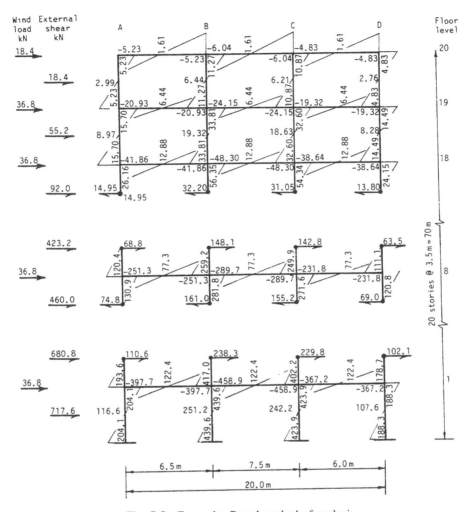

Fig. 7.8 Example: Portal method of analysis.

and moment are recorded and a bending moment diagram drawn on the diagram of the structure as the analysis progresses (Fig. 7.8). The bending moments are recorded on the girders above the left-hand end and below the right-hand end, and similarly on the columns as viewed from the right. The shears are written perpendicular to the columns and beams at the mid-heights and mid-spans, respectively. The bending moment diagram is drawn here on the tension side of the member.

If member forces are required only at a particular level in the structure, the horizontal row of modules at that level, consisting of the girders and half-columns above and below, can be analyzed separately by the above procedure without having to start the analysis at the top (Fig. 7.9c and d).

The consideration of vertical equilibrium of a joint module should give the increment of axial load picked up by a column at that level. However, the assumed

144 RIGID-FRAME STRUCTURES

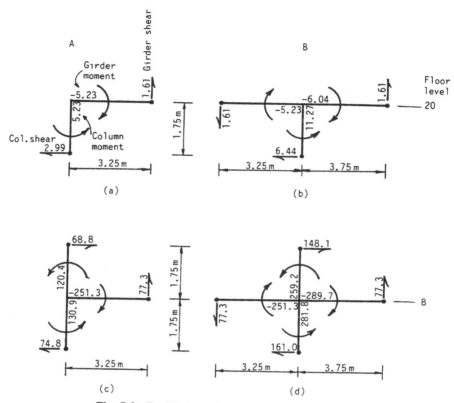

Fig. 7.9 Equilibrium of modules: portal method.

distribution of shear between the columns results in a zero increment for all except the two exterior columns. The axial force in the exterior columns in any story is equal, therefore, to the moment of the external loading about the mid-height level of that story, divided by the distance between the columns. The portal method tends to overestimate the axial force in the exterior columns and is incorrect in estimating zero axial force for the interior columns. However, when these forces are added to the gravity load axial forces, the effect of the discrepancies on the resultant axial force is generally negligible.

The simplicity of the portal method and the advantage that it allows a direct analysis of member forces at intermediate levels make it the most useful of the approximate methods for rigid-frame analysis. If, however, the frame is taller and more slender, so that overall bending of the structure by axial deformations of the columns becomes significant, it may be more appropriate to analyze it by the cantilever method.

Portal Method—Worked Example. It is required to determine the member forces in the 20-story frame of Fig. 7.8. The story height is typically 3.5 m, to

7.3 ANALYSIS OF MEMBER FORCES CAUSED BY HORIZONTAL LOADING

give a total height of 70 m. The bents are spaced at 7.0 m. The intensity of the wind loading is 1.5 kN/m² throughout the height.

Wind load per floor: At typical levels $1.5 \times 7.0 \times 3.5 = 36.8$ kN

At the roof level $1.5 \times 7.0 \times 1.75 = 18.4$ kN

Shear in the top story = 18.4 kN

Distributing this shear between the top-story columns in proportion to the widths of aisle supported:

For column A: $18.4 \times 3.25/20 = 2.99$ kN

For column B: $18.4(3.25 + 3.75)/20 = 6.44$ kN

The shear in columns C and D and in the columns of the stories below are allocated similarly. The values are recorded on Fig. 7.8.

Starting with the top-left module A20 (Fig. 7.9a) and considering its free-body equilibrium:

Moment at top of column = column shear × half-story height

$= 2.99 \times 1.75 = 5.23$ kNm

From moment equilibrium of the joint, the moment at left end of the first girder

$= -5.23$ kNm

Shear in girder = girder-end moment/half girder length

$= 5.23/3.25 = 1.61$ kN.

Because of the mid-length point of contraflexure, the moment at the right end of the girder has the same value as at the left end. Similarly, the column moments at the top and bottom of a story are equal. The sign convention for numerical values of the bending moment is that an anticlockwise moment applied by a joint to the end of a member is taken as positive.

The values of the moments and shears are recorded on Fig. 7.8. Continuing with the next module to the right, B20, in Fig. 7.9b:

Moment at top of column = column shear × half-story height

$= 6.44 \times 1.75 = 11.27$ kNm

From moment equilibrium of joint, moment at end of second girder

$= -(11.27 - 5.23) = -6.04$ kNm

Then shear in second girder = girder moment/half-girder length

$= 6.04/3.75 = 1.61$ kN

146   RIGID-FRAME STRUCTURES

The above procedure is repeated for successive modules to the right, and then continued on the floor below, starting again from left.

For the direct analysis of forces at an intermediate level, consider floor level 8 (Fig. 7.8).

Starting with the left module A8 (Fig. 7.9c):

$$\text{Moment in column above joint} = 68.8 \times 1.75 = 120.4 \text{ kNm}$$

$$\text{Moment in column below joint} = 74.8 \times 1.75 = 130.9 \text{ kNm}$$

From moment equilibrium of joint, moment at end of first girder

$$= -(120.4 + 130.9) = -251.3 \text{ kNm}$$

$$\text{Then shear in first girder} = 251.3/3.25 = 77.3 \text{ kN}$$

Continuing with the next-right module B8 (Fig. 7.9d):

$$\text{Moment in column above joint} = 148.1 \times 1.75 = 259.2 \text{ kNm}$$

$$\text{Moment in column below joint} = 161.0 \times 1.75 = 281.8 \text{ kNm}$$

From moment equilibrium of the joint, the moment at the end of the second girder

$$= -(259.2 + 281.8 - 251.3) = -289.7 \text{ kNm}$$

$$\text{Then the shear in the second girder} = 289.7/3.75 = 77.3 \text{ kN}$$

The above procedure is repeated for successive modules to the right, as in Fig. 7.8.

7.3.3 Approximate Analysis by Cantilever Method [7.5]

The cantilever method is based on the concept that a tall rigid frame subjected to horizontal loading deflects as a flexural cantilever (Fig. 7.2). The validity of this concept increases for taller, more slender frames, and for frames with higher girder stiffness. The method is recommended [7.4] as suitable for the analysis of structures of up to 35 stories high with height-to-width ratios of up to 5 : 1.

It is similar to the portal method in considering the equilibrium of joint modules in sequence. It differs, however, in starting by assuming values for the axial forces, rather than the shears, in the columns. It is less versatile than the portal method in not allowing a direct analysis of intermediate stories.

The assumptions for the cantilever method are as follows:

1. Horizontal loading on the frame causes double curvature bending of all the columns and girders with points of contraflexure at the mid-heights of columns and mid-spans of girders.

7.3 ANALYSIS OF MEMBER FORCES CAUSED BY HORIZONTAL LOADING

2. The axial stress in a column is proportional to its distance from the centroid of the column areas.

The procedure for the analysis is as follows:

1. Draw a line diagram of the frame and record on it the external moment M at each mid-story level (Fig. 7.10).
2. Find the centroid of the column areas and compute the second moment of the column areas about the centroid using

$$I = \sum A_j c_j^2 \qquad (7.4)$$

where c_j is the distance of column j from the centroid. In a case where the column areas A_j are not known, they are to be taken as unity. Calculate the column axial forces F_j in each story using

$$F_j = \frac{Mc_j}{I} A_j \qquad (7.5)$$

Record these on the diagram of the structure.

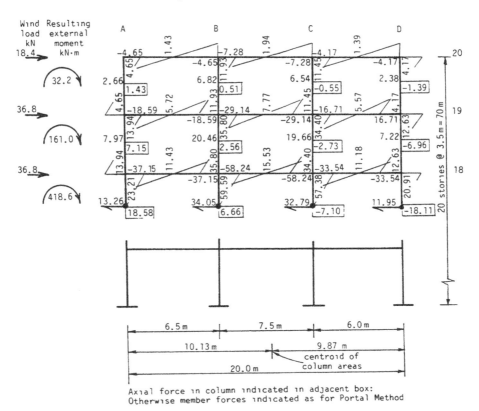

Fig. 7.10 Example: cantilever method of analysis.

148 RIGID-FRAME STRUCTURES

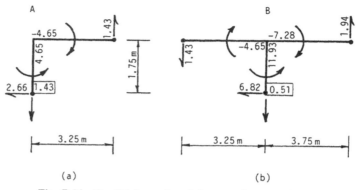

Fig. 7.11 Equilibrium of modules: cantilever method.

3. Starting with the top-left module (Fig. 7.11a) find the vertical shear in the girder from the vertical equilibrium of the module.
4. Compute the girder-end moments from the product of the girder shear and its half-span.
5. Compute the moment in the column just below the joint from the equilibrium of the girder and column moments at the joint.
6. Evaluate the column shear by dividing the column-top moment by half the story height.
7. Considering the next-right module (Fig. 7.11b) find the shear and moment in the second girder and column by repeating steps 3 to 6.

This is repeated for each module in turn, moving to the right across the top level, and then continuing from left to right in the level below. The values of shear and moment are recorded on the diagram of the structure (Fig. 7.10).

The convention for indicating forces in the members is the same as in Fig. 7.8, with the column axial forces written in boxes.

Cantilever Method—Worked Example. Analysis of the same 20-story, 70-m-high frame considered in the portal analysis. Referring to Fig. 7.10, external moments due to wind are

$$\text{At mid-height of story } 20 = 18.4 \times 1.75 = 32.2 \text{ kNm}$$
$$\text{At mid-height of story } 19 = 18.4 \times 5.25 + 36.8 \times 1.75$$
$$= 161.0 \text{ kNm}$$

Continue to calculate the external moment for each story down to the base and record the values on Fig. 7.10.

Assuming a unit sectional area for each column:
Location of centroid of areas = $1 \times (6.5 + 14.0 + 20)/4 = 10.13$ m from left

7.3 ANALYSIS OF MEMBER FORCES CAUSED BY HORIZONTAL LOADING

The second moment of area

$$= 1 \times (10.13^2 + 3.63^2 + 3.87^2 + 9.87^2)$$
$$= 228.2 \text{ m}^4$$

Column axial forces:

$$\text{Top story first column} = McA/I = 32.2 \times 10.13 \times 1/228.2$$
$$= 1.43 \text{ kN tension}$$
$$\text{second column} = McA/I = 32.2 \times 3.63 \times 1/228.2$$
$$= 0.51 \text{ kN tension}$$

Continue to find the axial forces in all the columns down to the base. The values are recorded (in boxes) on Fig. 7.10.

Starting with the top-left module, A20 (Fig. 7.11a):
From vertical equilibrium of module, shear in first girder

$$= 1.43 \text{ kN}$$

Moment at left end of girder = shear × half length of girder

$$= -(1.43 \times 3.25) = -4.65 \text{ kNm}$$

From moment equilibrium of joint, moment at top of column

$$= 4.65 \text{ kNm}$$

Shear in column = moment at top/half story height

$$= 4.65/1.75 = 2.66 \text{ kN}$$

The moments at opposite ends of the girders and columns are of the same value. The moments and shears and a bending moment sketch are recorded, as for the portal method.

Considering the next-right module, B20, Fig. 7.11b.
From vertical equilibrium of module, shear in second girder

$$= 1.43 + 0.51 = 1.94 \text{ kN}$$

Moment in left end of second girder = shear × half length of girder

$$= -(1.94 \times 3.75) = -7.28 \text{ kNm}$$

From moment equilibrium of joint, moment at top of column

$$= -(-4.65 - 7.28) = 11.93 \text{ kNm}$$

Shear in column = moment at column top/half story height

$$= 11.93/1.75 = 6.82 \text{ kN}$$

This procedure is repeated for successive modules to the right, then on the level below, working again from left to right.

7.3.4 Approximate Analysis of Rigid Frames with Setbacks

A rigid frame bent with setbacks, as shown in Fig. 7.12a, can be analyzed approximately by applying the cantilever method to the upper and lower parts as though they were two separate frames (Fig. 7.12b).

An analysis is made first of the upper part down to and including those parts of the setback girder that form the upper frame. A moment distribution is then carried out for the setback girder supported on the lower columns and subjected to the calculated vertical forces from the columns of the upper structure. Because the setback girder is so much stiffer in bending than the columns, it may be assumed for this part of the analysis that the girder rests on simple supports. The moment distribution yields the girder moments and shears and, hence, the vertical forces that the girder applies to the supporting columns; these forces are assumed to carry all the way to the foundation.

The lower structure, including the setback girder, may then be analyzed by the cantilever method applying, in addition to its story increments of wind load, a concentrated horizontal load at the setback level equal to the total horizontal force above that level. The column axial forces calculated from the moment of the external horizontal loading are added to those determined from the setback beam distribution to start the cantilever analysis for the lower part.

The total moments and shears in the setback girder due to wind forces are the superposed results of the three analyses: the cantilever analysis of the upper part, the cantilever analysis of the lower part, and the moment distribution of the girder.

If the complete setback structure has a low height-to-width ratio, it would be more appropriate to use the portal method of analysis. As described above, the two parts of the structure would be analyzed separately with the total shear from the upper structure applied as a concentrated load at the setback level for the analysis of the lower structure. The forces in the setback girder would be obtained by superposing the girder results from the two portal analyses, and the results of a moment distribution using vertical forces from the upper columns as for the cantilever method.

7.4 APPROXIMATE ANALYSIS FOR DRIFT

When the initial sizes of the frame members have been selected, an approximate check on the horizontal drift of the structure can be made. The drift in a nonslender rigid frame is mainly caused by racking (Fig. 7.1). The racking may be considered

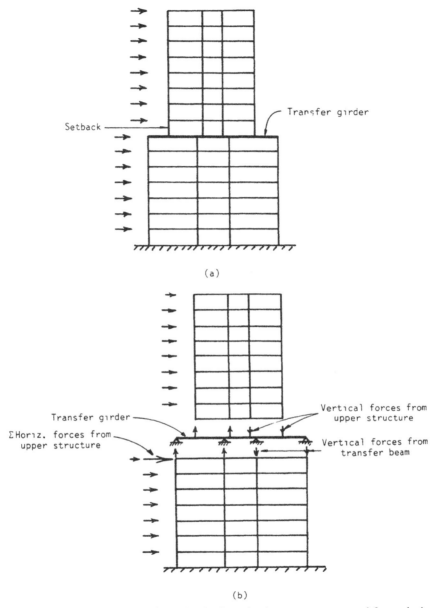

Fig. 7.12 (a) Rigid frame with setback; (b) setback structure separated for analysis.

as comprising two components: the first is due to rotation of the joints, as allowed by the double bending of the girders (Fig. 7.13a and b), while the second is caused by double bending of the columns (Fig. 7.13c). If a rigid frame is slender, a contribution to drift caused by the overall bending of the frame, resulting from axial deformations of the columns, may be significant (Fig. 7.2). If the frame has

152 RIGID-FRAME STRUCTURES

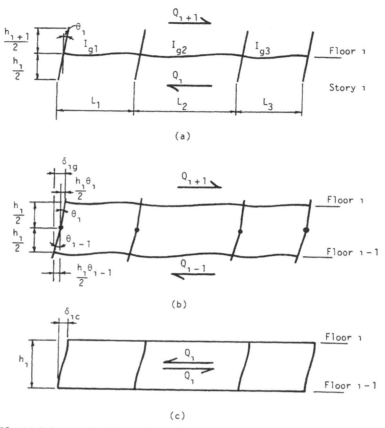

Fig. 7.13 (a) Joint rotation due to girder flexure; (b) story drift due to girder flexure; (c) story drift due to column flexure.

a height:width ratio less than 4:1, the contribution of overall bending to the total drift at the top of the structure is usually less than 10% of that due to racking.

The following method of calculation for drift allows the separate determination of the components attributable to beam bending, column bending, and overall cantilever action.

7.4.1 Components of Drift [7.6]

It is assumed for the drift analysis that points of contraflexure occur in the frame at the mid-story level of the columns and at the mid-span of the girders. This is a reasonable assumption for high-rise rigid frames for all stories except near the top and bottom.

Story Drift due to Girder Flexure. Consider a story-height segment of a frame at floor level i consisting of a line of girders and half-story-height columns above and below each joint (Fig. 7.13a). To isolate the effect of girder bending, assume the columns are flexurally rigid.

The average rotation of the joints can be expressed approximately as

$$\theta_{ig} \simeq \frac{\text{total moment carried by the joints}}{\text{total rotational stiffness of the joints}} \quad (7.6)$$

$$\text{The total moment} = \frac{Q_i h_i}{2} + Q_{i+1} \frac{h_{i+1}}{2} \quad (7.7)$$

and the total rotational stiffness

$$= 6E \left[\frac{I_{g1}}{L_1} + \left(\frac{I_{g1}}{L_1} + \frac{I_{g2}}{L_2} \right) + \left(\frac{I_{g2}}{L_2} + \frac{I_{g3}}{L_3} \right) + \frac{I_{g3}}{L_3} \right]_i$$

$$= 12E \sum \left(\frac{I_g}{L} \right)_i \quad (7.8)$$

From Eqs. (7.6) to (7.8)

$$\theta_{ig} = \frac{Q_i h_i + Q_{i+1} h_{i+1}}{24E \sum \left(\frac{I_g}{L} \right)_i} \quad (7.9)$$

A similar expression may be obtained for the average joint rotation in the floor ($i - 1$) below, but with subscripts ($i + 1$) replaced by i, and i by ($i - 1$).

Referring to Fig. 7.13b, the drift in story i due to the joint rotations is

$$\delta_{ig} = \frac{h_i}{2} (\theta_{i-1} + \theta_i) \quad (7.10)$$

that is

$$\delta_{ig} = \frac{h_i}{2} \left[\frac{Q_{i-1} h_{i-1} + Q_i h_i}{24E \sum \left(\frac{I_g}{L} \right)_{i-1}} + \frac{Q_i h_i + Q_{i+1} h_{i+1}}{24E \sum \left(\frac{I_g}{L} \right)_i} \right] \quad (7.11)$$

Assuming that the girders in floors $i - 1$ and i are the same, the story heights are the same, and the average of Q_{i+1} and Q_{i-1} is equal to Q_i,

$$\delta_{ig} = \frac{Q_i h_i^2}{12E \sum \left(\frac{I_g}{L} \right)_i} \quad (7.12)$$

154 RIGID-FRAME STRUCTURES

Story Drift due to Column Flexure. Referring to Fig. 7.13c, in which the drift due to bending of the columns is isolated by assuming the girders are rigid, the drift of the structure in story i is

$$\delta_u = \frac{Q_i h_i^3}{12E \sum I_{ci}} \qquad (7.13)$$

from which

$$\delta_{ic} = \frac{Q_i h_i^2}{12E \sum \left(\frac{I_c}{h}\right)_i} \qquad (7.14)$$

Story Drift due to Overall Bending. Although the component of total drift due to overall bending may be small relative to that caused by racking, the bending inclination increases cumulatively throughout the height. Consequently, in the upper stories, where the story shear drift tends to be less than in the lower region, the bending drift may become a significant part of the *story* drift. An estimate of the bending drift can be made by assuming the structure behaves as a flexural cantilever with a moment of inertia equal to the second moment of the column areas about their common centroid, that is $I_i = \sum (Ac^2)_i$ (Fig. 7.14a and b). If the moment diagram (Fig. 7.14c) is used to construct an M/EI diagram [in which $I = \sum (Ac^2)$] (Fig. 7.14d), the area of the diagram A'_0 between the base and the mid-height of story i gives the average slope of story i due to bending action, that is

$$\theta_{if} = A'_0 \qquad (7.15)$$

Then the bending component of drift in story i is given by

$$\delta_{if} = h_i \theta_{if} = h_i A'_0 \qquad (7.16)$$

Story Drift and Total Drift. The resulting drift in a single story i is the sum of the components,

$$\delta_i = \delta_{ig} + \delta_{ic} + \delta_{if} \qquad (7.17)$$

or

$$\delta_i = \frac{Q_i h_i^2}{12E \sum (I_g/L)_i} + \frac{Q_i h_i^2}{12E \sum (I_c/h)_i} + h_i A'_0 \qquad (7.18)$$

Denoting $\sum (I_g/L)_i$ by G_i, and $\sum (I_c/h)_i$ by C_i, this may be rewritten

$$\delta_i = \frac{Q_i h_i^2}{12E} \left(\frac{1}{G} + \frac{1}{C}\right)_i + h_i A'_0 \qquad (7.19)$$

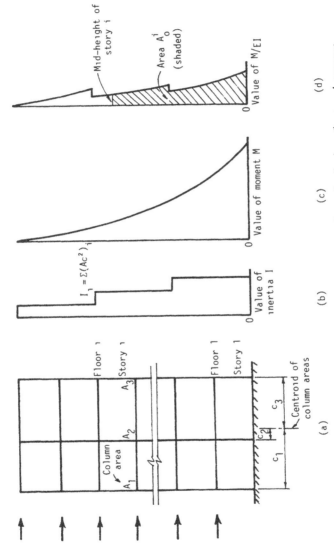

Fig. 7.14 (a) Frame structure; (b) distribution of inertia I; (c) distribution of external moment M; (d) M/EI diagram.

The assumption of a mid-story-height point of contraflexure is not valid for the first story of a rigid frame because of the fully fixed or hinged conditions at the base of the columns. Therefore, special expressions should be used for the first story drift attributable to column and girder bending [7.7]. If the columns have rigid base connections, the first story drift may be estimated by

$$\delta_{1,g+c} = \frac{Q_1 h_1^2}{12E} \frac{\left(\dfrac{2}{3G_1} + \dfrac{1}{C_1}\right)}{\left(1 + \dfrac{C_1}{6G_1}\right)} \tag{7.20}$$

If the columns have pinned base connections

$$\delta_{1,g+c} = \frac{Q_1 h_1^2}{12E}\left(\frac{3}{2G_1} + \frac{4}{C_1}\right) \tag{7.21}$$

The total drift at the nth floor of a building may then be found from

$$\Delta_n = \sum_{}^{n} \delta_i \tag{7.22}$$

A check on the story drift should be made for the top story and for intermediate stories where member size reductions occur. If, on the basis of the initially sized frame, the calculated drifts are well within the allowable values, these spot checks will probably be adequate.

7.4.2 Correction of Excessive Drift

The typical proportioning of member sizes in tall rigid frames is such that girder flexure is the major cause of drift, with column flexure a close second. Therefore, increasing the girder stiffness is usually the most effective and economical way of correcting excessive drift. If the girder in any single bay is substantially smaller than the others at that level, it should be increased first.

An estimate of the modified girder sizes required at level i to correct the drift in that story can be obtained by neglecting the contribution due to overall bending and rewriting Eq. (7.18) in the form

$$\sum (I_g/L)_i = \frac{Q_i h_i}{12E\left[\left(\dfrac{\delta_i}{h_i}\right) - \dfrac{Q_i h_i}{12E \sum (I_c/h)_i}\right]} \tag{7.23}$$

in which δ_i is assigned the value of the allowable story drift. If the frame is unusually proportioned so that column flexure contributes a major part of the drift,

7.4 APPROXIMATE ANALYSIS FOR DRIFT

Eq. (7.23) may be rewritten to allow an estimate of the required column sizes by interchanging $\Sigma (I_g/L)_i$ and $\Sigma (I_c/h)_i$.

A relatively simple check on whether girders or columns should be adjusted first has been proposed as follows [7.8]. Compute for each joint across the floor levels above and below the story whose drift is critical, the value of a parameter ψ where

$$\psi = \frac{I_c}{h} \bigg/ \Sigma \frac{I_g}{L} \qquad (7.24)$$

in which $\Sigma I_g/L$ refers to the girders connecting into the joint.

If a scan of the resulting values of ψ indicates that

1. $\psi \gg 0.5$, adjust the girder sizes;
2. $\psi \ll 0.5$, adjust the column sizes;
3. $\psi \approx 0.5$, adjust both column and girder sizes.

This test should preferably be accompanied by an inspection of the drift components of Eq. (7.18) to ascertain whether the allowable story drift is exceeded by any one component alone, as might occur in a grossly undersized initial design. If it is exceeded by any one component, whether as a result of undersized columns or of undersized beams, that component must be remedied first.

7.4.3 Effective Shear Rigidity (GA)

This parameter expresses the racking stiffness of a frame on a story-height average basis. It is a useful parameter when considering the allocation of loading between rigid frame bents, and the horizontal interaction of frames with walls. The composite symbol (GA) is used because it corresponds with the shear rigidity of an analogous shear cantilever of sectional area A and modulus of rigidity G. A story-height segment of such a cantilever may be compared (Fig. 7.15a) with a corresponding portion of a rigid frame (Fig. 7.15b).

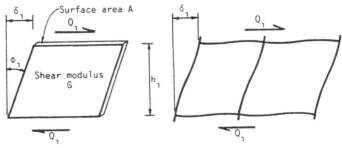

Fig. 7.15 (a) Story-height segment of analogous shear wall; (b) single story of rigid frame.

When the cantilever segment is subjected to a shear Q, its deflection is given by

$$\delta = \frac{Qh}{GA} \tag{7.25}$$

from which the shear rigidity is given by

$$(GA) = \frac{Qh}{\delta} = \frac{Q}{\phi} \tag{7.26}$$

where ϕ is the angle of inclination. That is, (GA) is the shear force necessary to cause unit inclination of the shear structure.

For the corresponding portion of frame, using Eq. (7.19) and neglecting drift caused by overall bending

$$(GA) = \frac{Qh}{\delta} = \frac{Qh}{\dfrac{Qh^2}{12E}\left(\dfrac{1}{G} + \dfrac{1}{C}\right)} \tag{7.27}$$

then

$$(GA) = \frac{12E}{h\left(\dfrac{1}{G} + \dfrac{1}{C}\right)} \tag{7.28}$$

If the value of (GA), at level i of a frame is known, the horizontal displacement in story i is given by

$$\delta_i = \left(\frac{Qh}{GA}\right)_i \tag{7.29}$$

7.5 FLAT PLATE STRUCTURE—ANALOGOUS RIGID FRAME

Flat plate structures, in which the columns are cast integrally with the floor slabs, behave under horizontal loading similarly to rigid frames. The lateral deflections of the structure are a result of simple double curvature bending of the columns, and a more complex three-dimensional form of double bending in the slab. If the columns are on a regular orthogonal grid (Fig. 7.16), the response of the structure can be studied by considering each bay-width replaced by an equivalent rigid frame bent. The slab is replaced for the analysis by an equivalent beam with the same double bending stiffness. The hand methods of estimating drift, outlined in Sections 7.4.1 to 7.4.3, or a computer analysis, can then be applied.

The flexural stiffness of the equivalent beam depends mainly on the width-to-length spacing of the columns and on the dimension of the column in the direction

7.5 FLAT PLATE STRUCTURE—ANALOGOUS RIGID FRAME

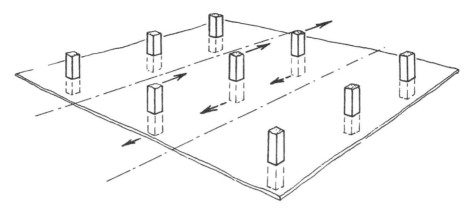

Fig. 7.16 Flat plate structure.

of drift. In Fig. 7.17, these parameters are used to present the effective width of the equivalent beam [7.9], that is, the width of the uniform-section beam having the same double curvature flexural stiffness as the slab, with the same depth, span, and modulus of elasticity as the slab. This equivalent beam may be used only in the lateral loading analysis of flat plate structures. It is not appropriate for gravity or combined loading analyses.

Figure 7.17 shows the equivalent beam stiffness to be very sensitive to the width of the column in the direction of drift. This is because of the "wide-column" effect that is demonstrated even more markedly by coupled shear walls (cf. Chapter 10). When the slab width-to-span ratio b/a exceeds 1.5, the effective width becomes virtually constant because the slab boundary regions parallel to the direction of drift deform negligibly and therefore contribute little to the stiffness. The apparent reduction in effective width shown by Fig. 7.17 as b/a increases is caused by plotting the effective width as a fraction of the transverse span. The curves in Fig. 7.17 were obtained for square section columns; however, they are equally applicable to rectangular section columns since additional analyses [7.9] have shown that variations in the column transverse dimension from one-half to two times the longitudinal dimension cause less than a 2% change in effective width.

7.5.1 Worked Example

A flat plate multistory structure consists of a regular rectangular grid of columns spaced at 8.0 m by 6.0 m ctrs. The columns are 0.6 m square and the slab is 0.2 m thick. For horizontal loading acting parallel to the 8 m dimension, determine the moment of inertia of an equivalent beam to replace the slab.

Referring to Fig. 7.17

$$a = 8.0 \text{ m}, \quad b = 6.0 \text{ m}, \quad u = 0.6 \text{ m}$$

$$\frac{u}{a} = \frac{0.6}{8.0} = 0.075; \quad \frac{b}{a} = \frac{6.0}{8.0} = 0.75$$

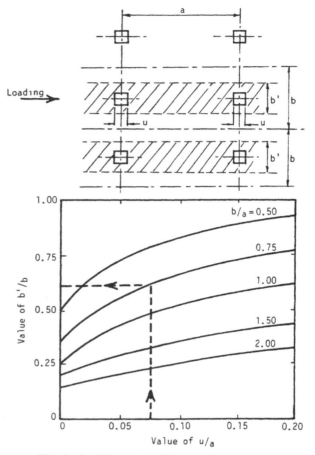

Fig. 7.17 Effective width of equivalent beam.

Referring to the graph, the above values give

$$\frac{b'}{b} = 0.61$$

Effective slab width $b' = 0.61 \times 6.0 = 3.66$ m. Therefore, moment of inertia of equivalent beam

$$I = \frac{3.66 \times 0.2^3}{12} = 0.0024 \text{ m}^4$$

This value would normally be reduced in the analysis by 50% to allow for the reduction in stiffness due to cracking as the slab bends.

7.6 COMPUTER ANALYSIS OF RIGID FRAMES

Although the previously described hand methods of determining deflections and forces in rigid frames have served engineers well for the design of rigid frames, and are still useful for preliminary analysis and checking, they have now been superseded for most practical purposes by computer analysis. A computer analysis is more accurate, and better able to analyze complex structures. A wide variety of commercial structural analysis programs, invariably based on the stiffness matrix method, are available.

Forming the model of the rigid frame for computer analysis has been described in Chapter 5. Briefly, it consists of an assembly of beam-type elements to represent both the beams and columns of the frame. The columns are assigned their principal inertias and sectional areas. The beams are assigned their horizontal axis inertia while their sectional areas are assigned to be effectively rigid. Torsional stiffnesses of the columns and beams are usually small and, therefore, neglected. Shear deformations of columns and beams are also usually neglected unless the member has a length-to-depth ratio of less than about 5, in which case a shear area is assigned.

If the frame is of reinforced concrete, reduced inertias are assigned to the members to allow for cracking: 50% of their gross inertia to the beams and 80% of their gross inertia to the columns.

Some analysis programs include the option of considering the slabs to be rigid in-plane, and some the option of including P-Delta effects. If a rigid slab option is not available, the effect can be simulated by interconnecting all vertical elements by a horizontal frame at each floor, adding fictitious beams where necessary, and assigning the beams to be effectively rigid axially and in flexure in the horizontal plane. Slabs are usually assumed to have a negligible transverse rigidity unless a flat plate or flat slab action is intended, in which case the slab is represented as a connecting grid of equivalent stiffness beams.

7.7 REDUCTION OF RIGID FRAMES FOR ANALYSIS

The reduction of a rigid frame to a simpler equivalent frame is a useful way of simplifying its analysis when it is not essential to obtain the exact member forces. The two techniques described below can be used separately, or in combination, to give large-scale reductions in the size of the computational problem.

7.7.1 Lumped Girder Frame

A repetitive floor system offers scope for the lumping of girders in successive floors to form a model with fewer stories. The lumped girder frame allows an accurate estimate of the drift and a good estimate of the member forces. The girders are usually lumped in threes or, if the frame is very tall, in fives. In the example of Fig. 7.18, three sets of three girders are lumped into single girders that converts

162 RIGID-FRAME STRUCTURES

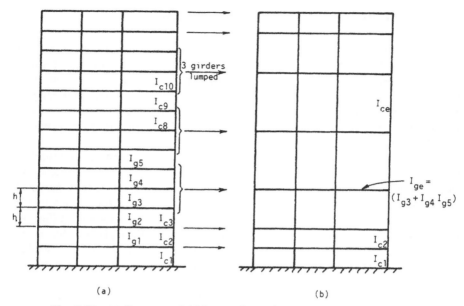

Fig. 7.18 (a) Prototype rigid frame; (b) equivalent lumped girder frame.

the 13-story frame into a 7-story equivalent frame. The first floor and roof girders must not be included in the lumping because the frame behavior near the top and the base differ significantly from that in middle regions. In Fig. 7.18b the second floor and next-to-roof girders are also left as in the original to give an even closer representation of the boundary conditions.

The requirement of a substitute frame is that, for horizontal loading, joint translations should be the same as those of the original structure. For translations caused by girder flexure, Eq. (7.12) shows this requirement to be satisfied by assigning the inertia of each equivalent girder to be equal to the sum of the lumped n-girder inertias in the original frame, that is,

$$I_{ge} = \sum_{}^{n} I_g \qquad (7.30)$$

To determine the properties of the columns in a lumped girder equivalent frame, reference is made to Eq. (7.14). Equating the component of drift caused by double curvature column bending in a story of height nh in the equivalent frame to the corresponding drift over n stories in the original frame

$$\frac{Q(nh)^2}{12E \sum I_{ce}/nh} = \frac{Qh^2}{12E} \sum_{}^{n} \frac{1}{\sum (I_c/h)_i} \qquad (7.31)$$

from which the equivalent column inertias are

$$I_{ce} = \frac{n^3}{\sum_{i}^{n}\left(\frac{1}{I_c}\right)_i} \qquad (7.32)$$

For example, in the structure shown in Fig. 7.18a, the vertical stack of three equal-height columns I_{c8}, I_{c9}, and I_{c10}, would be replaced in the equivalent lumped girder model (Fig. 7.18b) by a three-story-height equivalent column having an inertia

$$I_{ce} = \frac{3^3}{\dfrac{1}{I_{c8}} + \dfrac{1}{I_{c9}} + \dfrac{1}{I_{c10}}} \qquad (7.33)$$

If within each region to be lumped the inertia of a column in successive single stories is constant, the inertia of the equivalent column of height $3h$ would be nine times that of the original single-story-height columns, while the inertia of the equivalent column in an intermediate region (Fig. 7.18b), whose length is $2h$, would be four times that of the original single-story-height columns.

The columns' sectional areas, which control the cantilever component of deflection, must have the same second moment about their common centroid in the lumped and original structures. Consequently, the areas of the equivalent columns remain the same as those in the original frame. The horizontal loading on the equivalent frame is applied as equivalent concentrated loads at the lumped girder levels, taking the half new-story-height regions above and below the lumped girders as tributary areas.

When the lumped girder frame has been analyzed, the results must be transformed back to the original frame. The moments in the original girders at the lumped girder levels should be taken as $1/n$ of the resulting moment in the corresponding lumped girders. The moments in the original girders between the lumped girder levels should then be estimated by vertical interpolation. Girder shears are estimated by dividing the end moments by the half-span lengths.

For original column shears in a particular story, the actual external shear at the mid-height of the story should be distributed between the columns in the same ratio as that between the resulting shears at that level in the equivalent frame. The moments at the top and bottom of a column should be taken as the product of the column shear and the original half-story height.

7.7.2 Single-Bay Substitute Frame [7.10]

The reduction of a multibay rigid frame to a single-bay equivalent frame provides a model that closely simulates the response of the structure to horizontal loading. It is useful, therefore, in estimating deflections for stability analyses and for dynamic analyses of frames whose member forces are not required. It can also be used in a two-stage member force analysis of a large multibent, multibay frame,

164 RIGID-FRAME STRUCTURES

for which a first-stage overall analysis is made of the structure represented by an assembly of equivalent single-bay bents. The resulting shears in the bents are used to obtain the individual bent loadings, which are then used in second-stage analyses of the individual multibay bents to obtain their member forces (cf. Chapter 5, Section 5.1).

Figure 7.19a and b shows a multibay rigid frame and its single-bay analogy. Member sizes are assigned to the single-bay frame to cause it to deflect horizontally in the same manner as the prototype. First assume an arbitrary width l, for the single-bay frame. Then, equating the component of drift in story i of the prototype, caused by double bending of the girders, to that in the single-bay frame structure, and using Eq. (7.12)

$$\frac{Q_i h_i^2}{12E \sum (I_g/L)_i} = \frac{Q_i h_i^2}{12E(I_{ge}/l)_i} \qquad (7.34)$$

Therefore, the girder in floor i of the single-bay frame is assigned an inertia

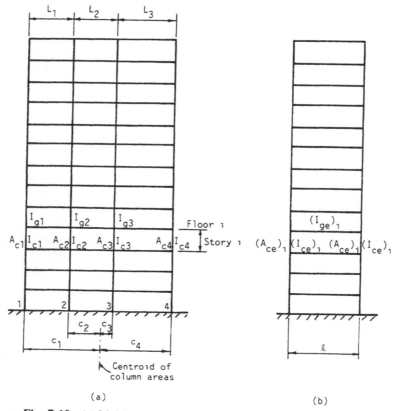

Fig. 7.19 (a) Multibay rigid frame; (b) equivalent single-bay frame.

$$(I_{gc})_t = l \sum (I_g/L)_t \qquad (7.35)$$

Considering similarly the component of drift due to double bending of the columns, and applying Eq. (7.14)

$$\frac{Q_t h_i^2}{12E \sum (I_c/h)_t} = \frac{Q_t h_i^2}{12E(2I_{cc}/h)_t} \qquad (7.36)$$

therefore, the moment of inertia of the equivalent single-bay column is

$$(I_{cc})_t = \frac{1}{2} \sum (I_c)_i \qquad (7.37)$$

Finally, equating the components of drift resulting from cantilever action in the prototype and the single-bay frames, and using Eq. (7.16)

$$h_t \left[\text{Area} \left(\frac{M}{E \sum A_c c^2} \right) \right]_0^t = h_t \left[\text{Area} \left(\frac{M}{EA_{cc} l^2/2} \right) \right]_0^t \qquad (7.38)$$

then

$$\left(\frac{A_{cc} l^2}{2} \right)_t = \sum (A_c c^2)_t \qquad (7.39)$$

therefore, the equivalent columns must be assigned sectional areas

$$(A_{cc})_t = \frac{2}{l^2} \sum (A_c c^2)_t \qquad (7.40)$$

Although the single-bay frame results for horizontal deflections will be fairly accurate, the resulting member forces for the single-bay frame are not transformable back to the multibay frame.

The lumped girder and single-bay frame techniques can also be used in combination to reduce an extremely large frame structure to one that is much more amenable to a first-stage, displacement and bent shear, analysis.

SUMMARY

The flexural continuity between the members of a rigid frame enables the structure to resist horizontal loading as well as to assist in carrying gravity loading. The probable worst combined effects of gravity and horizontal loading have to be estimated for the design of the frame.

Gravity loading causes regions of sagging moment near the mid-span of the girders and of hogging moment beside the columns. Pattern live loading must be used to estimate the worst effects of gravity loading. The girder maximum moments may be evaluated approximately from formulas or more accurately from conventional or shortened forms of moment distribution.

Horizontal loading causes racking of the frame due to double bending of the columns and girders, resulting in an overall shear mode of deformation of the structure. The portal and cantilever methods of analysis provide an estimate of the horizontal loading member forces that, when combined with the gravity loading member forces, allow a preliminary design of the frame members. The portal and cantilever methods may be used also for the analysis of rigid frames with setbacks.

The lateral displacement of rigid frames subjected to horizontal loading is due to three modes of member deformation: girder flexure, column flexure, and axial deformation of the columns. The horizontal displacements in each story attributable to these three components can be calculated separately and summed to give the total story drift. The sum of the story drifts from the base upward gives the horizontal displacement at any level. If the total drift, or the drift within any story, exceeds the allowable values, an inspection of the components of drift will indicate which members should be increased in size to most effectively control the drift.

A flat-plate structure responds to loading in a manner similar to a rigid frame but with the transversely varying vertical flexure of the floor slab replacing the single-plane vertical flexure of the rigid frame girder. A horizontal deflection analysis of a regular flat-plate structure can be made by considering the slabs replaced by equivalent girders, and treating it as a rigid frame.

When a rigid frame includes many repetitive stories it may be reduced for a horizontal loading analysis by lumping the girders in three, or five, successive floors to give an equivalent simpler structure. The properties of the girders and columns must be transformed initially in formulating the equivalent structure, and the resulting forces subsequently transformed back to give the forces in the members of the original structure. A multibay rigid frame may be reduced to an equivalent single-bay frame for a horizontal loading analysis. This model is useful for representing the horizontal response of the bent and for determining its horizontal deflections. The two reduction methods may be used, either separately or in combination, to simplify extremely large rigid frame structures for analysis.

REFERENCES

7.1 *Uniform Building Code 1988*. International Conference of Building Officials, Whittier, California 90601.

7.2 *Continuity in Concrete Building Frames*. Portland Cement Association, Skokie, Illinois 60076.

7.3 Smith, A. and Wilson, C. A. "Wind Stresses in the Steel Frames of Office Buildings." *J. Western Soc. Engineers* April 1915, 365–390.

7.4 *Wind Bracing in Steel Buildings*. Final Report of Sub-Committee No. 31 on Steel of the Structural Division, *Trans. ASCE* **105**, 1940, 1713–1739.

7.5 Wilson, A. C. "Wind Bracing with Knee-Braces or Gusset Plates." *Engineer. Rec.* September 1908, 227–274.

7.6 Goldberg, J. E. "Approximate Methods in Stress and Stability Analysis of Tall Building Frames." *Proc. IABSE, ASCE Regional Conference on Tall Buildings*, Bangkok, January 1974, 177-194.

7.7 Goldberg, J. E. "Analysis of Two-Column Symmetrical Bents and Vierendeel Trusses Having Parallel and Equal Chords." *J. Am. Conc. Inst.* **19**(3), November 1947, 225-234.

7.8 Cheong-Siat-Moy, F. "Stiffness Design of Unbraced Steel Frames." *AISC Engineer. Journal*, 1st Quarter, 1976, pp. 8-10.

7.9 Wong, Y. C. and Coull, A. "Effective Slab Stiffness in Flat Plate Structures." *Proc. Instn. Civ. Engineers* Part 2, **69,** September 1980, 721-735.

7.10 Goldberg, J. E. *Structural Design of Tall Steel Buildings*, Council on Tall Buildings and Urban Habitat, Monograph, Vol. SB, 1979, p. 53.

CHAPTER 8

Infilled-Frame Structures

The infilled frame consists of a steel or reinforced concrete column-and-girder frame with infills of brickwork or concrete blockwork (Fig. 8.1). In addition to functioning as partitions, exterior walls, and walls around stair, elevator, and service shafts, the infills may also serve structurally to brace the frame against horizontal loading. In nonearthquake regions where the wind forces are not severe, the masonry infilled concrete frame is one of the most common structural forms for high-rise construction. The frame is designed for gravity loading only and, in the absence of an accepted design method, the infills are presumed to contribute sufficiently to the lateral strength of the structure for it to withstand the horizontal loading. The simplicity of construction, and the highly developed expertise in building that type of structure have made the infilled frame one of the most rapid and economical structural forms for tall buildings.

In countries with stringently applied Codes of Practice the absence of a well-recognized method of design for infilled frames has severely restricted their use for bracing. It has been more usual in such countries, when designing an infilled-frame structure, to arrange for the frame to carry the total vertical and horizontal loading and to include the infills on the assumption that, with precautions taken to avoid load being transferred to them, the infills do not participate as part of the primary structure. It is evident from the frequently observed diagonal cracking of such infill walls that the approach is not always valid. The walls do sometimes attract significant bracing loads and, in so doing, modify the structure's mode of behavior and the forces in the frame. In such cases it would have been better to design the walls for the lateral loads, and the frame to allow for its modified mode of behavior.

In this chapter a design method is presented to allow the use of infilled frames as bracing. It is based on theoretical and experimental studies of interactive wall-frame behavior. Rather than being a method for the direct design of the frame members and the wall, it is intended for use more as a method of checking and adjusting an infilled frame that has already been designed to satisfy other criteria. The frame is sized initially to be adequate for gravity loading, while the thickness of the infill wall is probably decided on the basis of the acoustic, fire, and climatic requirements.

To brace a structure, the arrangement of infill walls within the three-dimensional frame must satisfy the same requirements as for the layout of bracing in a

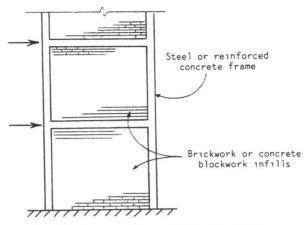

Fig. 8.1 Structural frame infilled with masonry.

steel structure. Within any story the infills must be statically capable of resisting horizontal shear in two orthogonal directions, as well as resisting a horizontal torque. To achieve this there must be at least three infills that may not be all parallel or all concurrent. They must, of course, also be able to satisfy the strength and stiffness requirements.

Certain reservations arise in the use of infilled frames for bracing a structure. For example, it is possible that as part of a renovation project, partition walls are removed with the result that the structure becomes inadequately braced. Precautions against this, either by including a generously excessive number of bracing walls, or by somehow permanently identifying the vital bracing walls, should be considered as part of the design. A reservation against their use where earthquake resistance is a factor is that the walls might be shaken out of their frames transversely and, consequently, be of little use as bracing in their own planes. On the basis of substantial field evidence this fear is well justified. Their use in earthquake regions, therefore, should be with the additional provision that the walls are reinforced and anchored into the surrounding frame with sufficient strength to withstand their own transverse inertial forces.

8.1 BEHAVIOR OF INFILLED FRAMES

The use of a masonry infill to brace a frame combines some of the desirable structural characteristics of each, while overcoming some of their deficiencies. The high in-plane rigidity of the masonry wall significantly stiffens the otherwise relatively flexible frame, while the ductile frame contains the brittle masonry, after cracking, up to loads and displacements much larger than it could achieve without the frame. The result is, therefore, a relatively stiff and tough bracing system.

The wall braces the frame partly by its in-plane shear resistance and partly by its behavior as a diagonal bracing strut in the frame. Figure 8.2a illustrates these

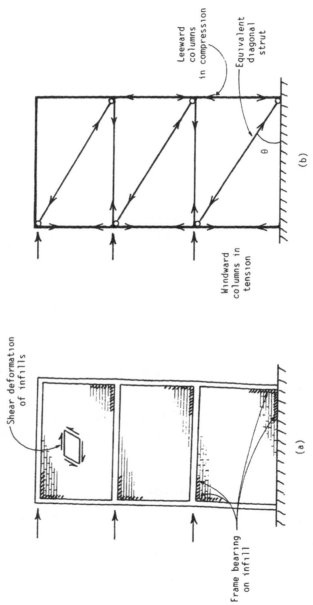

Fig. 8.2 (a) Interactive behavior of frame and infills; (b) analogous braced frame.

modes of behavior. When the frame is subjected to horizontal loading, it deforms with double-curvature bending of the columns and girders. The translation of the upper part of the column in each story and the shortening of the leading diagonal of the frame cause the column to lean against the wall as well as to compress the wall along its diagonal. It is roughly analogous to a diagonally braced frame (Fig. 8.2b).

Three potential modes of failure of the wall arise as a result of its interaction with the frame, and these are illustrated in Fig. 8.3a. The first is a shear failure stepping down through the joints of the masonry, and precipitated by the horizontal shear stresses in the bed joints. The second is a diagonal cracking of the wall through the masonry along a line, or lines, parallel to the leading diagonal, and

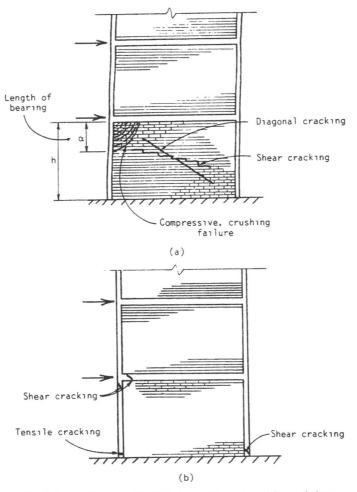

Fig. 8.3 (a) Modes of infill failure; (b) modes of frame failure.

caused by tensile stresses perpendicular to the leading diagonal. The "perpendicular" tensile stresses are caused by the divergence of the compressive stress trajectories on opposite sides of the leading diagonal as they approach the middle region of the infill. The diagonal cracking is initiated at and spreads from the middle of the infill, where the tensile stresses are a maximum, tending to stop near the compression corners, where the tension is suppressed. In the third mode of failure, a corner of the infill at one of the ends of the diagonal strut may be crushed against the frame due to the high compressive stresses in the corner.

The nature of the forces in the frame can be understood by referring to the analogous braced frame (Fig. 8.2b). The windward column is in tension and the leeward column is in compression. Since the infill bears on the frame not as a concentrated force exactly at the corners, but over short lengths of the beam and column adjacent to each compression corner, the frame members are subjected also to transverse shear and a small amount of bending. Consequently, the frame members or their connections are liable to fail by axial force or shear, and especially by tension at the base of the windward column (Fig. 8.3b).

8.2 FORCES IN THE INFILL AND FRAME

A concept of the behavior of infilled frames has been developed from a combination of results of tests [8.1–8.8], very approximate analyses [8.9], and more sophisticated finite element analyses [8.10]. An understanding of infilled-frame behavior is far from complete and further research needs to be done, especially with full-scale tests. Consequently, opinions about the approach to the design of infilled frames differ, especially as to whether it should be elastically or plastically based. The method presented here draws from a combination of test observations and the results of analyses. It may be classified as an elastic approach except for the criterion used to predict the infill crushing, for which a plastic type of failure of the masonry infill is assumed.

8.2.1 Stresses in the Infill

Relating to Shear Failure. Shear failure of the infill is related to the combination of shear and normal stresses induced at points in the infill when the frame bears on it as the structure is subjected to the external lateral shear. An extensive series of plane-stress membrane finite-element analyses [8.11] has shown that the critical values of this combination of stresses occur at the center of the infill and that they can be expressed empirically by

$$\text{Shear stress} \quad \tau_{xy} = \frac{1.43 Q}{Lt} \qquad (8.1)$$

$$\text{Vertical compressive stress} \quad \sigma_y = \frac{(0.8\, h/L - 0.2)Q}{Lt} \qquad (8.2)$$

where Q is the horizontal shear load applied by the frame to the infill of length L, height h, and thickness t.

Relating to Diagonal Tensile Failure.
Similarly, diagonal cracking of the infill is related to the maximum value of diagonal tensile stress in the infill. This also occurs at the center of the infill and, based on the results of the analyses, may be expressed empirically as

$$\text{Diagonal tensile stress} \quad \sigma_d = \frac{0.58Q}{Lt} \quad (8.3)$$

These stresses are governed mainly by the proportions of the infill. They are little influenced by the stiffness properties of the frame because they occur at the center of the infill, away from the region of contact with the frame.

Relating to Compressive Failure of the Corners.
Tests on model infilled frames have shown that the length of bearing of each story-height column against its adjacent infill is governed by the flexural stiffness of the column relative to the inplane bearing stiffness of the infill. The stiffer the column, the longer the length of bearing and the lower the compressive stresses at the interface. Tests to failure have borne out the deduction that the stiffer the column, the higher the strength of the infill against compressive failure. They have also shown that crushing failure of the infill occurs over a length approximately equal to the length of bearing of the column against the infill (Fig. 8.3a).

As a crude approximation, an analogy may be drawn with the theory for a beam on an elastic foundation [8.12], from which it has been proposed that [8.9] the length of column bearing α may be estimated by

$$\alpha = \frac{\pi}{2\lambda} \quad (8.4)$$

where

$$\lambda = \sqrt[4]{\frac{E_m t}{4EIh}} \quad (8.5)$$

in which E_m is the elastic modulus of the masonry and EI the flexural rigidity of the column.

The parameter λ expresses the bearing stiffness of the infill relative to the flexural rigidity of the column: the stiffer the column, the smaller the value of λ and the longer the length of bearing.

If it is assumed that when the corner of the infill crushes, the masonry bearing against the column within the length α is at the masonry ultimate compressive stress f'_m, then the corresponding ultimate horizontal shear Q'_c on the infill is given by

$$Q'_c = f'_m \alpha t \quad (8.6)$$

or

$$Q'_c = f'_m t \cdot \frac{\pi}{2} \cdot \sqrt[4]{\frac{4EIh}{E_m t}} \qquad (8.7)$$

Considering now the allowable horizontal shear Q_c on the infill, and assuming a value for E/E_m of 30 in the case of a steel frame and 3 in the case of a reinforced concrete frame, the allowable horizontal shear on a steel framed infill corresponding to a compressive failure is given by

$$Q_c = 5.2 f_m \sqrt[4]{Iht^3} \qquad (8.8)$$

and for a reinforced concrete framed infill

$$Q_c = 2.9 f_m \sqrt[4]{Iht^3} \qquad (8.9)$$

in which f_m is the masonry allowable compressive stress.

These semiempirical formulas indicate the significant parameters that influence the horizontal shear strength of an infill when it is governed by a compressive failure of one of its corners. The masonry compressive strength and the wall thickness have the most direct influence on the infill strength, while the column inertia and infill height exert control in proportion to their fourth roots. The infill strengths indicated by Eqn. (8.8) and (8.9) are very approximate. Experimental evidence has shown them to overestimate the real values; therefore, they will be modified before being used in the design procedure.

8.2.2 Forces in the Frame

Experiments on horizontally loaded model infilled frames, and finite-element stress analyses, have shown that the axial forces in the beams and columns of an infilled frame can be estimated reasonably well by a simple analysis of the analogous braced frame (Fig. 8.2b), assuming hinges at all joints. A conservative estimate of the shear in the columns is given by the horizontal component of the force in the diagonal strut and, similarly, an estimate of the shear in the beams is given by the vertical component of that force. The analyses have indicated that the bending moments in the columns and beams caused by the perpendicular thrust from the infill are small relative to the moments that would occur in a similarly loaded rigid frame without infills. A conservative nominal moment of $Q \cdot h/20$ is suggested as a maximum value.

8.3 DEVELOPMENT OF THE DESIGN PROCEDURE

The main factors to be provided for in the design method are as follows:

1. In the weakest of its three modes of failure (i.e., shear, diagonal tensile, and compressive) the infill must be capable of withstanding the stresses induced by the frame bearing on it under the action of the external shear.

2. The frame must be able to transmit to the infill the external shear imposed on it, as well as be strong enough to withstand the reactions it receives from the infill.

The following discussion concerns the development of a design procedure that attempts to satisfy the above criteria. It is assumed, conservatively, on the basis that the lateral stiffness of the infill is much greater than that of the frame, that the infill carries the total applied shear.

8.3.1 Design of the Infill

Shear Failure. Shear failure, which occurs along the masonry bed joints, is assumed to be initiated at the point in the infill where the ratio of horizontal shear stress to available shear strength is a maximum. As noted before, theoretical analyses have indicated, and tests have verified, that this occurs at the center of the infill.

The shear strength of masonry has commonly been represented in Codes of Practice [8.13, 8.14, 8.15] by a static friction type of equation

$$f_s = f_{bs} + \mu \sigma_y \tag{8.10}$$

together with a limiting maximum value.

The bond shear strength f_{bs} is similar in action to an adhesive shear strength while $\mu \sigma_y$ is a frictional component of resistance, in which μ is a coefficient of internal friction and σ_y is the vertical compressive stress across the horizontal joint. The allowable value of f_s depends on factored values of f_{bs} and μ, to allow for the type of masonry and a factor of safety.

Equating the shear stress at the wall center [Eq. (8.1)] to the allowable masonry shear stress [Eq. (8.10)], and substituting Eqn. (8.2) for the vertical stress σ_y, gives

$$\frac{1.43 Q_s}{Lt} = f_{bs} + \frac{\mu Q_s}{Lt} \left(\frac{0.8h}{L} - 0.2 \right) \tag{8.11}$$

Then, at any level of the structure, the allowable horizontal shear force based on the shear failure criterion is

$$Q_s = \frac{f_{bs} Lt}{1.43 - \mu \left(\frac{0.8h}{L} - 0.2 \right)} \tag{8.12}$$

Considering also the maximum allowable masonry shear stress

$$\frac{1.43 \, Q_s}{Lt} \not> f_{s,\max} \tag{8.13}$$

from which

176 INFILLED-FRAME STRUCTURES

$$Q_s \not> 0.7 Lt f_{s\,max} \qquad (8.14)$$

in which $f_{s\,max}$ is the specified maximum allowable shear stress.

Diagonal Tensile Failure. The diagonal tensile strength of masonry is somewhat uncertain in value. Tests [8.11] have shown, however, that it can be estimated conservatively as approximately equal to one-tenth of the mortar compressive strength. Codes of Practice give an allowable flexural tensile stress in masonry equal to approximately one-fortieth of the compressive strength of the weakest allowable mortar. Assuming a typical factor of safety of 4 for brickwork, it is reasonable to take the allowable diagonal tensile stress in masonry as equal to its allowable flexural tensile stress, that is

$$f_d = f_t \qquad (8.15)$$

Then, equating the maximum diagonal tensile stress [Eq. (8.3)] to the permissible diagonal tensile stress [Eq. (8.15)]

$$\frac{0.58Q}{Lt} = f_t \qquad (8.16)$$

from which the allowable horizontal shear Q_d, based on the diagonal tensile failure criterion, is given by

$$Q_d = 1.7 Lt f_t \qquad (8.17)$$

Comparing the allowable horizontal shear based on the maximum allowable shear stress criterion [Eq. (8.14)] with that based on diagonal tensile failure [Eq. (8.17)] by substituting values for unreinforced brick masonry from the *Uniform Building Code* [8.13], the latter always results in a higher allowable horizontal shear and, therefore, is less critical. Consequently, Eq. (8.17) may be dropped as a design consideration.

Comparing also the allowable horizontal force based on the friction-formula allowable shear stress [Eq. (8.12)], with that based on the maximum allowable shear stress [Eq. (8.14)], and substituting in the equations values of f_{bs}, μ, and $f_{s\,max}$, implicit in the *Uniform Building Code*, the friction-formula expression always gives lower values of allowable horizontal force for practically proportioned frames and is therefore the more critical condition. Consequently, the maximum allowable shear stress condition [Eq. (8.14)] can also be dropped. Therefore Eq. (8.12) for the shear failure remains as one of the design criteria for a satisfactory infill.

Compressive Failure. Equation (8.7) demonstrates how the relative stiffness of the column and infill influence the magnitude of the shear load required to cause compressive failure of the infill. This influence occurs because of the effect that the column stiffness has on its length of bearing against the infill.

Tests to failure of model masonry infilled frames [8.16] have shown that compressive failure shear loads may be represented more accurately by

$$Q'_c = 1.12(\lambda h)^{-0.88} f'_m ht \cos^2 \theta \tag{8.18}$$

in which θ is the angle of the infill diagonal to the horizontal (Fig. 8.2b).

Substituting for λ from Eq. (8.5) yields

$$Q'_c = 1.12 \left(\frac{4EI}{E_m t h^3} \right)^{0.22} f'_m ht \cos^2 \theta \tag{8.19}$$

Then, assuming for E/E_m a value of 30 in the case of a steel frame, and 3 for a reinforced concrete frame, and using the allowable compressive stress f_m, Eq. (8.19) gives the allowable shear on a masonry infilled steel frame approximately as

$$Q_c = 3.2 f_m \cos^2 \theta \sqrt[4]{Iht^3} \tag{8.20}$$

and that on a reinforced concrete infilled frame approximately as

$$Q_c = 1.9 f_m \cos^2 \theta \sqrt[4]{Iht^3} \tag{8.21}$$

Equations (8.20) and (8.21), which are more conservative than the theoretically deduced Eqs. (8.8) and (8.9), will be used in the design procedure.

8.3.2 Design of the Frame

Because an infilled frame behaves under horizontal loading in approximately the same way as the analogous diagonally braced truss, the members can be designed directly on the basis of the dead, live, and wind loading.

Columns. The design axial forces in the columns will be the worst combinations of the forces from gravity and wind loading acting on the analogous braced frame. In addition, on the basis of the results of the stress analyses, columns should be assigned to have a bending moment with a conservative value of $Q \cdot h/20$. The shear force in the ends of a column should be assumed equal to the horizontal component of the infill force at that level, that is Q.

Beams. The axial force in a beam may also be obtained from the analysis of the analogous frame. Theoretically, this will be a tensile or compressive force equal to the external shear at that level; however, this will be a conservative value because in reality the force will be shared with the floor slab.

If a beam has an infill above and below, it will be restrained against bending in the vertical plane. If, however, a beam does not have an infill above, or it does not have one below, the vertical thrust from the infill will cause a bending moment in the beam. As for the columns, this bending moment may be taken conservatively to be equal to $Q \cdot h/20$.

The shear force in the ends of a beam may be taken to be equal to the vertical component of the infill diagonal force, that is $Q \cdot h/L$.

In designing for beam moments and shears caused by an infill above, when there is no infill below, these moments and shears must be added to those from gravity loading on the beam.

Connections. These should be designed to carry the axial and shear forces in the connected members. Since moment resistance of the joints has been found to make only a small difference in the overall behavior of the structure, it is not necessary for the beam-to-column connections to be designed for moment.

8.3.3 Horizontal Deflection

In contrast to the shear configuration of a laterally loaded rigid frame without infills, an infilled frame deflects in a flexural shape. This difference in deflected shape occurs because the infill greatly reduces the shear mode deformations. Attempts have been made to estimate the diagonal stiffness of the infill for the purpose of including it in a deflection analysis of the analogous truss. It has been proposed by some to take a stiffness based on an equivalent width of strut equal to a fraction of the diagonal length of the infill, while others have modified this concept to assess the equivalent width as a function of the column stiffness. Unfortunately, the correlation between predicted and observed experimental deflections has been poor, probably because of unpredictable factors such as differences in the tightness of fit of the infill in the frame, or the possible adherence of the infill to the frame at low load levels.

In braced and infilled bents, the more slender the structure, the relatively greater the influence of the column axial stiffness on the horizontal top deflection compared with that of the diagonal bracing stiffness. In view of this, together with the uncertainties about the diagonal stiffness of the infill, it is proposed here that the deflection should be calculated as for the analogous diagonally braced frame, taking the area of the equivalent diagonal struts as the product of one-tenth of the infills' diagonal length and their thickness. Assuming an elastic modulus of 7×10^3 N/mm^2, (1×10^6 lb/in.^{2}) for the equivalent diagonal strut, an analysis of the analogous braced frame will yield a conservative, that is excessive, estimate of the deflection.

8.4 SUMMARY OF THE DESIGN METHOD

On the basis of the previous discussion, procedures for checking the strengths of the frame and infills of an infill-braced structure, as well as the drift, can be formulated.

In the case of a tall infill-braced building structure, the initial design of the frame would probably be on the basis of the gravity loading, and the design of the infills on the basis of their acoustic and fire requirements. The number and direction of infills in each story must be arranged so they at least equal, and preferably exceed significantly, the minimum requirements for static stability of the structure. The loads carried by the individual bents should then be assessed so that the most heavily loaded bents can be checked for the strength of their infills and frames.

The recommended design procedure for an individual bent would be as follows.

8.4.1 Provisions

1. The axis of the frame member sections should lie within the middle third of the thickness of the infill to ensure the effective interaction of the frame and infill.
2. The height-to-length ratio of the wall should be within the range of 0.3 to 3.
3. Care should be taken during construction to ensure a tight fit of the infill in the frame.
4. The slenderness ratio of the wall should conform with the relevant Masonry Code, assuming an effective height equal to the height of the infill.
5. Openings should not be allowed in the infill except at the edges, within the middle third of the length of the sides. The maximum dimension of such openings must not exceed one-tenth of the height or length of the infill, whichever is the lesser value.

8.4.2 Design of the Infill

Two modes of infill failure may cause collapse of the structure. The first is a shear failure, stepping down diagonally through the bed joints of the masonry, and the second is by spalling and crushing of the masonry in the corners of the infill. The lesser of the two strengths should be taken as the critical value.

Shear Failure. The shear strength of the structure based on the shear failure of the infill should be estimated from:

$$Q_s = \frac{f_{bs}Lt}{1.43 - \mu[0.8(h/L) - 0.2]} \quad (8.12)$$

in which f_{bs} and μ are the allowable values of the bond shear stress and the coefficient of internal friction, respectively, as given in the relevant Code formula for the allowable shear stress in masonry [see Eq. (8.10)].

Compressive Failure. If the infill is bounded by a steel frame, the shear strength of the structure relating to a compressive failure of the infill should be estimated from

180 INFILLED-FRAME STRUCTURES

$$Q_c = 3f_m \cos^2 \theta \sqrt[4]{Iht^3} \tag{8.22}$$

and, if it is bounded by a reinforced concrete frame, from

$$Q_c = 2f_m \cos^2 \theta \sqrt[4]{Iht^3} \tag{8.23}$$

8.4.3 Design of the Frame

Axial Forces. The gravity load forces in the columns should be calculated from the tributary areas, applying reduction factors to the live load forces as appropriate. Axial forces in the columns and beams resulting from the horizontal loading should be estimated by a simple static analysis of the analogous braced frame, considering each infill as a diagonal strut.

Bending Moments and Shear Forces

Columns. In addition to the axial forces determined as above, columns should be able to withstand a design bending moment of $Q \cdot h/20$ and a shear of Q.

Beams. The beams and their connections should be designed to carry an upward shear force of $Q \cdot h/L$, less the shear force due to dead load, and a downward shear force of $Q \cdot h/L$, plus the shear force due to dead and live load.

Where an upper beam of an infilled panel is not restrained by an infill above, it should be designed to carry a negative (i.e., "hogging") moment of $Q \cdot h/20$, less the moment due to dead load. Where a lower beam of an infilled panel is not restrained by an infill below, it should be designed to carry a mid-span positive moment of $Q \cdot h/20$, in addition to the moment caused by vertical dead and live load.

8.4.4 Deflections

A conservative estimate of the horizontal deflection of an infilled frame would be given by the calculated deflection of the equivalent pin-jointed braced frame, assuming each infill to be replaced by a diagonal strut with a cross-sectional area equal to the product of one-tenth of its diagonal length and its thickness, and taking an elastic modulus equal to 7×10^3 N/mm² (1×10^6 lb/in.²). Methods for calculating the lateral deflection of a braced frame are given in Section 6.4.2.

8.5 WORKED EXAMPLE—INFILLED FRAME

A reinforced concrete, rigid-frame structure consists of a system of parallel three-bay bents, as shown in Fig. 8.4, at 20 ft centers. The outer bays of each bent are infilled by 8-in.-thick walls of 10,000 lb/in.² clay brickwork.

It is required to assess the adequacy of the walls' strength to serve as bracing for a horizontal wind pressure of 30 lb/ft².

8.5 WORKED EXAMPLE—INFILLED FRAME

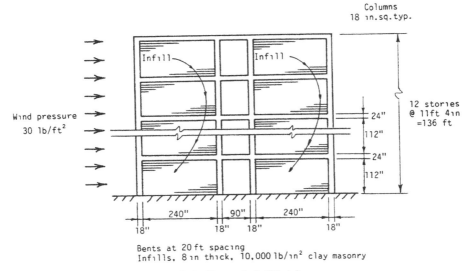

Fig. 8.4 Example infilled frame.

Properties of the Frame

$$\text{Inertia of columns} \quad I = \frac{18 \times 18^3}{12} \text{ in.}^4 = 8748 \text{ in.}^4$$

Reduced by 20% for cracking $I = 6998$ in.^{4}

For slope θ of infill diagonal $\cos \theta = 0.906$

Properties of Masonry Infill.
Taking, for example, values of 10,000 lb/in.^{2} clay masonry properties given implicitly in the UBC [8.13]

$$\text{Allowable compressive stress} \quad f_m = 3300 \text{ lb/in.}^2$$

$$\text{Allowable coefficient of friction} \quad \mu = 0.2$$

$$\begin{aligned}\text{Allowable bond shear strength} \quad f_{bv} &= 0.3(f'_m)^{1/2} \\ &= 0.3(10{,}000)^{1/2} \\ &= 30 \text{ lb/in.}^2\end{aligned}$$

Wind Shear at Base of Structure

$$Q_1 = 20 \times 12 \times 11.33 \times 30 = 81576 \text{ lb}$$
$$= 81.6 \text{ kip}$$

Structure Shear Strength—Infill Shear Failure. Using Eqn. (8.12): strength for two infills

$$Q_s = \frac{2 \times 30 \times 240 \times 8}{1.43 - 0.2[0.8 \times (112/240) - 0.2]}$$

$$= 82{,}561 \text{ lb} = 82.6 \text{ kip}$$

Structure Shear Strength—Infill Compressive Failure. Using Eqn. (8.23): strength for two infills

$$Q_c = 2 \times 2 \times 3300 \times 0.906^2 \times \sqrt[4]{6998 \times 112 \times 8^3}$$

$$= 1{,}533{,}542 = 1533.5 \text{ kip}$$

Conclusion. The infill is just adequate to carry the external shear on the basis of the shear failure criterion (strength = 82.6 kip compared with load of 81.6), and more than adequate on the basis of the compressive failure criterion (strength = 1533.5 kip). In addition to these calculations for the strength of the infill, the members of the frame should be checked to see that they are adequate to carry the forces described in Section 8.4.3. This is not included here, however, because the procedure would be the same as for the members of a low-rise structure, as well as being particular to the local Code-recommended method.

SUMMARY

An infilled frame consists of a steel or reinforced concrete frame of columns and beams containing panels of brickwork or concrete blockwork. When an infilled frame is subjected to horizontal loading, the infills behave as diagonal struts to brace the structure and restrain its lateral deflection.

A method of design is developed that considers three possible modes of failure of the infill: shear along the bedding planes of the masonry, diagonal cracking through the masonry, and crushing of a corner of the infill against a column. The estimated strengths of the three modes are based on a combination of experimental evidence and the results of theoretical stress analyses.

The forces in the frame are estimated by a simple static analysis of the analogous braced frame, considering the infills to be diagonal bracing struts. Failure of the columns at the base of the structure, by tension on the windward side and by shear on the leeward side, are of particular concern.

It is proposed that a conservative estimate of the lateral deflection of an infilled frame would be made by calculating the deflection of the analogous braced frame, assuming the equivalent diagonal strut to have an effective width equal to one-tenth of the diagonal length of the infill.

REFERENCES

8.1 Polyakov, S. V. *Masonry in Infilled Framed Buildings*, 1956, G. L. Cairns, (trans.). B.R.S. (U.K.) Publication, 1963.

8.2 Thomas, F. G. "The Strength of Brickwork." *Struct. Engineer* (U.K.), February 1953, 35–46.

8.3 Benjamin, J. R. and Williams, H. A. "The Behavior of One-Story Brick Shear Walls." *Proc. A.S.C.E.* **84,** ST4, 1958, 1723-1–1723-30.

8.4 Esteva, L. "Behavior under Alternating Loads of Masonry Diaphragms Framed by Reinforced Concrete Members." *Int. Symp. on the Effects of Repeated Loading of Materials and Structures*, Vol. V, RILEM, Mexico City, September 15–17, 1966.

8.5 Stafford Smith, B. "Model Tests Results of Vertical and Horizontal Loading of Infilled Frames." *J. A.C.I.* August 1968, 618–624.

8.6 Meli, R. and Salgardo, G. *Comportamiento de muros de mamposteria sujetos a cargo lateral* (in Spanish). National University of Mexico, September 1969.

8.7 Mainstone, R. J. and Weeks, G. A. "Influence of a Bounding Frame on the Racking Stiffness and Strengths of Brick Walls." *Proc. 2nd. Int. Brick Masonry Conference*, Stoke-on-Trent, April 1970, pp. 165–171.

8.8 Fiorato, A. E., Sozen, M. A., and Gamble, W. L. *Investigation of the Interaction of Reinforced Concrete Frames with Masonry Filler Walls*. Technical Report to the Department of Defense, University of Illinois, November 1970.

8.9 Stafford Smith, B. "The Composite Behaviour of Infilled Frames." *Proc. Symp. on Tall Buildings*, University of Southampton, 1966, pp. 481–492.

8.10 Riddington, J. R. and Stafford Smith, B. "Analysis of Infilled Frames Subject to Racking with Design Recommendations." *The Struct. Engineer* (U.K.), June 1977, 263–268.

8.11 Riddington, J. R. "Composite Behaviour of Walls Interacting with Flexural Members." Ph.D. Thesis, University of Southampton, 1974.

8.12 Hetenyi, M. *Beams on Elastic Foundations*, Vol. XVI. University of Michigan Studies, Scientific Series, 1946.

8.13 *Uniform Building Code*. International Conference of Building Officials, Whittier, California, 1976.

8.14 *Code of Practice for Use of Masonry*, BS 5628, Part 1 Structural Use of Unreinforced Masonry, British Standards Institution, 1978.

8.15 *Masonry Design for Buildings*. National Standard of Canada, CAN3-S304-M84, Canadian Standards Association, 1984.

8.16 Mainstone, R. J. "Supplementary Note on the Stiffness and Strengths of Infilled Frames." *Current Paper CP 13/74*, Building Research Establishment, U.K., February 1974.

CHAPTER 9
Shear Wall Structures

A shear wall structure is considered to be one whose resistance to horizontal loading is provided entirely by shear walls. The walls may be part of a service core or a stairwell, or they may serve as partitions between accommodations (Fig. 9.1). They are usually continuous down to the base to which they are rigidly attached to form vertical cantilevers. Their high inplane stiffness and strength makes them well suited for bracing buildings of up to about 35 stories, while simultaneously carrying gravity loading. It is usual to locate the walls on plan so that they attract an amount of gravity dead loading sufficient to suppress the maximum tensile bending stresses in the wall caused by lateral loading. In this situation, only minimum wall reinforcement is required. The term "shear wall" is in some ways a misnomer because the walls deform predominantly in flexure. Shear walls may be planar, but are often of L-, T-, I-, or U-shaped section to better suit the planning and to increase their flexural stiffness.

This chapter is concerned with the behavior of single walls and "linked-wall" systems, that is, walls that are connected by floor slabs or beams with negligible bending resistance, so that only horizontal interactive forces are transmitted. Walls connected by bending members, termed "coupled walls," are considered separately in Chapter 10.

9.1 BEHAVIOR OF SHEAR WALL STRUCTURES

A tall shear wall building typically comprises an assembly of shear walls whose lengths and thicknesses may change, or which may be discontinued, at stages up the height. The effects of such variations can be a complex redistribution of the moments and shears between the walls, with associated horizontal interactive forces in the connecting girders and slabs. As an aid to understanding the behavior of shear wall structures, it is useful to categorize them as proportionate or nonproportionate systems.

A proportionate system is one in which the ratios of the flexural rigidities of the walls remain constant throughout their height, as in Fig. 9.2a. For example, a set of walls whose lengths do not change throughout their height, but whose changing wall thicknesses are the same at any level, is proportionate. Proportionate systems of walls do not incur any redistribution of shears or moments at the change

9.1 BEHAVIOR OF SHEAR WALL STRUCTURES   185

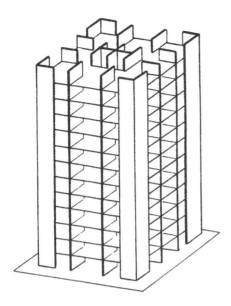

Fig. 9.1 Shear wall structure.

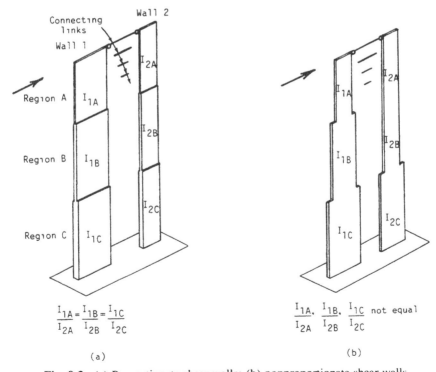

Fig. 9.2 (a) Proportionate shear walls; (b) nonproportionate shear walls.

186 SHEAR WALL STRUCTURES

levels. The statical determinacy of proportionate systems allows their analysis to be made by considerations of equilibrium, with the external moment and shear on nontwisting structures distributed between the walls simply in proportion to their flexural rigidities.

A nonproportionate system is one in which the ratios of the walls' flexural rigidities are not constant up the height (Fig. 9.2b). At levels where the rigidities change, redistributions of the wall shears and moments occur, with corresponding horizontal interactions in the connecting members and the possibility of very high local shears in the walls. Nonproportionate structures are statically indeterminate and therefore much more difficult to visualize in behavior, and to analyze.

In this chapter, hand methods of analysis are described for proportionate nontwisting and twisting structures. A hand method of analysis is presented for nonproportionate, nontwisting structures also. However, it is generally more expedient for nonproportionate, nontwisting structures, and essential for nonproportionate twisting structures, to be analyzed by computer.

9.2 ANALYSIS OF PROPORTIONATE WALL SYSTEMS

The problem of analyzing a proportionate wall system is relatively uncomplicated because of its statical determinacy. It will be considered in two subcategories of structure—those that do not twist and those that twist.

9.2.1 Proportionate Nontwisting Structures

A structure that is symmetrical on plan about the axis of loading, as in Fig. 9.3, will not twist. At any level i, the total external shear Q_i, and the total external moment M_i, will be distributed between the walls in the ratio of their flexural rigidities. The resulting shear and moment in a wall j at a level i can be expressed as

$$Q_{ji} = Q_i \frac{(EI)_{ji}}{\sum (EI)_i} \qquad (9.1)$$

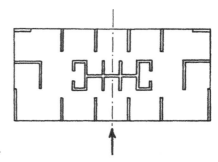

Fig. 9.3 Symmetric shear wall structure.

9.2 ANALYSIS OF PROPORTIONATE WALL SYSTEMS

and

$$M_{ji} = M_i \frac{(EI)_{ji}}{\Sigma (EI)_i} \qquad (9.2)$$

where $(EI)_{ji}$ is the flexural rigidity of wall j at level i and $\Sigma(EI)_i$ represents the summation of the flexural rigidities of all the walls at level i.

In such a proportionate nontwisting structure, there is no redistribution of shear or moment at the change levels, and no redistributive interactive forces between the walls.

9.2.2 Proportionate Twisting Structures

A structure that is not symmetric on plan about the axis of loading will generally twist as well as translate. In a proportionate shear wall structure that twists under the action of horizontal loading (Fig. 9.4) the resulting horizontal displacement of any floor is a combination of a translation and a rotation of the floor about a center of twist, which, in a proportionate structure, is located at the "centroid" of the flexural rigidities of the walls. Referring to the asymmetric cross-wall structure in Fig. 9.5, and assuming that the stiffness of a planar wall transverse to its plane is negligible, the X-location of the center of twist from an arbitrary origin is

$$\bar{x} = \frac{\Sigma (EIx)_i}{\Sigma (EI)_i} \qquad (9.3)$$

in which $(EI)_i$ and $(EIx)_i$ are, respectively, the sum of the flexural rigidities and the sum of the first moments of the flexural rigidities about the origin, for all the walls parallel to the Y axis at level i.

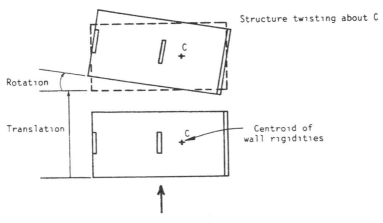

Fig. 9.4 Displacements of asymmetric structure.

188 SHEAR WALL STRUCTURES

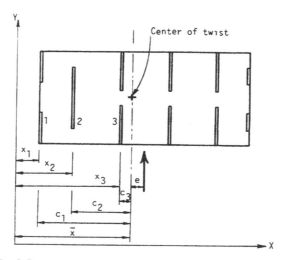

Fig. 9.5 Asymmetric structure with walls parallel to loading.

In a proportionate structure, the center of twist and the shear center axis of the structure coincide. Consequently, the effect of horizontal loading on the structure is to produce at level i a resultant shear Q_i, and a resultant horizontal torque, which is equal to the product of the resultant shear Q_i and its eccentricity e from the shear center, that is $Q_i e$. The resultant shear in any wall j at level i is a combination of its share of the external shear and the shear due to resisting its share of the external torque at that level, which may be expressed as

$$Q_{ji} = Q_i \cdot \frac{(EI)_{ji}}{\Sigma (EI)_i} + Q_i e \frac{(EIc)_{ji}}{\Sigma (EIc^2)_i} \tag{9.4}$$

in which c_{ji} is the distance of wall j from the shear center.

Noting that the moment in a wall can be obtained by integrating the shear ($M = \int_z^H Q dz$), integrating Eq. 9.4 leads to an expression for the moment in wall j at level i,

$$M_{ji} = M_i \cdot \frac{(EI)_{ji}}{\Sigma (EI)_i} + M_i e \frac{(EIc)_{ji}}{\Sigma (EIc^2)_i} \tag{9.5}$$

The first terms on the right-hand sides of Eqs. 9.4 and 9.5 are the shear and moment, respectively, associated with bending translation of the structure, while the second terms are associated with bending of the walls as the structure twists.

In Eqs. 9.4 and 9.5, c_{ji} is taken as positive when on the same side of the center of twist as the eccentricity e. Consequently, walls on the same side of the center

9.2 ANALYSIS OF PROPORTIONATE WALL SYSTEMS   189

of twist as the resultant loading will have their shears and moments increased by the twisting behavior, while those on the opposite side will have their shears and moments reduced.

If a proportionate structure also includes walls perpendicular to the direction of external loading, that is, aligned in the X direction, as in Fig. 9.6, the Y location of the center of twist can be defined by

$$\bar{y} = \frac{\Sigma (EIy)_t}{\Sigma (EI)_t} \qquad (9.6)$$

in which the flexural rigidities refer to only the "perpendicular" walls.

As the structure twists under the action of horizontal loading, the total set of orthogonally oriented walls will rotate about the axis of twist.

The effect of the "perpendicular" walls will be to stiffen the structure in torsion, to reduce the twist, and, in doing so, to influence the contributions to the "parallel" walls' shears and moments that result from the structure's twisting. The denominator of the second terms in Eqs. (9.4) and (9.5), for the shears and moments in the "parallel" walls, must then be modified to $\Sigma(EIc^2) + \Sigma(EId^2)$, in which EIc^2 is the second moment of the "parallel" walls' flexural rigidities about the center of twist while EId^2 refers correspondingly to the "perpendicular" walls.

Shears and moments will result in the "perpendicular" walls only from twisting of the structure. The shear at level i in a "perpendicular" wall r will be

$$Q_{ri} = Q_i e \frac{(EId)_{ri}}{[\Sigma (EIc^2) + \Sigma (EId^2)]_i} \qquad (9.7)$$

and the moment will be

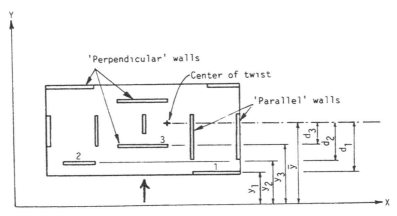

Fig. 9.6 Asymmetric structure including "perpendicular" walls.

190 SHEAR WALL STRUCTURES

$$M_{ri} = M_t e \frac{(EId)_{ri}}{\left[\sum (EIc^2) + \sum (EId^2)\right]_t} \tag{9.8}$$

Shear walls that are not aligned with the structure axes can be incorporated into such an analysis by resolving their rigidities into components along the axes at the shear centers of the walls, and treating the components of rigidity as those of walls aligned parallel or perpendicular to the direction of loading.

9.3 NONPROPORTIONATE STRUCTURES

Nonproportionate structures consist of walls whose flexural rigidity ratios are not constant throughout the height, and that consequently have different load-deflection characteristics. When the system of walls is subjected to horizontal loading, so that the structure deflects and possibly twists, the rigidity of the floor slabs constrains the dissimilar walls to deflect with similar configurations thereby inducing horizontal interactive forces between them. The horizontal interactions play a significant role in redistributing the horizontal shears and moments between the walls.

9.3.1 Nonproportionate Nontwisting Structures

A nonproportionate, plan symmetric, nontwisting structure, such as shown in Fig. 9.7a, could be analyzed using a plane frame analysis program by assembling half of the walls in a single plane, representing them by column elements, and connecting them at floor levels by axially rigid links (Fig. 9.7b) and then subjecting them to half of the loading.

A hand method exists, however, that is accurate and can also be easily programmed for a small computer [9.1]. It is an iterative relaxation method, somewhat similar in its derivation to the well-known moment distribution method. The iterations are reduced, however, by a series expression to make it a relatively concise two-step operation. Rather than presenting the derivation, which is lengthy, the principles of the method are explained, and the resulting procedure illustrated with a worked example.

Referring again to Fig. 9.7a, in which a plan symmetric set of shear walls changes flexural rigidity ratios at levels A and B, the external moment at each floor level is allocated initially between the individual walls in proportion to their flexural rigidities. At each of the change levels A and B two allocations are made, one above the change and one below. Considering any single wall, because of the nonproportional system the moments that have been allocated just above and below level A are out of balance. Equilibrium of the wall is then restored at A by applying a correcting moment that is shared above and below A in proportion to the respective wall rigidities. This is repeated at each change level in each wall.

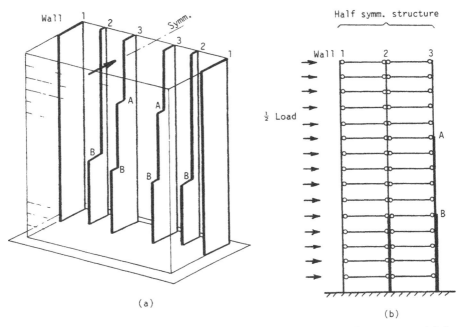

Fig. 9.7 (a) Nonproportionate, plan-symmetric structure; (b) half-structure model for computer analysis.

Now returning to level A, the correcting moments applied to the set of walls just above A will not sum to zero. This means that at that level the wall moments are not in equilibrium with the total external moment. The same situation exists just below A with a residual moment of the same amount. A correcting moment is then applied to the set of walls just above A, and similarly just below A, and distributed between the walls in proportion to their rigidities. In any one wall at level A, because of nonproportionality, the distributed moments just above and below A are again out of balance, and so the cycle of balancing begins over.

The final moment in a wall just above level A consists of the sum of the initially allocated "primary" moment, and a secondary moment that comprises all the allocations of correcting moments from the vertical and horizontal distributions. The iteration converges to a solution in which the moments in each wall just above and below each change level balance, while the sum of moments in the walls at each change level balance with the external moment. Fortunately, the steps of the iteration can be written as a mathematical series that can be represented by a simple expression.

In each wall the moments at levels other than the change points receive carryover moments from the change level correction moments; however, these diminish so rapidly with each story further from a change level that it is necessary to calculate them only for two stories above and two below each change level. This reveals the interesting information that, in nonproportionate structures at lev-

192 SHEAR WALL STRUCTURES

els more than two stories away from change levels, the external moment is shared between the walls almost exactly in proportion to their flexural rigidities, as in proportionate structures.

A procedure with a worked example for illustration is now presented. The procedure as given provides for the determination of wall moments at all levels of the structure; such a complete analysis, however, would be lengthy and tedious. Therefore, a shortened form of the analysis, which would be adequate for most design purposes, is used in the accompanying worked example. In this, the moments are found in the walls at the change levels, at one story above and one below the change levels, and at the base.

Procedure and Worked Example. Consider the structure shown in Fig. 9.8a and b, which consists of 20 3.5-m stories with a total height of 70 m. The five shear walls include two symmetrical pairs (Types 1 and 2) and a central core

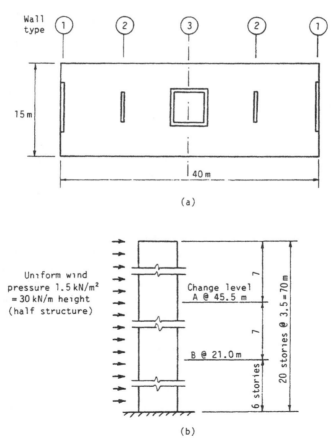

Fig. 9.8 (a) Example structure-plan; (b) example structure-end view.

9.3 NONPROPORTIONATE STRUCTURES

TABLE 9.1 Wall Dimensions and Inertias

	Wall 1		Wall 2		Wall 3		
	Dimensions (m)	Inertia I_1 (m^4)	Dimensions (m)	Inertia I_2 (m^4)	Dimensions (m)	Total Inertia I_3 (m^4)	Half Inertia $I_3/2$ (m^4)
Top region, 45.5–70 m	8 × 0.2	8.533	5 × 0.2	2.083	Outside 6 × 6, walls 0.2 m thick	26.046	13.023
Middle region, 21–45.5 m	8 × 0.3	12.800	5 × 0.3	3.125	Outside 6 × 6, walls 0.2 m thick	26.046	13.023
Bottom region, 0–21 m	8 × 0.45	19.200	7 × 0.5	14.292	Outside 6 × 6, walls 0.4 m thick	47.070	23.535

(Type 3). Two change levels, A and B, divide the structure into three regions. The wall dimensions and inertias are given in Table 9.1. Making use of plan symmetry, one-half of the structure, comprising one Type 1 wall, one Type 2 wall, and a half of the core, will be analyzed. Each stage of the worked example consists of a procedural step in algebraic terms, together with a corresponding numerical step for the example structure.

Procedure. For a nontwisting system of nonproportionate shear walls numbered 1, 2, 3, ..., j, ..., n, and with change levels A, B, ..., x.

1. Determine for each wall j at each change level x, the parameters

   a.
   $$k^t_{xj} = \frac{I^t_{xj}}{\sum_{j=1}^{n} I^t_{xj}} \quad \text{and} \quad k^b_{xj} = \frac{I^b_{xj}}{\sum_{j=1}^{n} I^b_{xj}} \qquad (9.9)$$

   where I^t_{xj} and I^b_{xj} are, respectively, the inertias of wall j just above and just below change point x. For example, for wall 1 at change level A

   $$k^t_{a1} = \frac{8.533}{8.533 + 2.083 + 13.023} = 0.361$$

   and

   $$k^b_{a1} = \frac{12.800}{12.800 + 3.125 + 13.023} = 0.442$$

   The other values, for change level B and for other walls, are obtained similarly.

b.
$$\Delta k_{\nu j} = k_{\nu j}^b - k_{\nu j}^t \qquad (9.10)$$

For example, for wall 1 at change level A

$$\Delta k_{a1} = 0.442 - 0.361 = 0.081$$

The other values, for change level B and for other walls, are obtained similarly.

c.
$$\rho_{xj}^t = \frac{-I_{\nu j}^t}{I_{\nu j}^t + I_{xj}^b} \quad \text{and} \quad \rho_{\nu j}^b = \frac{I_{\nu j}^b}{I_{\nu j}^t + I_{\nu j}^b} \qquad (9.11)$$

For example, for wall 1 at change level A

$$\rho_{a1}^t = \frac{-8.533}{8.533 + 12.800} = -0.400$$

$$\rho_{a1}^b = \frac{12.800}{8.533 + 12.800} = 0.600$$

The other values for change level B and for other walls are obtained similarly. It should be noted that as a check, at a change level, $\Sigma k_{xj}^t = \Sigma k_{\nu j}^b = 1$ for the set of walls, and $|\rho_{\nu j}^t| + |\rho_{\nu j}^b| = 1$ for each wall.

The values of k, Δk, and ρ, obtained from steps 1a, 1b, and 1c are entered in Table 9.2.

2. Determine also for each change level

$$\alpha_x = \sum_{j=1}^{n} \rho_{\nu j}^t \, \Delta k_{\nu j} \qquad (9.12)$$

For example, considering change level A, and using values from Table 9.2.

TABLE 9.2 Parameters for Analysis

Change Level	Wall	$k_{\nu j}^t$	$k_{\nu j}^b$	$\Delta k_{\nu j}$	$\rho_{\nu j}^t$	$\rho_{\nu j}^b$	$\beta_{\nu j}^t$	$\beta_{\nu j}^b$
A	1	0.361	0.442	0.081	−0.400	0.600	−0.036	0.045
	2	0.088	0.108	0.020	−0.400	0.600	−0.009	0.011
	3	0.551	0.450	−0.101	−0.500	0.500	0.045	−0.056
				$\alpha_a = 0.0101$				
B	1	0.442	0.336	−0.106	−0.400	0.600	0.030	−0.076
	2	0.108	0.251	0.143	−0.179	0.821	−0.030	0.113
	3	0.450	0.413	−0.037	−0.356	0.644	0	−0.037
				$\alpha_b = 0.0296$				

9.3 NONPROPORTIONATE STRUCTURES   195

$$\alpha_a = (-0.400)(0.081) + (-0.400)(0.020) + (-0.500)(-0.101)$$
$$= 0.0101$$

The other value for change level B is obtained and entered in Table 9.2.

3. Using the parameters evaluated in Steps 1 and 2, determine for each wall j at change level x:

$$\beta'_{xj} = \frac{1}{1 - \alpha_x} (\rho'_{xj} \Delta k_{xj} - \alpha_x k'_{xj})$$

and

$$\beta^b_{xj} = \frac{1}{1 - \alpha_x} (\rho^b_{xj} \Delta k_{xj} - \alpha_x k^b_{xj}) \qquad (9.13)$$

For example, for wall 1 at change level A, and using the values from Table 9.2

$$\beta'_{a1} = \frac{1}{1 - 0.0101} (-0.400 \times 0.081 - 0.0101 \times 0.361) = -0.036$$

$$\beta^b_{a1} = \frac{1}{1 - 0.0101} (0.600 \times 0.081 - 0.0101 \times 0.442) = 0.045$$

The other values, for change level B and for other walls, are obtained similarly and the results entered in Table 9.2.

4. Calculate the total external moment M_i on the structure at each level i, designating as $M_a, M_b, \ldots, M_x$, the external moments at change levels, $A, B, \ldots, X$.

For example, at levels $A + 1$, A, and $A - 1$

$$M_{a+1} = 30(70 - 49)^2/2 = 6615 \text{ kNm}$$
$$M_a = 30(70 - 45.5)^2/2 = 9004 \text{ kNm}$$
$$M_{a-1} = 30(70 - 42)^2/2 = 11760 \text{ kNm}$$

The other values, for change level B, plus and minus one story, and for the base, are obtained similarly. The results for this step and for all subsequent steps are entered in Table 9.3.

5. Determine the primary moments in each wall j
   a. just above and below each change level X, using respectively

$$M'_{pxj} = k'_{xj} M_x \quad \text{and} \quad M^b_{pxj} = k^b_{xj} M_x \qquad (9.14)$$

TABLE 9.3 Bending Moments in Shear Walls (in kNm)

Floor Level	External Moment M_t	Wall 1			Wall 2			Half Wall 3		
		Primary Moment M_{pt}	Second Moment M_{vt}	Final Moment M_{fit}	Primary Moment M_{pt}	Second Moment M_{vt}	Final Moment M_{fit}	Primary Moment M_{pt}	Second Moment M_{vt}	Final Moment M_{fit}
$A + 1$	6615	2388	−87	2301	582	−22	560	3645	+109	3754
A^t	9004	3250	+324	3574	792	+81	873	4961	−405	4556
A^b	9004	3980	−405	3575	972	−99	873	4052	+504	4556
$A − 1$	11760	5198	+109	5307	1270	+27	1297	5292	−135	5157
$B + 1$	31054	13726	+289	14015	3354	−289	3065	13974	0	13974
B^t	36015	15919	−1080	14839	3890	+1080	4970	16207	0	16207
B^b	36015	12101	+2737	14838	9040	−4070	4970	14874	+1333	16207
$B − 1$	41344	13892	−734	13158	10377	+1091	11468	17075	−357	16718
Base	73500	24696	0	24696	18449	0	18449	30356	0	30356

9 3 NONPROPORTIONATE STRUCTURES

For example, at change level A in wall 1

$$M'_{pa1} = 0.361 \times 9004 = 3250 \text{ kNm}$$

$$M^b_{pa1} = 0.442 \times 9004 = 3980 \text{ kNm}$$

The other values, for change level B and for other walls, are obtained similarly.

b. at all other floor levels i using

$$M_{pij} = k_{ij} M_i \qquad (9.15)$$

For example, in wall 1 at levels $a + 1$ and $a - 1$

$$M_{p,a+1} = 0.361 \times 6615 = 2388 \text{ kNm}$$

$$M_{p,a-1} = 0.442 \times 11760 = 5198 \text{ kNm}$$

The other values, for levels just above and below floor levels B, for the base and for other walls, are obtained similarly.

6. Determine the secondary moments in each wall j at the following levels:
   a. Just above and below each change level X using, respectively,

$$M'_{xxj} = -\beta'_{xj} M_x \quad \text{and} \quad M^b_{xxj} = -\beta^b_{xj} M_x \qquad (9.16)$$

For example, in wall 1 at change level A

$$M'_{xa1} = -(-0.036) \times 9004 = 324 \text{ kNm}$$

$$M^b_{xa1} = -(0.045) \times 9004 = -405 \text{ kNm}$$

The other values, above and below change level B and for other walls, are obtained similarly.

b. At two levels above and two levels below change levels using

$$M_{x,x+1,j} = -0.268\, M'_{xxj}$$

$$M_{x,x+2,j} = (-0.268)^2 M'_{xxj} \qquad (9.17)$$

$$M_{x,x-1,j} = -0.268\, M^b_{xxj}$$

$$M_{x,x-2,j} = (-0.268)^2 M^b_{xxj}$$

If further refinement in results is sought for stories beyond these, the progressions in Eq. (9.17) should be extended.

In the example, at one level above and one level below change level A in wall 1

$$M_{x,a+1,1} = -0.268 \times 324 = -87 \text{ kNm}$$

$$M_{x,a-1,1} = -0.268 \times (-405) = 109 \text{ kNm}$$

7. The final moments are then obtained from the sum of the primary moment and the corresponding secondary moment, thus
   a. at change level x in wall j using

$$M_{fxj} = M'_{pxj} + M'_{sxj} \tag{9.18}$$

which, as a check on the result, should equal

$$M_{fxj} = M^b_{pxj} + M^b_{sxj} \tag{9.19}$$

For example, at change level A in wall 1

$$M'_{fa1} = 3250 + 324 = 3574 \text{ kNm}$$

and, as a check

$$M^b_{fa1} = 3980 - 405 = 3575 \text{ kNm}$$

b. at intermediate floors i in wall j using

$$M_{fij} = M_{pij} + M_{sij} \tag{9.20}$$

As an example, at level $a + 1$ in wall 1, and using Eq. (9.17) the final moment

$$M_{f,a+1,1} = 2388 + (-0.268)324 = 2301 \text{ kNm}$$

All other primary moments and their corresponding secondary moments are summed to obtain the final moments.

8. The shear within a story-height region of a wall is then given by the difference in moments in the wall at the top and bottom of the story, divided by the story height. For example, the shear in wall 1 in story 14, that is between levels A and $A + 1$ is given by

$$Q_{a+1,1} = \frac{M_{a,1} - M_{a+1,1}}{h}$$

$$= \frac{3574 - 2301}{3.5} = 364 \text{ kN} \tag{9.21}$$

These values have not been included in Table 9.3.

9.3.2 Nonproportionate Twisting Structures

Structures that are asymmetric in plan, as in Fig. 9.9a, generally twist when subjected to horizontal loading. The complication of z axis rotation, as an additional variable, makes the problem even less tractable, and a computer analysis is the only practical method.

A convenient model for the computer analysis of a shear wall structure uses column elements along the centroidal axes of the walls to represent the shear walls (Fig. 9.9b), with either the assignment of a "rigid-floor" option or with constraining members in the horizontal plane to represent the inplane rigidity of the floor slabs. The columns are assigned the flexural and shear rigidities of the corresponding walls. For walls that change only in thickness, or symmetrically in width, so that their axes are vertically continuous, the column elements are stacked in a vertical line. For a wall that changes width asymmetrically, the column element above the change should be connected to the offset column element below by a horizontal rigid beam (Fig. 5.2b).

If the available structural analysis program does not have a "rigid-floor" option, the constraint exerted by the inplane rigidity of the slabs on the relative horizontal displacements of the walls can be represented by incorporating at each floor level a horizontal rigid frame interconnecting the centroidal axes of the walls. The beams are assigned to be axially rigid and flexurally rigid in the horizontal plane, but to have negligible stiffness against vertical plane bending. If the line of the resultant load does not coincide with one of the columns it may be applied at each floor as a statically equivalent pair of forces to two of the columns.

An alternative to the three-dimensional model described above would be an equivalent planar model, as explained in Chapter 5. Such a model would allow a much reduced, two-dimensional analysis, which would be more amenable to analysis by a microcomputer.

9.4 BEHAVIOR OF NONPROPORTIONATE STRUCTURES

Considering the case of a nonproportionate shear wall structure that does not twist, as represented by the equivalent planar structure in Fig. 9.10a, the links constrain the walls to have the same curvature in the uniform regions away from the change levels. Consequently, in those regions, the external moment is distributed between the walls in the same ratio as their flexural rigidities, as would be the case if the structure were proportionate. In the transition from above to below a change level, a redistribution of the wall moments must take place to satisfy the change in the ratio of the wall rigidities. Because the only mechanism allowing a force transfer between the walls is by horizontal forces in the connecting links, the moment redistribution must occur by couples consisting of horizontal forces and reverse forces in the links at successive levels around the exchange level, as in Fig. 9.10b. The size of the moments transferred is usually large enough to cause the interactive forces at the change level to be very large, with the result that, locally, the shear

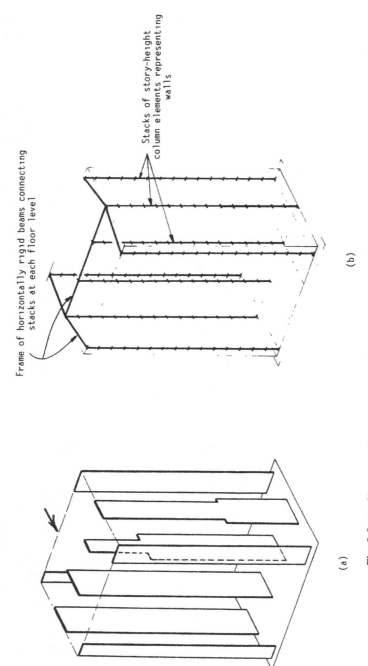

Fig. 9.9 (a) Nonproportionate, plan-asymmetric structure; (b) model for computer analysis.

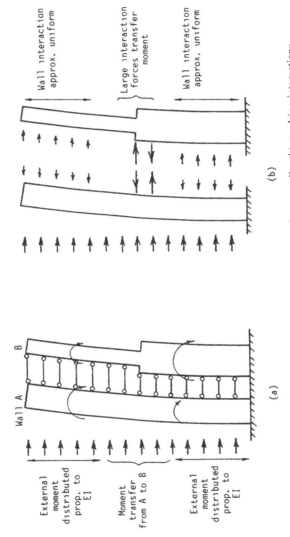

Fig. 9.10 (a) Allocation of moment between nonproportionate walls; (b) resulting interactions.

in a wall and the reverse shear in another wall may easily exceed the total external shear at that level. The severe local effects on the walls' loading close to the change level due to the moment transfer give rise to carryover effects above and below, which diminish within one or two stories before becoming negligible. The shear force diagrams of the walls are, therefore, significantly disturbed by the change levels. The wall moment diagrams in changing to a different distribution across a change level are, however, much less disturbed. These effects are exemplified by the shear and moment diagrams in Fig. 9.11a and 9.11b, respectively, which are plotted from the results of a full computer analysis of the structure in the worked example of Section 9.3 (Fig. 9.8).

Nonproportionate shear wall structures that twist and translate under horizontal loading also behave similarly to proportionate structures in regions away from the change levels, with the walls' resulting moments being a combination of the moments from their flexural and flexural torsion rigidity effects, as expressed by Eq. (9.5). At the change levels, transfers of moment occur with severe disturbances in the walls' interactions and shears. Because of the twist, the transfers of moment at the change levels have to accommodate the effects of both the walls' flexural resistance to twisting of the structure, which involves the plan locations of the walls, as well as the walls' resistance to the structure's bending.

9.5 EFFECTS OF DISCONTINUITIES AT THE BASE

In medium high-rise apartment blocks that depend for their horizontal resistance on cross walls, it is not uncommon for some of the walls to be partially discontinued at the base to provide for lobby space. To discuss the different types of force interactions that may occur, two extreme cases of discontinuity will be considered. Case 1 is illustrated by Fig. 9.12a and Case 2 by Fig. 9.13a.

In Case 1 the inner pair of walls have openings in the ground story, which leave each wall standing, in effect, on a pair of edge columns. In Case 2, the inner walls are cut back in the ground story to leave each wall supported on a much shorter central wall.

Consider for Case 1 the equivalent half-structure planar model, as shown in Fig. 9.12b. When the structure is subjected to horizontal loading, the flexibility of the columns supporting the right-hand wall causes the ground story of that wall to be very much less transversely stiff than that of the left-hand wall. The flexural stiffness of the right-hand wall, however, has been reduced by a proportionately much lesser amount because of the edge location of the columns.

The resulting effect in Case 1, therefore, is a heavy transfer of shear from the discontinuous wall to the continuous wall, with a relatively smaller transfer of moment. As an approximate illustration, the resulting forces on the walls are as shown in Fig. 9.12c.

Referring to Fig. 9.13b, which is the equivalent planar model for Case 2, the flexural rigidity of the cut-back wall is very much reduced in the ground story, whereas its transverse rigidity has suffered by a proportionately lesser amount.

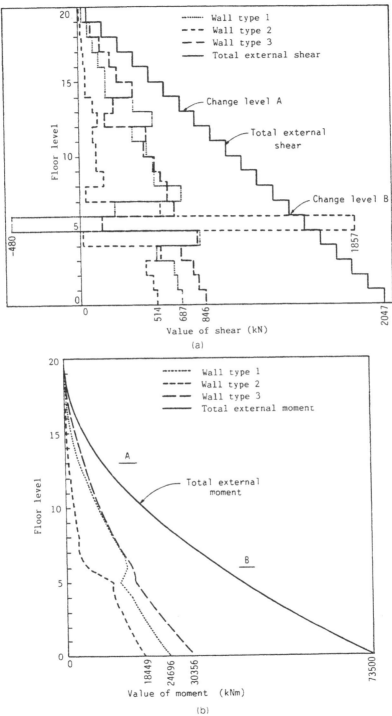

Fig. 9.11 (a) Resulting shear in walls; (b) resulting moments in walls.

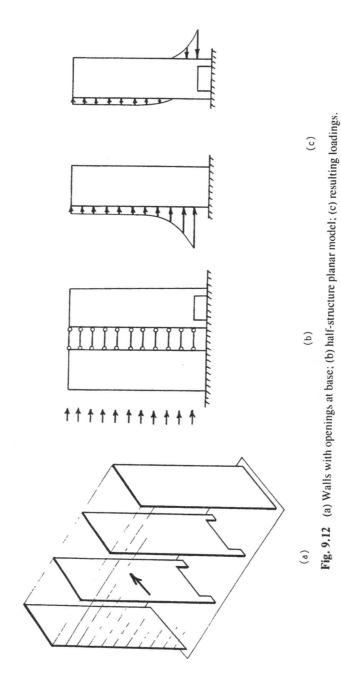

Fig. 9.12 (a) Walls with openings at base; (b) half-structure planar model; (c) resulting loadings.

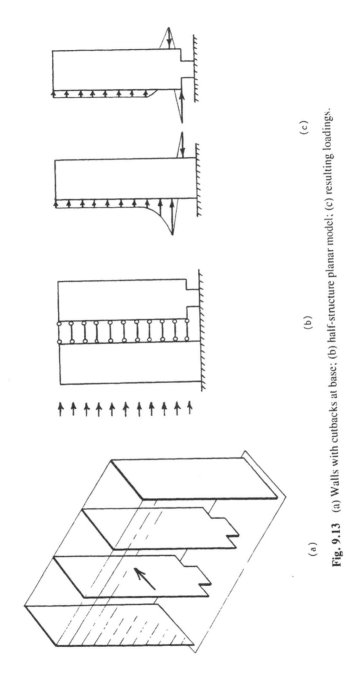

Fig. 9.13 (a) Walls with cutbacks at base; (b) half-structure planar model; (c) resulting loadings.

Consequently, there is a very large transfer of moment from the cut-back wall to the continuous wall in the levels just above the ground story, with correspondingly large horizontal interactions between the walls, and high forward and reverse shears.

As an approximate illustration, the forces on the walls are as in Fig. 9.13c. It is quite possible, in Case 2-type structures with severe reductions in length of certain walls, for the remaining walls to be subjected to a shear of twice, or more than twice, the total external shear on the structure [9.2].

This discussion of the effects of discontinuities is intended mainly to explain the modes of the resulting actions, and to serve notice about their potential significance. The desirability of a detailed analysis in such a case, rather than an approximate intuitive estimate of the forces, is evident.

9.6 STRESS ANALYSIS OF SHEAR WALLS

When the loads acting on an individual shear wall have been determined, the next stage of the design process is to use the loads to determine the wall stresses. If the wall is rectangular in elevation and has a height-to-width ratio greater than 5, a close estimate of the axial stresses is given by simple bending theory. If, however, the aspect ratio is less than 5:1, or if it is irregular with changes in width or openings, or if beams or other walls connect to it, a more detailed analysis is necessary. For this, the finite element technique is convenient, versatile, and accurate.

The major structural analysis programs usually include a selection of elements suitable for problems of walls bending in their planes. If only a frame analysis program is available a shear wall can be analyzed alternatively by an analogous frame consisting of beam members. Although not able to compete with finite elements in representing nonrectangular shapes, the analogous frame has some virtue in its simpler mathematical concept and in its amenability to solution by an ordinary frame analysis program.

In this section, a discussion of the use of finite elements for shear wall analysis is followed by a description of an analogous frame particularly appropriate to shear wall analysis.

9.6.1 Membrane Finite Element Analysis

The predominantly inplane action of shear walls allows them to be satisfactorily represented in most cases by plane stress membrane elements. The simplest types of rectangular and quadrilateral elements are usually adequate. In some wall analyses, greater accuracy could be achieved by using a smaller number of complex elements. Typically, however, shear wall analysis, with its need for a refined stress description only in local regions, is better satisfied by a larger number of simpler elements.

A detailed discussion of modeling techniques for the computer analysis of shear walls is given in Chapter 5. If a refined analysis is needed in certain regions of a shear wall, either because the wall has irregularities such as openings, or because of anticipated high stress gradients in certain regions, such as close to the base, a refined mesh should be used in these regions. The refined mesh may be included as part of the primary model, with a transition mesh connecting between the coarse and fine mesh regions, as described in Chapter 5. If, however, the required stress detail calls for the mesh in a region to be very much finer than elsewhere it may be inadvisable to analyze the complete model simultaneously, even if the computer capacity allows it. Large differences in element size can lead to computational errors. Such cases may be better treated by a preliminary analysis with the region in question represented relatively coarsely. A separate more detailed analysis of the region can then be made using a very refined mesh with imposed boundary loads or point displacements taken from the first analysis. Caution should be applied in accepting detailed stress results immediately adjacent to sharp internal corners where, theoretically, the stresses are infinite. The more refined the mesh used in such regions, the larger and more alarming the stress results become.

9.6.2 Analogous Frame Analysis

When only a frame analysis program is available, and the shape of the shear wall can be divided into a mesh of rectangular segments, an analogous frame may be used for analysis. Earlier analogous frames [9.2] were developed for plate stretching and compression, without particular reference to inplane bending. Only if used in a refined mesh across the wall will these satisfactorily represent bending. More recently, analogous frames have been developed for shear wall analysis that accommodate inplane bending.

One of these which has proved to be superior in efficiency is termed the braced-frame analogy [9.3]. Its concept, derivation and application are considered next.

Braced-Frame Analogy. The analogy was conceived originally with a symmetric module (Fig. 9.14a). It consisted of two columns joined by rigid beams at the top and bottom, and diagonal braces. This was satisfactory as a full-width unit for plane walls and for orthogonal shear wall assemblies with not more than two walls joining at any corner. In general cases of shear walls with fractional-width meshes, or assemblies with more than two walls at a corner, the symmetric module is inadequate because the adjacent columns of intersecting modules cannot easily be arranged to bend independently, as they must. Consequently, the module was modified to be asymmetric (Fig. 9.14b), with a column on the left-hand side connecting to the rigid beams, a hinged-end link on the right-hand side, and diagonal braces. The left-hand ends of the beams and the ends of the column rotate with the nodes, while the right-hand ends of the beams and the link are rotationally released from the nodes. Although physically asymmetric about its vertical center line, the module behaves in the same way as the symmetric module.

The requirement of the frame module is that it should simulate the bending, shear, and vertical axial stiffnesses of the corresponding wall segment. The properties of the module members are derived as follows.

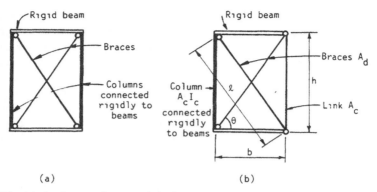

Fig. 9.14 (a) Analogous frame: original symmetric module; (b) analogous frame: improved module.

Bending Stiffness. The bending stiffness of the wall segment (Fig. 9.15a) must be matched by the bending stiffness of the module (Fig. 9.15b). The latter is given by the sum of the flexural stiffness of the column, and the bending resistance of the column and link sectional areas acting about the center line of the module. Assuming the wall and module to have the same elastic modulus, E.

$$EI_c + 2EA_c \left(\frac{b}{2}\right)^2 = E\frac{tb^3}{12} \qquad (9.22)$$

Shear Stiffness. The shear stiffnesses of the wall segment (Fig. 9.16a) and the frame module (Fig. 9.16b) should be equal. This is provided in the module by the sum of the transverse stiffness of the column in double curvature bending and the horizontal components of the axial stiffness of the diagonals.

$$\frac{12EI_c}{h^3} + \frac{2EA_d \cos^2 \theta}{l} = \frac{Gbt}{h} = \frac{Ebt}{2(1+\mu)h} \qquad (9.23)$$

in which G is the shear modulus, μ is Poisson's ratio, and $G = E/2(1+\mu)$.

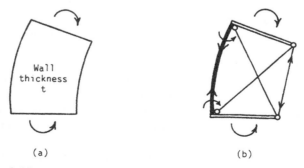

Fig. 9.15 (a) Wall segment: flexure; (b) analogous frame: flexure.

9.6 STRESS ANALYSIS OF SHEAR WALLS

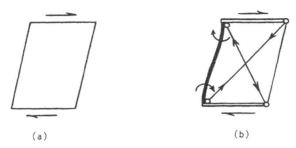

Fig. 9.16 (a) Wall segment: shear; (b) analogous frame: shear.

Axial Stiffness. The axial stiffnesses of the wall segment (Fig. 9.17a) and the frame module (Fig. 9.17b) should be the same. The axial stiffness of the module comprises the sum of the axial stiffnesses of the column and link, and the vertical components of the axial stiffness of the diagonals.

$$\frac{2EA_c}{h} + \frac{2EA_d \sin^2 \theta}{l} = \frac{Ebt}{h} \tag{9.24}$$

Solving Eqs. (9.22) to (9.24) simultaneously gives the properties of the frame members:

COLUMN:

$$\text{Moment of inertia} \quad I_c = \frac{tb^3}{12}(6B - 0.5) \tag{9.25}$$

$$\text{Sectional area} \quad A_c = tb(0.25 - B) \tag{9.26}$$

LINK:

$$\text{Sectional area} \quad A_c = tb(0.25 - B) \tag{9.27}$$

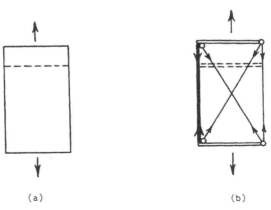

Fig. 9.17 (a) Wall segment: axial deformation; (b) analogous frame: axial deformation.

210 SHEAR WALL STRUCTURES

DIAGONAL BRACES:

$$\text{Sectional area} \quad A_d = \frac{tb(0.25 + B)}{\sin^3 \theta} \quad (9.28)$$

in which

$$B = \frac{h^2}{16b^2(1 + \mu)} \quad (9.29)$$

The other properties of the braced frame members are assigned to be rigid. Equation (9.25) can be used to show that I_c is negative for segment height-to-width ratios less than $2[(1 + \mu)/3]^{1/2}$, and Eq. (9.26) can be used to show that A_c is negative for height-to-width ratios greater than $2(1 + \mu)^{1/2}$. Although negative property members are a fictitious concept, frame analysis programs will usually accept and process them provided the resulting direct stiffness coefficients of the structure are all positive.

Application of the Analogous Frame. An approximate analysis of a shear wall structure can be made by dividing each wall into a coarse mesh of wall-width story height modules (Fig. 9.18). For a more detailed and accurate solution, a refined mesh of fractional wall-width and story-height modules can be used. The components of each module are assigned properties according to the dimensions of the corresponding wall segment. The two rigid beams between each pair of vertically adjacent modules are replaced by a single rigid beam. Beams connecting to the wall from other parts of the building are joined rigidly to the ends of the rigid beams. The resulting analogous frame may be analyzed by any standard frame analysis program.

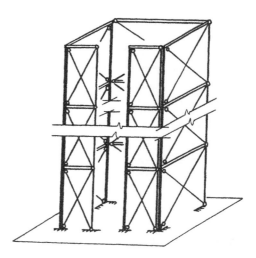

Fig. 9.18 Analogous frame model for elevator core.

Conversion of Analogous Frame Forces to Wall Stresses. The stresses in a wall segment are obtained by applying the resultant moment, axial force, and shear in the corresponding frame module as evaluated from the results of the analysis, to the horizontal section of the segment.

The resultant bending moment in a module is obtained by summing algebraically the average of the moments at the top and bottom of the column and the couples given by multiplying the axial forces in the column and link by the half-width of the module. The resultant axial force is the algebraic sum of the axial force in the column and link, and the vertical components of the axial forces in the diagonals. The resultant shear force is the algebraic sum of the shear in the column and the horizontal components of the axial forces in the diagonals. The resulting moment and axial force is applied to the segment section to obtain the wall axial stresses, and the resulting shear is applied to obtain the wall shear stresses.

The wall stresses are referred to the mid-height of the segment. The vertical stresses vary linearly over the width of the segment and are constant over its height. The shear stress is uniform over the whole segment. Horizontal direct stresses are not evaluated because the axial rigidity of the arms causes the values of the resulting internal horizontal forces to be meaningless.

The stresses from shear wall analyses using this analogous frame have shown to compare closely, that is within 1%, with results from finite element analyses having a similar sized mesh.

SUMMARY

The chapter is concerned with tall building structures that consist of assemblies of shear walls, connected only by members of low flexural resistance such as thin slabs. The walls interact, therefore, primarily through horizontal forces. The chapter is also concerned with the analysis of individual shear walls.

Assemblies of shear walls may be categorized as proportionate, when the ratios of the walls' flexural stiffnesses are constant throughout the height, or nonproportionate. Each category may be divided further for consideration into nontwisting or twisting structures.

The allocation of external shear and external moment between the walls of a proportionate system can be made on the basis of each wall's proportionate contribution to the overall flexural stiffness, and the overall flexural-torsional stiffness of the structure.

The allocation of shear and moment between the walls of a nonproportionate system is more complex because of the horizontal interaction between the walls. For a nontwisting system, a hand method based on a type of moment distribution gives an accurate solution. The method is amenable to programming for use on a very small microcomputer.

For a twisting, nonproportionate system, it is recommended that a stiffness matrix computer analysis is carried out with the walls represented by column elements.

The stress analysis of rectangular walls with a height-to-length ratio greater than five can be made with acceptable accuracy by simple bending theory. Shorter walls, irregular walls, and walls with openings should be analyzed by using either membrane finite element or analogous frame models. Plane stress membrane elements of rectangular or quadrilateral shape are recommended, with refinements of the mesh in regions of special interest.

If only a frame analysis computer program is available, and if the wall can be divided into a rectangular mesh, an equivalent frame may be used to obtain a reasonably accurate solution.

REFERENCES

9.1 Gluck, J. "Lateral Load Analysis of Multi-story Structures Comprising Shear Walls with Sudden Changes in Stiffness." *J. Am. Conc. Inst.* **66,** September 1969, 729-736.

9.2 Lerner, E. and Stafford Smith, B. "Severe Interaction Effects between Plain and Irregular Shear Walls." *Proc. 4th Canadian Conf. on Earthquake Engineering*, University of British Columbia, Vancouver, June 1983, 220-230.

9.3 McCormick, C. W. "Plane Stress Analysis." *ASCE J. Struct. Div.* **89**(St 4), August 1963, 37-54.

9.4 Stafford Smith, B. and Girgis, A. "Simple Analogous Frames for Shear Wall Analysis." *ASCE J. Struct. Engineer.* **110**(11), November 1984, 2655-2666.

CHAPTER 10

Coupled Shear Wall Structures

The previous chapter considered shear wall structures in which lateral loads on the building are resisted by the independent actions of the individual walls. In many practical situations, however, walls are connected by moment-resisting members. For example, walls in residential buildings will be perforated by vertical rows of openings that are required for windows on external gable walls or for doorways or corridors in internal walls. In the design of slab residential blocks consisting of walls and floor slabs only, self-contained apartment units are generally arranged on opposite sides of a central corridor along the length of the building. This arrangement naturally results in parallel assemblies of division walls running perpendicular to the face of the building, with intersecting longitudinal walls along the corridor and facade enclosing the living spaces (Fig. 10.1). In addition to serving the functional requirements of dividing and enclosing space, and providing fire and acoustic insulation between dwellings, the cross walls are employed as load bearing walls, since their disposition favors an efficient distribution of both gravity and lateral loads to the structural elements. If the floor slabs are rigidly connected to the walls, they serve in effect as connecting beams to produce a shear interaction between the two inplane cross walls. Such structures, which consist of walls that are connected by bending-resistant elements, are termed ''coupled shear walls,'' in which the presence of the moment-resisting connections greatly increases the stiffness and efficiency of the wall system.

10.1 BEHAVIOR OF COUPLED SHEAR WALL STRUCTURES

If a pair of inplane shear walls is connected by pin-ended links that transmit only axial forces between them, any applied moment will be resisted by individual moments in the two walls, the magnitudes of which will be proportional to the walls' flexural rigidities. The bending stresses are then distributed linearly across each wall, with maximum tensile and compressive stresses on opposite edges (Fig. 10.10d). If, on the other hand, the walls are connected by rigid beams to form a dowelled vertical cantilever, the applied moment will be resisted by the two walls acting as a single composite unit, bending about the centroidal axis of the two walls. The bending stresses will then be distributed linearly across the composite unit, with maximum tensile and compressive stresses occurring at the opposite

214 COUPLED SHEAR WALL STRUCTURE

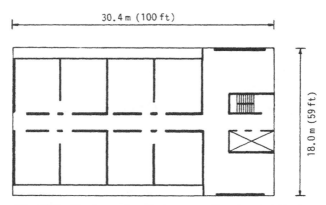

Fig. 10.1 Planform of typical cross-wall residential block.

extreme edges (Fig. 10.10c). The practical situation of a pair of walls connected by flexible beams will lie between these two extreme cases, which may be regarded as bounds on the structural behavior of a coupled wall system. The stiffer the connecting beams, the closer the structural behavior will approach that of a fully composite cantilever. The efficiency of the system may be assessed by the degree to which it approaches the optimum behavior of a composite cantilever.

When the walls deflect under the action of the lateral loads, the connecting beam ends are forced to rotate and displace vertically, so that the beams bend in double curvature and thus resist the free bending of the walls (Fig. 10.2). The bending action induces shears in the connecting beams, which exert bending moments, of opposite sense to the applied external moments, on each wall. The shears also induce axial forces in the two walls, tensile in the windward wall and compressive in the leeward wall. The wind moment M at any level is then resisted by the sum of the bending moments M_1 and M_2 in the two walls at that level, and the moment of the axial forces Nl, where N is the axial force in each wall at that level and l is the distance between their centroidal axes.

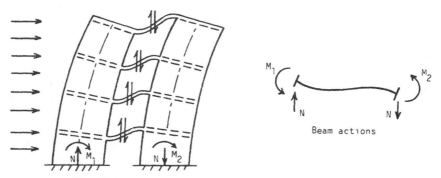

Fig. 10.2 Behavior of laterally loaded coupled shear walls.

$$M = M_1 + M_2 + Nl \tag{10.1}$$

The last term Nl represents the reverse moment caused by the bending of the connecting beams which opposes the free bending of the individual walls. This term is zero in the case of linked walls, and reaches a maximum when the connecting beams are infinitely rigid.

The action of the connecting beams is then to reduce the magnitudes of the moments in the two walls by causing a proportion of the applied moment to be carried by axial forces. Because of the relatively large lever arm l involved, a relatively small axial stress can give rise to a disproportionally larger moment of resistance. The maximum tensile stress in the concrete may then be greatly reduced. This makes it easier to suppress the wind load tensile stresses by gravity load compressive stresses.

10.2 METHODS OF ANALYSIS

As with other structural forms considered, it is possible to analyze coupled shear wall structures by either approximate or more exact techniques. The former are simpler and more amenable to hand calculation, but tend to be restricted to regular or quasiregular structures and load systems. The latter can deal with irregular structures and complex loadings, but require the services of a digital computer. The method to be employed will generally depend on the structural layout and on the degree of accuracy required.

The most important approximate method is termed the continuous medium technique. (In the literature it has also been variously termed the "continuous connection method," the "continuum method," or the "shear connection method.") As the name suggests, the structure is simplified by making the assumption that all horizontal connecting elements are effectively smeared over the height of the building to produce an equivalent continuous connecting medium between the vertical elements. This can be achieved with reasonable accuracy only for a uniform system of connecting beams or floor slabs. The two-dimensional plane structure is thereby transformed into an essentially one-dimensional one, in which all major actions depend on the height coordinate. This enables the behavior of the structure to be expressed in the form of an ordinary linear differential equation, allowing closed form solutions to be obtained.

In many practical situations, the building layout will involve walls that are not uniform over their height, but have changes in width or thickness, or in the disposition of the openings. In addition, the base support conditions may be complex due to either a discontinuation of the walls at the first story level, or the form of substructure employed. Such discontinuities do not lend themselves to a uniform smeared representation, and the continuous medium approach cannot be used with any confidence. Such irregular systems are most conveniently and accurately analyzed by using an equivalent frame approach, in conjunction with standard frame analysis programs based on the stiffness matrix method of analysis. The analyst

216 COUPLED SHEAR WALL STRUCTURE

must use his skill and experience to replace the coupled wall structure by an equivalent plane framework of beams and columns. As discussed in Chapter 5, relatively wide walls may be modeled by a line column situated at the centroidal axis, with rigid horizontal elements connecting the centroidal axis to the outer fibers at each floor level, to transmit the rotational and vertical displacement effects at the edge of the wall to the connecting beams. As a consequence, the method is frequently referred to as the "wide-column frame method." Practical shear wall structures are generally analyzed by this method.

If the wall contains irregular openings or has a complex support system, it may prove difficult to represent the structure by a plane frame model with any degree of confidence. In that case, it is better to use a finite element model with an assembly of plane stress elements (cf. Chapter 5, Section 5.5).

Previously, it was considered uneconomical to use the finite element technique for the complete analysis of such structures. However, with the availability of general purpose computer programs that include a wide variety of line and surface elements, a complete analysis is now a more reasonable proposition.

This chapter considers in detail the continuous medium and the analogous frame methods of analysis of coupled walls. Since the analysis of plane structures by the finite element method is well documented elsewhere in the literature, only a brief discussion of the method is given with particular reference to its use in the analysis of coupled walls.

10.3 THE CONTINUOUS MEDIUM METHOD

This approximate method allows a broad look at the behavior of coupled wall structures and, simultaneously, gives a good qualitative and quantitative understanding of the relative influences of the walls and the connecting beams or slabs in resisting lateral forces.

10.3.1 Derivation of the Governing Differential Equations

Consider the plane coupled-wall structure shown in Fig. 10.3a subjected to distributed lateral loading of intensity w per unit height. A general form of loading is used to illustrate the derivation of the governing differential equation, before solutions are derived for common standard design load cases.

The basic assumptions made in the analysis are as follows:

1. The properties of the walls and connecting beams do not change over the height, and the story heights are constant.
2. Plane sections before bending remain plane after bending for all structural members.
3. The discrete set of connecting beams, each of flexural rigidity EI_b, may be replaced by an equivalent continuous connecting medium of flexural rigidity EI_b/h per unit height, where h is the story height (Fig. 10.3b). Strictly

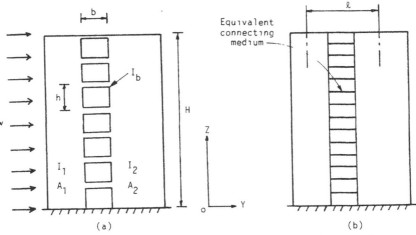

Fig. 10.3 Representation of coupled shear walls by continuum model.

speaking, for this analogy to be correct, the inertia of the top beam should be half of the other beams.

4. The walls deflect equally horizontally, as a result of the high inplane rigidity of the surrounding floor slabs and the axial stiffness of the connecting beams. It follows that the slopes of the walls are everywhere equal along the height, and thus, using a straightforward application of the slope-deflection equations, it may be shown that the connecting beams, and hence the equivalent connecting medium, deform with a point of contraflexure at mid-span. It also follows from this assumption that the curvatures of the walls are equal throughout the height, and so the bending moment in each wall will be proportional to its flexural rigidity.

5. The discrete set of axial forces, shear forces, and bending moments in the connecting beams may then be replaced by equivalent continuous distributions of intensity n, q, and m, respectively, per unit height.

In particular, if the connecting medium is assumed cut along the vertical line of contraflexure, the only forces acting there are a shear flow of intensity $q(z)$ per unit height and an axial force of intensity $n(z)$ per unit height, as in Fig. 10.4. The axial force N in each wall at any level z will then be equal to the integral of the shear flow in the connecting medium above that level, that is,

$$N = \int_z^H q\, dz \qquad (10.2)$$

or, on differentiating

$$q = -\frac{dN}{dz} \qquad (10.3)$$

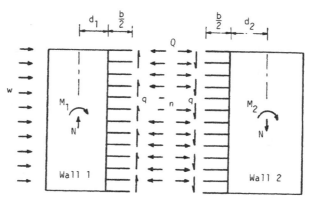

Fig. 10.4 Internal forces in coupled shear walls.

Consider now the condition of vertical compatibility along the cut line of contraflexure of Fig. 10.4. Relative vertical displacements will occur at the cut ends of the cantilevered laminas due to the following four basic actions. [In the derivation, positive relative displacements are taken to mean that the end of the left-hand lamina (1) moves downward relative to the end of the right-hand lamina (2).]

1. Rotations of the wall cross sections due to bending (Fig. 10.5a). Under the action of a bending moment, the wall will deflect, and cross sections will rotate as shown in Fig. 10.5a. Two forms of bending action occur: first, the free bending of the walls due to the applied external moments, and second, the reverse bending caused by the shear forces and axial forces in the connecting beams.

The relative vertical displacement δ_1 is given by (Fig. 10.5a)

$$\delta_1 = \left(\frac{b}{2} + d_1\right)\frac{dy}{dz} + \left(\frac{b}{2} + d_2\right)\frac{dy}{dz} = l\frac{dy}{dz} \tag{10.4}$$

where dy/dz is the slope of the centroidal axes of the walls at level z due to the combined bending actions.

2. Bending and shearing deformations of the connecting beams under the action of the shear flow (Fig. 10.5b). Consider a small element of the connecting medium of depth dz, which may be assumed cantilevered from the inner edge of the wall. The flexural rigidity of this small lamina is $(EI_b/h)\,dz$, and the cantilever is subjected to a tip shear force of $q\,dz$.

Due to bending only, the relative displacement δ_2 is given by

$$\delta_2 = -2\frac{q\,dz}{3(EI_b/h)\,dz}\left(\frac{b}{2}\right)^3 = -\frac{qb^3h}{12EI_b} \tag{10.5}$$

where b is the clear span of the beams.

The effects of shearing deformations in the connecting beams may readily be

10.3 THE CONTINUOUS MEDIUM METHOD

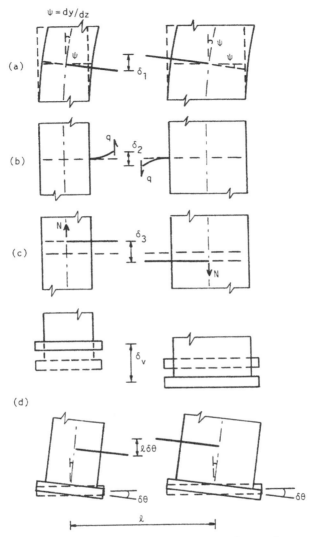

Fig. 10.5 Relative displacements at line of contraflexure.

included by replacing the true flexural rigidity EI_b by an equivalent flexural rigidity EI_e [10.1], where

$$I_e = \frac{I_b}{1+r} \tag{10.6}$$

and

$$r = \frac{12EI_b}{b^2 GA}\lambda$$

in which GA is the shearing rigidity and λ is the cross-sectional shape factor for shear, equal to 1.2 in the case of rectangular sections. The correction is necessary only in the case of connecting beams with a span-to-depth ratio less than about 5.

The evaluation of δ_2 has assumed that the connecting beam is rigidly connected to the wall, and thus ignores the effects of local elastic deformations at the beam–wall junction that will increase the flexibility of the lamina. Both elasticity and finite element studies have shown that the additional flexibility that arises may be included by the simple expedient of extending the beam length by a further quarter-beam depth into the wall at each end. The length b in Eq. (10.5) should thus be taken as the true length $b + \frac{1}{2}$ beam depth.

Equations (10.3) and (10.6) enable Eq. (10.5) to be expressed in terms of the axial forces N, as

$$\delta_2 = +\frac{b^3 h}{12 E I_c} \frac{dN}{dz} \qquad (10.7)$$

3. Axial deformations of the walls under the actions of the axial forces N (Fig. 10.5c). The action of the shear forces in the connecting beams will be to induce tensile forces in the windward wall 1 and compressive forces in the leeward wall 2. Consequently, the relative displacement, δ_3 at level z will be

$$\delta_3 = -\frac{1}{E}\left(\frac{1}{A_1} + \frac{1}{A_2}\right) \int_0^z N\, dz \qquad (10.8)$$

where A_1 and A_2 are the cross-sectional areas of walls 1 and 2, respectively.

4. Any vertical or rotational relative displacements at the base (Fig. 10.5d). Vertical or rotational deformations of the base may occur as a result of displacements of the foundations (proportional to the modulus of subgrade reaction, for example) or as a result of the flexibility of the supporting substructure. Such foundation displacements will induce rigid body movements of the superstructure above, and will give rise to displacements that are constant over the height as shown in Figs. 10.5d.

Assuming relative displacements (δ_r) and rotations (δ_θ) occur in the same senses as the internal axial forces and moments, the relative vertical displacement δ_4 is,

$$\delta_4 = -\delta_r + l\delta_\theta = \delta_b, \text{ say} \qquad (10.9)$$

In the original deflected structure (Fig. 10.2) there can be no relative vertical displacement on the line of contraflexure of the connecting beams. Consequently, the condition of vertical compatibility at this position is

$$\delta_1 + \delta_2 + \delta_3 + \delta_4 = 0$$

or, using the appropriate expressions for each,

$$I\frac{dy}{dz} + \frac{b^3 h}{12EI_c}\frac{dN}{dz} - \frac{1}{E}\left(\frac{1}{A_1} + \frac{1}{A_2}\right)\int_0^z N\,dz + \delta_b = 0 \qquad (10.10)$$

The last term will be zero in the common case of a rigid base.

On considering both the free bending due to the externally applied moment M and the reverse bending due to the shears and axial forces in the connecting medium (Fig. 10.4), the moment–curvature relationships for the two walls are, at any level,

$$EI_1\frac{d^2y}{dz^2} = M_1 = M - \left(\frac{b}{2} + d_1\right)\int_z^H q\,dz - M_a \qquad (10.11)$$

$$EI_2\frac{d^2y}{dz^2} = M_2 = -\left(\frac{b}{2} + d_2\right)\int_z^H q\,dz + M_a \qquad (10.12)$$

where M_a is the moment caused by the axial forces in the connecting beams.

The addition of Eqs. (10.11) and (10.12) yields the overall moment–curvature relationship for the coupled walls,

$$E(I_1 + I_2)\frac{d^2y}{dz^2} = M - l\int_z^H q\,dz = M - lN \qquad (10.13)$$

Differentiating Eq. (10.10) with respect to z, and combining with Eq. (10.13) to eliminate the curvature d^2y/dz^2 gives

$$\frac{d^2N}{dz^2} - (k\alpha)^2 N = -\frac{\alpha^2}{l}M \qquad (10.14)$$

This is the governing equation for coupled walls expressed in terms of the axial force N.

The parameters in the equation are defined as

$$\alpha^2 = \frac{12I_c l^2}{b^3 h I}$$

$$k^2 = 1 + \frac{AI}{A_1 A_2 l^2}$$

and

$$I = I_1 + I_2, \qquad A = A_1 + A_2$$

As usual, the left-hand side of Eq. (10.14) describes the inherent physical properties of the structure, and the right-hand side involves the form of applied loading.

Alternatively, eliminating the axial force N from Eqs. (10.10) and (10.13) gives

$$\frac{d^4y}{dz^4} - (k\alpha)^2 \frac{d^2y}{dz^2} = \frac{1}{EI}\left(\frac{d^2M}{dz^2} - (k\alpha)^2 \frac{k^2 - 1}{k^2} M\right) \quad (10.15)$$

The is the governing equation for coupled walls expressed in terms of the lateral deflection y.

10.3.2 General Solutions of Governing Equations

The general solution of Eq. (10.14) will always be of the form

$$N = C_1 \cosh k\alpha z + C_2 \sinh k\alpha z - \frac{1}{(k\alpha)^2}\left|1 + \frac{D^2}{(k\alpha)^2} + \frac{D^4}{(k\alpha)^4} + \cdots\right|\left|\frac{\alpha^2 M}{l}\right| \quad (10.16)$$

in which D is the operator d/dz and C_1 and C_2 are integration constants that must be determined from the appropriate boundary conditions at the top and bottom expressed in terms of the variable N.

The general solution of Eq. (10.15) is, similarly

$$y = C_3 + C_4 z + C_5 \cosh k\alpha z + C_6 \sinh k\alpha z - \frac{1}{EI(k\alpha)^2}\left|\frac{1}{D^2} + \frac{1}{(k\alpha)^2} + \frac{D^2}{(k\alpha)^4} + \frac{D^4}{(k\alpha)^6} + \cdots\right|\left|\frac{d^2M}{dz^2} - (k\alpha)^2 \frac{k^2 - 1}{k^2} M\right| \quad (10.17)$$

where C_3 to C_6 are constants to be determined from the boundary conditions expressed in terms of the variable y.

Boundary Conditions. By considering conditions of compatibility and equilibrium at the top and bottom of the structure, appropriate boundary conditions may be derived for a range of base conditions.

For example, for a structure that is free at the top and rigidly built in at the base, the two boundary conditions for Eq. (10.16) will be

$$\text{At } z = H, \quad N = 0 \quad (10.18)$$

At the base, the first term in Eq. (10.10) is the base slope, dy/dz, which is zero. The third term also is zero, and hence the boundary condition becomes

$$\text{At } z = 0, \quad \frac{dN}{dz} = 0 \qquad (10.19)$$

In Eq. (10.17), the four boundary conditions will be

$$\text{At } z = 0, \quad y = 0 \qquad (10.20)$$

$$\frac{dy}{dz} = 0 \qquad (10.21)$$

At the top, the axial force and moment are zero, and, hence, from Eq. (10.13), $d^2y/dz^2 = 0$.

The second boundary condition at the top may readily be derived by substituting for N and its first derivative dN/dz from Eq. (10.13) into the compatibility Eq. (10.10), and making use of Eq. (10.21). The required boundary conditions are then

$$\text{At } z = H, \quad \frac{d^2y}{dz^2} = 0 \qquad (10.22)$$

$$\frac{d^3y}{dz^3} - (k\alpha)^2 \frac{dy}{dz} = \frac{1}{EI}\left(\frac{dM}{dz} - \alpha^2(k^2 - 1) \int_0^H M \, dz\right) \qquad (10.23)$$

Corresponding conditions may be derived for other practical cases such as walls supported on elastic foundations and walls supported on different types of portal frames. In the latter case, it may prove necessary to make further simplifying assumptions regarding the mode of behavior of the support structure in order to derive the appropriate number of conditions in terms of the variable used. This usually involves a reduction in the degree of statical indeterminancy of the support frame by the insertion of hinges at assumed points of contraflexure, or at the junctions of relatively slender columns with relatively stiff beams. These situations are discussed further in Section 10.3.7.

10.3.3 Solution for Standard Load Cases

A complete solution is now obtained for a uniformly distributed lateral load that is often used to simulate wind loading. Solutions are also given in Appendix 1 for two other standard load cases, a triangularly distributed load and a point load at the top, which, as discussed in Chapter 3, are used to simulate earthquake loading.

Uniformly Distributed Lateral Loading. Consider the case of a pair of coupled shear walls on a rigid base, subjected to a uniformly distributed load of intensity w per unit height. The external moment is

$$M = w(H - z)^2/2 \qquad (10.24)$$

Axial Forces in Walls. The particular integral part of the solution may be determined from Eq. (10.16) and, on evaluating the integration constants C_1 and C_2 from Eqs. (10.18) and (10.19), the complete solution becomes

$$N = \frac{wH^2}{k^2 l} \left\{ \frac{1}{2}\left(1 - \frac{z}{H}\right)^2 + \frac{1}{(k\alpha H)^2}\left[1 - \frac{\cosh k\alpha z + k\alpha H \sinh k\alpha (H-z)}{\cosh k\alpha H}\right]\right\} \quad (10.25)$$

Equation (10.25) shows that the distribution of the axial force throughout the height depends on two nondimensional variables only, the relative height z/H and the stiffness parameter $k\alpha H$, since $k\alpha z = k\alpha H \cdot z/H$.

Equation (10.25) may be expressed as

$$N = w\frac{H^2}{k^2 l} F_1(z/H, k\alpha H) \quad (10.26)$$

where F_1 is the expression in the braces in Eq. (10.25). The variation of the axial force factor F_1 with the two parameters z/H and $k\alpha H$ is shown in Fig. 10.6. The curves demonstrate how the axial forces increase with an increase in $k\alpha H$.

Shear in Connecting Members. The shear flow in the connecting medium then follows from Eq. (10.3) as

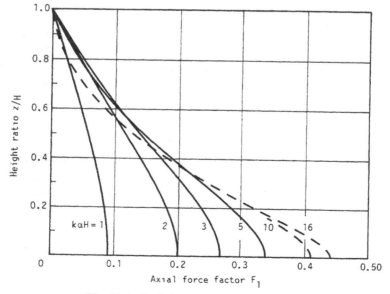

Fig. 10.6 Variation of axial force factor F_1.

10.3 THE CONTINUOUS MEDIUM METHOD

$$q = \frac{wH}{k^2 l}\left\{\left(1 - \frac{z}{H}\right) + \frac{1}{k\alpha H}\left|\frac{\sinh k\alpha z - k\alpha H \cosh k\alpha (H-z)}{\cosh k\alpha H}\right|\right\} \quad (10.27)$$

which may be expressed in a form similar to (10.26) as

$$q = \frac{wH}{k^2 l} F_2(z/H, k\alpha H) \quad (10.28)$$

The variation of the shear flow factor F_2 with the two parameters z/H and $k\alpha H$ is shown in Fig. 10.7.

The maximum value $F_2(\max)$ of the factor F_2 may be obtained by differentiation of Eq. (10.27), and the resulting curve of $F_2(\max)$ is indicated by the broken line on Fig. 10.7. The maximum intensity of the shear flow in the connecting medium is then given by

$$q_{\max} = \frac{wH}{k^2 l} F_2(\max) \quad (10.29)$$

Figure 10.7 shows that as the parameter $k\alpha H$ increases, corresponding mainly to an increase of stiffness of the connecting members, the position of maximum shear flow in the connecting medium, and, hence, the position of the most heavily stressed connecting beam, moves progressively down the height of the structure.

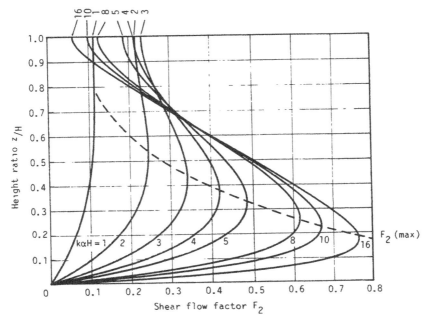

Fig. 10.7 Variation of shear flow factor F_2.

Wall Moments. Since the moments are proportional to the flexural rigidities, the bending moment at any level in wall 1 will be

$$M_1 = \frac{I_1}{I}\left[\frac{1}{2}wH^2(1 - z/H)^2 - Nl\right] \quad (10.30)$$

where N is given by Eq. (10.26). It follows that the moment may also be expressed in terms of the function F_1 as

$$M_1 = \frac{I_1}{I}\frac{1}{2}wH^2\left[\left(1 - \frac{z}{H}\right)^2 - \frac{2}{k^2}F_1\right] \quad (10.31)$$

Similarly, the bending moment in wall 2 will be

$$M_2 = \frac{I_2}{I}\frac{1}{2}wH^2\left[\left(1 - \frac{z}{H}\right)^2 - \frac{2}{k^2}F_1\right] \quad (10.32)$$

The terms in the brackets indicate the proportion of the wind moment that is resisted by bending actions in the walls. The greater the value of N and F_1, the smaller will be the proportion resisted by wall bending.

Deflections. The lateral deflection y may be obtained from Eqs. (10.17) and (10.24) with the aid of Eqs. (10.20) to (10.23) as

$$y = \frac{wH^4}{EI}\left|\frac{1}{24}\left\{\left(1 - \frac{z}{H}\right)^4 + 4\frac{z}{H} - 1\right\} + \frac{1}{k^2}\left\{\left(\frac{1}{2(k\alpha H)^2}\right)\left|2\frac{z}{H} - \left(\frac{z}{H}\right)^2\right|\right.\right.$$

$$\left.\left. - \frac{1}{24}\left|\left(1 - \frac{z}{H}\right)^4 + 4\frac{z}{H} - 1\right| - \frac{1}{(k\alpha H)^4 \cosh k\alpha H}\right.\right.$$

$$\left.\left. \cdot [1 + k\alpha H \sinh k\alpha H - \cosh k\alpha z - k\alpha H \sinh k\alpha(H - z)]\right\}\right| \quad (10.33)$$

The expression (10.33) consists of two terms in braces, the first representing the pure cantilever action that would occur if the walls were connected by pin-ended axially rigid links, and the second representing the modification to the independent cantilever action that occurs because of the reverse bending of the walls induced by the coupling beams. The deflection is now dependent on three independent parameters, the height z/H and the two structural parameters k and αH. It is therefore not possible to produce single generalized curves, such as those for N and q, which show the variation of the deflection profile with the different parameters involved. One particular quantity, however, which is of fundamental significance in design, is the maximum lateral deflection y_H at the top of the structure since, as discussed in Chapter 2, the designer will wish to contain the ratio y_H/H within a particular value such as $\frac{1}{500}$.

10.3 THE CONTINUOUS MEDIUM METHOD

At the top, at $z/H = 1$, the maximum deflection is

$$y_H = \frac{wH^4}{8EI} F_3(k, \alpha H)$$

where

$$F_3 = 1 - \frac{1}{k^2}\left[1 - \frac{4}{(k\alpha H)^2} + \frac{8}{(k\alpha H)^4 \cosh k\alpha H}(1 + k\alpha H \sinh k\alpha H - \cosh k\alpha H)\right] \quad (10.34)$$

The variation of the deflection factor F_3 may be expressed most conveniently in terms of the parameters k and $k\alpha H$, as shown in Fig. 10.8. The curves show that the maximum lateral deflection is reduced by more than 60% for values of $k\alpha H$ greater than 4.

Forces in the Discrete Structure. The results that have been obtained relate to the equivalent continuous system, and it is necessary to transform them to the real coupled wall structure.

The shear force Q_i in any particular connecting beam i at level z_i may be obtained from the difference in values of axial force N at levels $h/2$ above and below the level concerned, that is,

$$Q_i = N(z_i - h/2) - N(z_i + h/2) \quad (10.35)$$

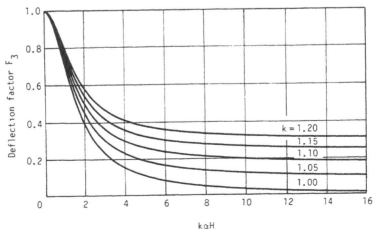

Fig. 10.8 Variation of top deflection factor F_3.

COUPLED SHEAR WALL STRUCTURE

Alternatively, Q_i may be obtained by integrating the shear flow q over the height concerned

$$Q_i = \int_{z_i - h/2}^{z_i + h/2} q \, dz$$

which becomes, on substituting for q from Eq. (10.27),

$$Q_i = \frac{wH}{k^2 l} \left\{ \left(1 - \frac{z_i}{H}\right) h + \frac{2H \sinh k\alpha(h/2)}{(k\alpha H)^2 \cosh k\alpha H} \left[\sinh k\alpha z_i - k\alpha H \cosh k\alpha (H - z_i) \right] \right\} \quad (10.36)$$

If the shear flow curve (Fig. 10.7) has already been plotted, Q_i may be obtained by evaluating the area under the curve over the story height concerned, or by multiplying the estimated average value of q by the story height. The shear in the top beam will be derived from the shear flow over half a story height at the top.

The maximum moment in any connecting beam may be obtained from the shear flow curve F_2 where an inspection will indicate the position of the most heavily loaded beams. The value of $Q_i(\max)$ may be evaluated quickly by considering the two beam positions on either side of the maximum value $F_2(\max)$.

The maximum value will be given approximately by

$$Q_i(\max) \simeq w \frac{H}{k^2 l} F_2(\max) h \quad (10.37)$$

The maximum bending moment in any connecting beam i, at the junction with the wall, is equal to $Q_i b/2$.

Equations (10.31) and (10.32) will give the continuous distribution of bending moments in the two walls. In the real discrete structure, the distribution is discontinuous as a component $Q_i b/2$ is fed into the wall at each beam position. If desired, the discrete distribution may then be obtained from the known values of shear force in the beams. It may be noted that although there is no bending moment or axial force at the top of each wall in the continuous system, discrete values, obtained by integrating the continuous forces over half a story height at the top of the building, exist in the real system. Although the distributions of bending moments and axial forces in the real structure may apparently be determined more accurately by adding up the effects of individual beam forces, it is very doubtful if the additional laborious work would be warranted. The continuous distributions may be regarded as sufficiently accurate for the practical purposes of determining the wall stresses, unless one wall is very much smaller than the other.

In the above context, it is worth bearing in mind that the bending stresses and shear stresses at the ends of the connecting beams must be diffused into the edges of the walls over the depth of the beam. In most cases, the semiwidth of the wall is much greater than the depth of the beam, and diffusion of stresses into the interior of the wall takes place over some distance from the junction. Such discontinuities in the wall's bending moment and axial force, although statically correct in relation to the centroidal axes of the members, may have in reality little meaning for relatively wide walls when using simple bending and axial load theories to determine the wall stresses. Consequently, the continuous bending moment distributions in the walls may well be as accurate as the discontinuous distributions in many practical situations.

Wall Shears. By considering the equilibrium conditions for a small element of wall and the associated connecting medium in the continuous structure, (Fig. 10.9) it may be shown that the shear forces S_1 and S_2 in the two walls are given by

$$S_1 = \left(\frac{I_1}{I} l - \frac{b}{2} - d_1\right) \frac{dN}{dz} - \frac{I_1}{I} \frac{dM}{dz} \qquad (10.38)$$

$$S_2 = \left(\frac{I_2}{I} l - \frac{b}{2} - d_2\right) \frac{dN}{dz} - \frac{I_2}{I} \frac{dM}{dz} \qquad (10.39)$$

in which M is the total external moment.

Hence, on using Eqs. (10.3) and (10.24), the shears in the walls are

$$S_1 = wH \frac{I_1}{I} \left(1 - \frac{z}{H}\right) - \left(\frac{I_1}{I} l - \frac{b}{2} - d_1\right) q \qquad (10.40)$$

$$S_2 = wH \frac{I_2}{I} \left(1 - \frac{z}{H}\right) - \left(\frac{I_2}{I} l - \frac{b}{2} - d_2\right) q \qquad (10.41)$$

in which the function q is given by Eq. (10.27) and represented graphically in Fig. 10.7. The first term in each gives the shear that would exist if the walls were pin-connected. It may be noted that since $q = 0$ at a fixed base, the base shears will always be given by the first terms in the equations.

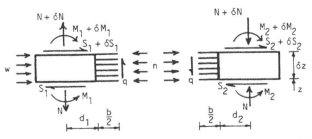

Fig. 10.9 Forces on small element of wall-continuum model.

In the case of two equal walls, Eqs. (10.40) and (10.41) reduce, as required by symmetry, to

$$S_1 = S_2 = \frac{1}{2} wH \left(1 - \frac{z}{H}\right) \qquad (10.42)$$

Since

$$\frac{I_2}{I} l - \frac{b}{2} - d_2 = \frac{b}{2} + d_1 - \frac{I_1}{I} l$$

it follows that the second terms in Eqs. (10.40) and (10.41) are equal in magnitude but of opposite sign. This is essential for horizontal equilibrium since the sum of the two shears must equal the applied shear $wH(1 - z/H)$.

It follows that, at the top of the structure, when $dM/dz = 0$,

$$S_1(H) = -S_2(H) = -\left(\frac{I_1}{I} l - \frac{b}{2} - d_1\right) q(H)$$

This situation will be true for any distributed loading for which the static shear force is zero at the top. It must therefore be deduced that, in order to produce a shear force in each wall at the top of the continuous structure, there must exist at the top of the connecting medium a concentrated horizontal interactive force Q between the walls, of magnitude

$$Q = -\left(\frac{I_1}{I} l - \frac{b}{2} - d_1\right) q(H) \qquad (10.43)$$

The force Q will be tensile or compressive according to whether the top shear force $S_1(H)$ is positive or negative, but will vanish in the case of two equal walls. The magnitude of the top interactive force depends on the structural parameters involved, and is an indication of the heavy interactions that can occur at the tops of such structures.

Axial Forces in Beams. Consideration of the horizontal equilibrium of the small wall element of Fig. 10.9 gives the distribution of axial forces n in the connecting medium as

$$n = w \left[\frac{I_2}{I} - \frac{1}{k^2 l} \left(\frac{I_1}{I} l - \frac{b}{2} - d_1\right) F_4 \right] \qquad (10.44)$$

in which the function F_4 is

$$F_4 = \frac{1}{\cosh k\alpha H} \left[\cosh k\alpha z + k\alpha H \sinh k\alpha (H - z)\right] - 1$$

10.3.4 Graphic Design Method

In the detailed derivation of the solution for a uniformly distributed load, it was shown that the forces in the structure are dependent on two nondimensional parameters, the relative height z/H and the stiffness parameter $k\alpha H$. This allowed a series of curves to be derived showing the forms of distributions of axial force N, bending moment M_1 or M_2, and shear force intensity q, with height z/H, for a range of values of $k\alpha H$. The lateral deflection is dependent on three variables, but a similar generalized curve was obtained for the particular case of the maximum top deflection. These curves may be used directly as design aids to give the internal forces at any required level.

A simple alternative technique has been devised [10.2] to give directly the stresses in the walls. It has the added advantage that it is devised in such a way that the influence of the different parameters on the structural behavior is readily seen, and the designer can check the efficiency of the wall with flexible connecting beams relative to the optimum case of a dowelled system.

The technique is described in detail here for the case of a pair of coupled shear walls on a rigid base subjected to a uniformly distributed load. Design curves are also presented in Appendix 1 for the other standard load cases of a triangularly distributed lateral load and a point load at the top.

Consider the pair of coupled shear walls shown in plan in Fig. 10.10a. The

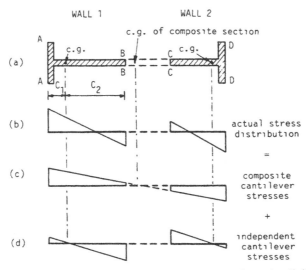

Fig. 10.10 Superposition of stress distributions due to composite and individual cantilever actions to give true stress distribution in walls.

stress distribution at any section, under the action of the bending moments M_1 and M_2, and the axial force N, will be as shown in Fig. 10.10b. Taking tensile stresses as positive, the maximum extreme fiber stresses in wall 1 will be given by

$$\sigma_A = \frac{M_1 c_1}{I_1} + \frac{N}{A_1}$$

$$= \left[\frac{1}{2} w(H-z)^2 - Nl\right]\frac{c_1}{I} + \frac{N}{A_1} \quad (10.45)$$

$$\sigma_B = -\frac{M_1 c_2}{I_1} + \frac{N}{A_1}$$

$$= -\left[\frac{1}{2} w(H-z)^2 - Nl\right]\frac{c_2}{I} + \frac{N}{A_1} \quad (10.46)$$

Similar expressions hold for the stresses in wall 2.

The complete stress distribution, which consists in reality of a superposition of uniform axial stresses on linear bending stress distributions in each wall, may be considered derived from an alternative superposition of two hypothetical pure bending stress distributions, namely (1) a bending stress distribution based on the assumption that the wall system acts as a single composite cantilever, the neutral axis being situated at the centroid of the two wall elements, as shown in Fig. 10.10c, and (2) two linear stress distributions obtained on the assumption that the walls act as completely independent cantilevers, with the neutral axis at the centroid of each wall, as shown in Fig. 10.10d. The real stress distribution of Fig. 10.10b is then obtained by superposition of appropriately sized hypothetical distributions of Fig. 10.10c and d. By this procedure, the solution of the complex coupled structure is reduced to the combination of solutions for two separate simple cantilever problems.

Suppose that K_1 is the percentage of the total wind moment that is resisted by independent cantilever action, and that K_2 is the percentage resisted by composite cantilever action, and consider the two component stress distributions.

1. Composite cantilever action (cf. Fig. 10.10c). The total bending moment at any section which is carried by composite action is

$$M_c = \frac{K_2}{100}\frac{1}{2} w(H-z)^2 \quad (10.47)$$

By taking moments about one edge, it is found that the centroid of the pair of walls is located at a distance of $(A_2 l/A) + c_1$ from the edge A, c_1 being the distance from A to the centroid of wall 1 (cf. Fig. 10.10a). The second moment of area I_g of the two walls about their centroidal axis becomes

$$I_g = I_1 + I_2 + \frac{A_1 A_2}{A} l^2 \quad (10.48)$$

10.3 THE CONTINUOUS MEDIUM METHOD

Consequently, the extreme fiber stresses in wall 1 of the composite cantilever will be

$$\sigma_A = \frac{w(H-z)^2}{2I_g}\left(\frac{A_2 l}{A} + c_1\right)\frac{K_2}{100} \quad (10.49)$$

and

$$\sigma_B = \frac{w(H-z)^2}{2I_g}\left(\frac{A_2 l}{A} - c_2\right)\frac{K_2}{100} \quad (10.50)$$

Similar expressions again hold for the stresses in wall 2.

2. Individual cantilever action (cf. Fig. 10.10d). Since both walls are assumed to deflect equally, the individual moments carried by the walls will be proportional to their second moments of area. The total moment carried by individual cantilever action is

$$\frac{1}{2}w(H-z)^2\frac{K_1}{100}$$

and the bending moments in walls 1 and 2 will then be

$$M_1 = \frac{1}{2}w(H-z)^2\frac{I_1}{I}\frac{K_1}{100} \quad (10.51)$$

and

$$M_2 = \frac{1}{2}w(H-z)^2\frac{I_2}{I}\frac{K_1}{100} \quad (10.52)$$

The extreme fiber stresses in wall 1 are given by

$$\sigma_A = \frac{M_1 c_1}{I_1} = \frac{1}{2}w(H-z)^2\frac{c_1}{I}\frac{K_1}{100} \quad (10.53)$$

$$\sigma_B = -\frac{M_1 c_2}{I} = -\frac{1}{2}w(H-z)^2\frac{c_2}{I}\frac{K_1}{100} \quad (10.54)$$

Similar expressions again hold for the stresses in wall 2.

The individual and composite cantilever factors K_1 and K_2 must vary throughout the height in such a way that when the linear stress distributions defined by Eqs. (10.49) and (10.50) and (10.53) and (10.54) are superimposed, they give a distribution defined by Eqs. (10.45) and (10.46). The accordance may be defined by equating the corresponding stresses at the four extreme wall fiber positions.

Hence, on substituting in these equations for the axial force N from Eq. (10.25), and equating stresses, the composite cantilever factor K_2 may be shown to be

$$K_2 = \frac{200}{(k\alpha H)^2 [1 - (z/H)]^2} \left[1 + \frac{\sinh k\alpha H - k\alpha H}{\cosh k\alpha H} \sinh k\alpha (H - z) \right.$$

$$\left. - \cosh k\alpha (H - z) + \frac{1}{2} (k\alpha H)^2 \left(1 - \frac{z}{H}\right)^2 \right] \quad (10.55)$$

and, by definition,

$$K_1 = 100 - K_2 \quad (10.56)$$

The proportions of composite and individual cantilever action required to produce the true stress distribution at any position are thus functions only of the structural parameter $k\alpha H$ and the height ratio z/H. The form of functions K_1 and K_2 are shown in Fig. 10.11 for a range of values of the parameter $k\alpha H$, covering the range of all practical situations.

The curves for F_2 and F_3 in Figs. 10.7 and 10.8 may be used directly as corresponding design curves for the shear flow q and the top deflection y_H in conjunction with Eqs. (10.28) and (10.34).

Significance of Parameter kαH. It has been shown that at any height z/H, the distributions of the axial forces, and thus bending moments, in the walls, and the shear flow in the connecting medium, depend on the relative stiffness parameter $k\alpha H$, which, using expressions for k and α as defined for Eq. (10.14), is given by

$$k\alpha H = \left[\frac{12 I_c l^2}{b^3 h I} \left(1 + \frac{AI}{A_1 A_2 l^2}\right) H^2 \right]^{1/2}$$

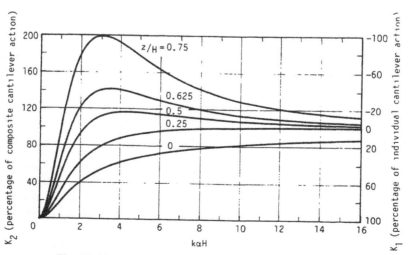

Fig. 10.11 Variation of wall moment factors K_1 and K_2.

For a given set of walls, with fixed dimensions, the value of $k\alpha H$ will be a measure of the stiffness of the connecting beams, and it will increase if either I_c is increased or the clear span b is decreased.

If the connecting beams have negligible stiffness ($k\alpha H = 0$) then the applied moment M will be resisted entirely by bending moments in the walls, and the axial forces N will be negligible. That is, the structure behaves as a pair of linked walls.

If the connecting beams are rigid ($k\alpha H = \infty$), the structure will behave as a single composite dowelled beam, with a linear bending stress distribution across the entire section, and zero stress at the neutral axis, which is situated at the centroid of the two wall elements.

The stress distributions for these two limiting cases are shown in Fig. 10.10d and c, and correspond to the cases of $K_1 = 100\%$ and $K_2 = 100\%$, respectively.

For practical situations involving beams of finite stiffness the stress distribution will lie between these two extremes (Fig. 10.10b). As the beams increase in stiffness the induced axial forces in the walls increase, increasing the component of uniform tensile or compressive stresses in the walls, and reducing the wall bending moments, and hence the component of bending stress in each wall.

The value of $k\alpha H$ will thus define the degree of composite action and will indicate the mode of resistance to applied moments. If $k\alpha H$ is small, say less than about 1, the beams may be regarded as flexible and the walls tend to act as independent linked cantilevers. If $k\alpha H$ is large, say greater than about 8, the beams are classed as stiff and the structure tends to act like a composite cantilever. In between these values, the mode of action will vary with the level concerned, as indicated in Fig. 10.11. Particular attention must be given to the most heavily loaded section at the base.

It is shown in Chapter 15 that corresponding general relative stiffness parameters may also be used to describe the lateral load behavior of rigid-frames, braced-frames, and wall-frame structures. Further consideration is given to the significance of the general governing parameters k and αH, and their relationship to the shearing and bending stiffnesses of a bent.

10.3.5 Coupled Shear Walls with Two Symmetrical Bands of Openings

Shear walls with two symmetrical bands of openings are frequently encountered in specific situations, such as on the flank walls of residential buildings. They may also be conveniently analyzed by the procedures developed for a single row of openings.

Consider the structure shown in Fig. 10.12a. Because of symmetry, there will be no axial force in the central wall, the moments in the outer walls and the shear flows in the two sets of connecting laminas will be equal, while the axial forces N_1 in the outer walls will be equal and opposite. The wind moment will then be resisted by the wall moments and axial forces N_1 in the outer walls, as,

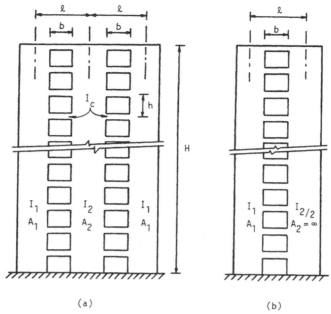

Fig. 10.12 Representation of coupled shear walls with two symmetric rows of openings by equivalent single bay structure.

$$M = 2M_1 + M_2 + 2N_1 l$$

in which M_1 and M_2 are the moments in the outer and central walls, respectively.

The simplest way of treating the structure is to consider a half-structure, consisting of a wall type 1 and a half-thickness wall type 2, joined by connecting beams, and subjected to half the applied loading. Since there is no axial force in the real wall 2, it will not undergo any axial deformation, and this may be achieved by assigning to it a very large area A_2. Similarly, the second moment of area of wall 2 should be taken as $\frac{1}{2} I_2$ (Fig. 10.12b). The formulas established previously for a pair of dissimilar walls may then be used directly.

10.3.6 Worked Example of Coupled Shear Wall Structure

The theory and design curves in the preceding sections provide a practical method for the rapid analysis of coupled shear walls of any cross-sectional shape. Although they are accurate only for structures with uniform properties over the height, they can, by the judicial use of average properties, serve as a useful guide to the forces in nonuniform structures. The analysis allows an assessment of how much of the applied moment is resisted by bending moments in the walls, and how much by axial forces.

To illustrate the practical use of the design curves, the typical system of plane coupled shear walls shown in Fig. 10.13 is considered. It is assumed that the

10.3 THE CONTINUOUS MEDIUM METHOD   237

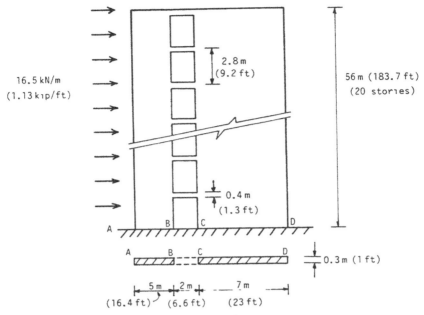

Fig. 10.13 Example structure.

system forms one of a series spaced at 6.1 m (20 ft) centers in a 20-story building, and is subjected to a uniformly distributed wind loading of intensity 16.5 kN/m (1.13 kip/ft) height.

It is required to determine the stresses in the walls at base level, the maximum shear, and hence the maximum moment, in any connecting beam, and the maximum lateral deflection at the top.

In addition, the distributions of axial forces and bending moments in the walls are evaluated to illustrate the general forms of distribution of such forces.

The required procedures are first described, and then illustrated numerically for the considered structure.

Step 1. Determine the areas and second moments of area of the walls, and the second moment of area of the connecting beams. Depending on the design code used, the latter may be taken as gross values, calculated from the beam section's full dimensions, or reduced values to take account of cracking in the concrete.

Wall properties:

$$I_1 = \tfrac{1}{12} \times 5^3 \times 0.3 = 3.125 \text{ m}^4 \ (362.1 \text{ ft}^4)$$

$$I_2 = \tfrac{1}{12} \times 7^3 \times 0.3 = 8.575 \text{ m}^4 \ (993.5 \text{ ft}^4)$$

$$\therefore \ I = I_1 + I_2 = 11.700 \text{ m}^4 \ (1355.6 \text{ ft}^4)$$

COUPLED SHEAR WALL STRUCTURE

$$A_1 = 5 \times 0.3 = 1.5 \text{ m}^2 \text{ (16.15 ft}^2\text{)}$$
$$A_2 = 7 \times 0.3 = 2.1 \text{ m}^2 \text{ (22.60 ft}^2\text{)}$$
$$\therefore A = A_1 + A_2 = 3.6 \text{ m}^2 \text{ (38.75 ft}^2\text{)}$$
$$l = 8 \text{ m (26.25 ft)}$$

For the connecting beams, assuming that the entire cross section is effective

$$I_b = \tfrac{1}{12} \times 0.4^3 \times 0.3 = 1.6 \times 10^{-3} \text{ m}^4 \text{ (0.185 ft}^4\text{)}$$

The second moment of area is reduced to include shearing deformations. Assuming a Poisson's ratio ν of 0.15 for concrete, the shear modulus G is

$$G = \frac{E}{2(1 + \nu)} = \frac{E}{2.3}$$

$\therefore$ From Eq. (10.6)

$$r = \frac{12 \times E \times 1.6 \times 10^{-3} \times 1.2}{2^2 \times 0.4 \times 0.3 \times E/2.3} = 0.1104$$

$\therefore$ Effective second moment of area $I_e = \dfrac{1.6 \times 10^{-3}}{1.1104}$

$$= 1.441 \times 10^{-3} \text{ m}^4 \text{ (0.167 ft}^4\text{)}$$

Taking account of the wall-beam flexibility, effective length = true length + $\tfrac{1}{2}$ beam depth = 2.2 m (7.22 ft).

Step 2. Determine the structural parameters k, α, and $k\alpha H$ from Eq. (10.14).

$$k^2 = 1 + \frac{Al}{A_1 A_2 l^2} = 1 + \frac{3.6 \times 11.7}{1.5 \times 2.1 \times 8^2} = 1.2089$$

$\therefore k = 1.0995$

$$\alpha^2 = \frac{12 I_e l^2}{b^3 h I} = \frac{12 \times 1.441 \times 10^{-3} \times 8^2}{2.2^3 \times 2.8 \times 11.7} = 3.1725 \times 10^{-3}$$

$\therefore \alpha = 0.05633 \text{ m}^{-1} \text{ (0.0172 ft}^{-1}\text{)}$

$\therefore k\alpha H = 1.0995 \times 0.05633 \times 56 = 3.468$

This value indicates that the beams are of intermediate relative stiffness.

Step 3. For the particular level z considered, calculate the wind moment $[\tfrac{1}{2} w(H - z)^2]$. Generally, this will be initially at the base where the applied moment is greatest.

10 3 THE CONTINUOUS MEDIUM METHOD

Determine from Fig. 10.11 the percentage of the moment at this level carried by individual cantilever action (K_1) and the percentage carried by composite cantilever action (K_2). The individual moment acting on each wall will be proportional to its second moment of area.

From Fig. 10.11. the values of K_1 and K_2 at base level ($z/H = 0$) are

$K_1 = 42\%$ $K_2 = 58\%$

Total base moment $= \frac{1}{2} \times 16.5 \times 56^2$
$= 25,872$ kNm (19,083 kipft)

∴ Portion of base moment due effectively to individual cantilever action is $0.42 \times 25,872 = 10,866$ kNm (8015 kipft)

Moment on wall 1, $M_1 = \dfrac{3.125}{11.70} \times 10,866 = 2902$ kNm (2141 kipft)

Moment on wall 2, $M_2 = \dfrac{8.575}{11.70} \times 10,866 = 7964$ kNm (5874 kipft)

Portion of base moment due effectively to composite cantilever action is $0.58 \times 25,872 = 15,006$ kNm (11,068 kipft)

Step 4. Calculate the second moment of area I_g of the composite cross section. Hence calculate the stresses at the extreme fibers of the walls, using ordinary beam theory, due to the individual and composite moments, and add these to obtain the true stresses at these positions. The bending stress distribution is linear across each wall.

From Eq. (10.48), effective composite second moment of area of cross section is

$$I_g = 3.125 + 8.575 + \frac{1.5 \times 2.1}{3.6} \times 8^2 = 67.70 \text{ m}^4 \text{ (7843 ft}^4\text{)}$$

The position of the center of gravity (c.g.) of the cross section and the distances from it to the extreme wall fibers are shown in Fig. 10.14.

Using ordinary beam theory, the stresses at the salient points A, B, C, and D are, on adding the stresses due to individual and composite cantilever stresses taking tensile stresses as positive,

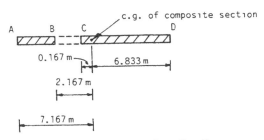

Fig. 10.14 Cross section of walls.

$$\sigma_A = \frac{2902 \times 2.5}{3.125} + \frac{15{,}006 \times 7.167}{67.70} = +3910 \text{ kN/m}^2 \ (+567 \text{ lb/in.}^2)$$

$$\sigma_B = -\frac{2902 \times 2.5}{3.125} + \frac{15{,}006 \times 2.167}{67.70} = -1841 \text{ kN/m}^2 \ (-267 \text{ lb/in.}^2)$$

$$\sigma_C = \frac{7964 \times 3.5}{8.575} + \frac{15{,}006 \times 0.167}{67.70} = +3288 \text{ kN/m}^2 \ (+477 \text{ lb/in.}^2)$$

$$\sigma_D = -\frac{7964 \times 3.5}{8.575} - \frac{15{,}006 \times 6.83}{67.70} = -4765 \text{ kN/m}^2 \ (-691 \text{ lb/in.}^2)$$

It is of interest to note that if the walls were uncoupled and behaved as independent cantilevers, the corresponding base stresses would be as follows:

$$\sigma_A = -\sigma_B = 5528 \text{ kN/m}^2 \ (802 \text{ lb/in.}^2)$$

$$\sigma_C = -\sigma_D = 7739 \text{ kN/m}^2 \ (1122 \text{ lb/in.}^2)$$

illustrating the considerable reduction in stresses that results from the coupling action.

Step 5. From Fig. 10.7, determine the maximum shear force factor $F_2(\max)$ and hence the shear flow $q_{\max}$ at the most heavily loaded beam [Eq. (10.29)]. The maximum possible shear in any beam is equal to $q_{\max} h$, and the maximum possible beam moment is $q_{\max} h \cdot b/2$.

This procedure will overestimate the beam shears and moments, by an amount which depends on the story height. If a more accurate estimate is required, the actual beam locations may be superimposed on Fig. 10.7 to indicate the position of the most heavily loaded one. The maximum shear may then be obtained by finding the area under the shear flow curve over the story height concerned.

From Fig. 10.7, the value of the maximum shear force factor for $k\alpha H = 3.468$, $F_2(\max) = 0.381$ at a level $z/H = 0.39$.

The maximum shear flow $q_{\max}$ becomes, from Eq. (10.29),

$$q_{\max} = 16.5 \times \frac{56}{8} \times \frac{1}{1.209} \times 0.381 = 36.39 \text{ kN/m} \ (2.49 \text{ kip/ft})$$

Thus, the maximum possible shear in any connecting beam is

$$Q_{\max} = q_{\max} h = 36.39 \times 2.8 = 101.9 \text{ kN} \ (22.91 \text{ kip})$$

and the maximum possible moment in any connecting beam is

$$M_{\max} = (q_{\max} h) b/2 = 101.9 \times 1 = 101.9 \text{ kNm} \ (75.16 \text{ kipft})$$

Alternatively, if the positions of the 20 connecting beams are superimposed on the shear flow diagram of Fig. 10.7, the most heavily loaded beam is that at the eighth floor level, where an average value over the story height of the shear force function F_2 is 0.380. The maximum shear is then

$$Q_{max} = 36.30 \times 2.8 = 101.6 \text{ kN } (22.84 \text{ kip})$$

and the maximum beam moment is

$$M_{max} = 101.6 \times 1 = 101.6 \text{ kNm } (74.93 \text{ kipft})$$

In this case, the position of the maximum value of F_2 coincides almost exactly with a beam position, and there is insignificant difference between the two values of Q_{max}. Provided the number of beams exceeds 10, the percentage error obtained in using F_2(max) rather than the true average beam value should not exceed more than a few percent since the shear flow curve is fairly flat near the most heavily loaded region.

Step 6. Determine the deflection factor F_3 from Fig. 10.8. The maximum lateral deflection at the top of the structure is then obtained from Eq. (10.34).

From Fig. 10.8, for $k\alpha H = 3.468$ and $k = 1.0995$, the value of the maximum deflection factor F_3 is 0.333. That is, since F_3 is equal to unity if no coupling beams are present, the effect of the coupling is to increase the stiffness by 300%.

Assuming that the dynamic modulus of elasticity of the concrete employed is estimated to be 36 kN/mm² (5.22 × 10⁶ lb/in.²), the maximum top deflection becomes, from Eq. (10.34),

$$y_{max} = \frac{1}{8} \times \frac{16.5 \times 56^4}{36 \times 10^6 \times 11.7} \times 0.333$$

$$= 0.016 \text{ m or } 16 \text{ mm } (0.052 \text{ ft or } 0.63 \text{ in.})$$

With no coupling beams, the maximum top deflection would have been

$$y_{max} = 0.048 \text{ m } (0.157 \text{ ft})$$

Step 7. If desired, the variation of axial force N in each wall throughout the height may be determined from Eq. (10.26) and Fig. 10.6 from which the value of F_1 may be determined directly at any level.

The bending moment in each wall then follows from Eq. (10.31) and (10.32).

For completeness, in order to demonstrate the influence of the axial forces in reducing the wind moments in the walls, these have been evaluated and are shown in Fig. 10.15a and b. The curves illustrate the considerable degree of negative

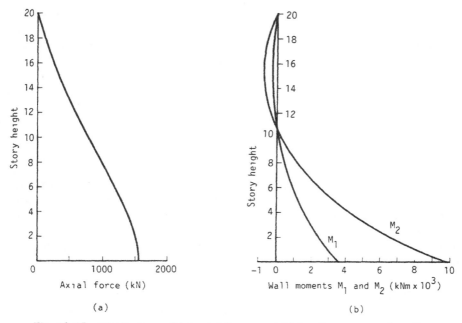

Fig. 10.15 Distributions of (a) axial forces and (b) bending moments in walls.

bending that takes place in the walls in the upper levels, due to the resisting moment induced by the connecting beams. In the lower levels, the amount of deformation of the connecting beams is reduced, and the relative influence of the resulting axial forces is diminished. The proportion of the wind moment that is resisted by axial forces diminishes toward the base, but it still accounts for 48% of the total moment at the base. A redistribution of forces takes place continuously throughout the height, and the behavior is much more complex than that of ordinary cantilevered walls. The beneficial effects that arise from the coupling action are clear, but these, of course, must be balanced against the cost of providing moment connections between the beams and walls.

Step 8. If required, the shear forces in the walls may be calculated from Eqs. (10.40) and (10.41). At the most heavily loaded section, at the base, the total shear force is shared between the walls in proportion to their flexural rigidities.

At the base, the wall shears become

$$S_1 = 16.5 \times 56 \times \frac{3.125}{11.7} = 246.8 \text{ kN } (55.5 \text{ kip})$$

$$S_2 = 16.5 \times 56 \times \frac{8.575}{11.7} = 677.2 \text{ kN } (152.2 \text{ kip})$$

10.3.7 Coupled Shear Walls with Different Support Conditions

The previous sections have considered the analysis of laterally loaded walls supported on rigid foundations. However, architectural requirements may dictate that shear walls are discontinued in the first story to provide large open areas for entrance foyers. The coupled walls may then be supported at first floor level on a set of columns or on a portal frame allowing relative deformations to occur at the base of the walls. Similarly, if the coupled walls are supported on individual footings or other foundation systems that interact directly with the underlying soil, relative vertical or rotational deformations may occur at the base. These effects are frequently modeled by linear or rotational springs, whose stiffness is proportional to the modulus of subgrade reaction.

The governing equations and the general form of solution are not affected by the boundary conditions. Only the constants of integration C_1 to C_6 in Eqs. (10.16) and (10.17) will be affected. The boundary conditions for different base support systems are most conveniently derived in terms of the fundamental condition of compatibility (Eq. 10.10). Provided that the relative base displacement δ_b in Eq. (10.10) can be expressed in terms of the variable in the governing equation, sufficient boundary conditions may be derived at the base to allow a solution to be obtained.

Two particular cases are considered to illustrate the method of dealing with different forms of base conditions: walls supported on elastic foundations and walls supported on a portal frame.

Walls Supported on Separate Elastic Foundations. In this case, the walls are supported on individual foundations (Fig. 10.16) that deflect elastically, both vertically and rotationally, under the actions of the imposed axial forces and moments at the bases. Since both walls are assumed to deflect horizontally equally at all levels, it follows that the slopes and curvatures of the walls at the base are also equal.

The relative vertical and rotational displacements will then be

$$\delta_v = \left(\frac{1}{k_{v1}} + \frac{1}{k_{v2}}\right) N_0 \qquad (10.57)$$

and

$$\delta_\theta = \frac{M_{10} + M_{20}}{k_{\theta 1} + k_{\theta 2}} \qquad (10.58)$$

in which N_0, M_{10}, and M_{20} are the axial forces and wall moments at the base, and k_{v1} and k_{v2} are the vertical stiffnesses and $k_{\theta 1}$ and $k_{\theta 2}$ are the rotational stiffnesses of the elastic foundations under walls 1 and 2. The wall base moments are equal to $M_0 - N_0 l$, where M_0 is the applied moment at that position.

If the walls are fixed to footings on an elastic foundation with modulus of subgrade reaction λ, the stiffnesses are given by

244 COUPLED SHEAR WALL STRUCTURE

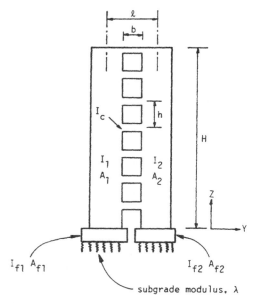

Fig. 10.16 Coupled shear walls on individual elastic foundations.

$$k_{v1} = \lambda A_{t1} \quad k_{v2} = \lambda A_{t2}$$

$$k_{\theta 1} = \lambda I_{t1} \quad k_{\theta 2} = \lambda I_{t2}$$

where A_{t1} and A_{t2} are the cross-sectional areas, and I_{t1} and I_{t2} are the second moments of area of the footings.

Consequently, if Eqs. (10.57) and (10.58) are substituted into Eq. (10.9) to give δ_b, the compatibility condition [Eq. (10.10)] at the base may be expressed in terms of the axial force N_0, and the required boundary condition obtained. A corresponding condition in terms of the deflection y may be achieved by the use of the moment–curvature relationships [10.3].

Walls Supported on Columns or Portal Frames. For supports of this nature, it is convenient to assume that the base ($z = 0$) of the system of coupled shear walls is situated at the axis of the portal beam or at the top of the columns, at a height h_0 above the rigid foundations. The total height of the building is then $H + h_0$. The governing differential equation holds over the normal range of the height variable z from 0 to H, and the lower boundary condition must then be expressed at the level $z = 0$.

It will again be possible to obtain general closed form solutions for any support system provided that the lower boundary condition can be expressed explicitly in terms of the axial force N (or the shear flow q), or the lateral deflection y, at the

10.3 THE CONTINUOUS MEDIUM METHOD

level concerned, depending on the particular form of governing differential equation employed. To do so, it is frequently necessary to make certain simplifying assumptions regarding the behavior of the support system in order to obtain the appropriate load deformation characteristics at level $z = 0$.

To illustrate the procedure involved, consider the general case of a support system that consists of a trapezoidal portal frame with pinned column bases, as shown in Fig. 10.17. The forces transmitted from the walls to the frame are shown in Fig. 10.18. In view of the continuity within the beam system, it must be assumed that the portal beam also deflects with a point of contraflexure at the midspan position. Figure 10.18 shows the portal frame with a hypothetical cut at the point of contraflexure at which only vertical shear forces Q_0 and axial forces H_0 occur.

Since the structure is statically determinate, it is possible to calculate the relative displacement between points B and C on the two walls in terms of the base forces M_0 and N_0. The appropriate boundary condition then follows again from the compatibility condition at the base.

Provided that the supporting structure can be rendered statically determinate by the insertion of pins or moment releases at appropriate positions, such as those indicated "X" in Fig. A1.9, it will generally be possible to carry out an independent analysis of the substructure to allow the relative base displacement to be evaluated. A wide range of support structures may be considered by this technique [10.3].

The solutions for a pair of coupled shear walls, supported either on elastic foundations or on different forms of portal frame, are tabulated in Appendix 1. Results are given for the three standard load cases of a uniformly distributed load, a triangularly distributed load, and a point load at the top.

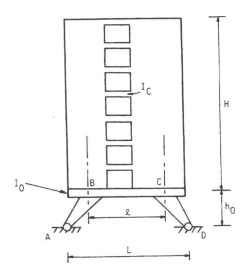

Fig. 10.17 Coupled shear walls on supporting frame.

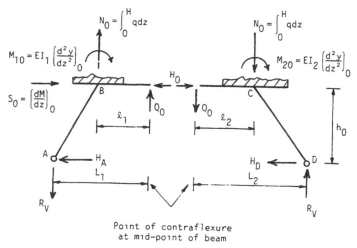

Fig. 10.18 Forces in supporting frame.

10.4 COMPUTER ANALYSIS BY FRAME ANALOGY

The approximate method of analysis described in Section 10.3 is most appropriate for uniform or quasiuniform structures. For other than simple regular systems, it is necessary to use a more sophisticated model to obtain an accurate analysis. The most convenient method is by the use of a frame analogy, which is a very versatile and economic approach and may be used for the majority of practical situations.

The analysis requires the modeling of the interaction between the vertical shear walls and the horizontal connecting beams. Over the height of a single story, a wall panel may appear to be very broad, but when viewed in the context of the entire height it will appear as a slender cantilever beam. When subjected to lateral forces, the wall will be dominated by its flexural behavior, and shearing effects will generally be insignificant.

In the simplest analogous frame model, the wall can then be represented by an equivalent column located at the centroidal axis, to which is assigned the axial rigidity EA and flexural rigidity EI of the wall. The condition that plane sections remain plane may be incorporated by means of stiff arms located at the connecting beam levels, spanning between the effective column and the external fibers as shown in Fig. 10.19. The rigid arms ensure that the correct rotations and vertical displacements are produced at the edges of the walls. The connecting beams may be represented as line elements in the conventional manner, and assigned the correct axial, flexural, and if necessary, shearing rigidities. Generally, shearing deformations should be included if the beam length/depth ratio is less than about 5. The coupled shear wall structure of Fig. 10.19a may then be represented by the analogous wide-column plane frame of Fig. 10.19b.

10.4 COMPUTER ANALYSIS BY FRAME ANALOGY

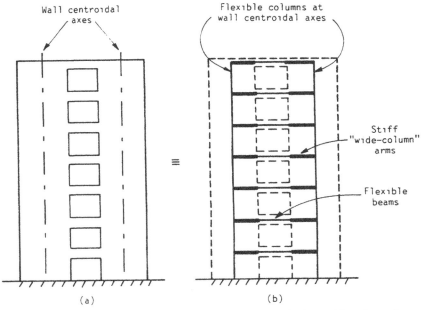

Fig. 10.19 Representation of coupled shear walls by equivalent wide-column frame.

10.4.1 Analysis of Analogous Frame

The analogous frame may be analyzed most conveniently by the conventional stiffness method, which has been extensively developed over the years, and is now well documented in the literature [e.g., 10.4, 10.5]. There is no need to treat the subject in detail, and only its specific use in the analysis of coupled wall structures is considered.

General purpose frame analysis programs are now widely available to carry out the matrix operations required on both micro- and main frame computers. These require no more of the engineer than a specification of the geometric and structural data, and the applied loading.

Different approaches are possible for modeling the rigid-ended connecting beams in the analytical model, depending on the facilities and options available in the program used. The most important techniques are as follows.

Direct Solution of the Analytical Model (Fig. 10.19b). A direct application of the stiffness method will require a series of nodes at the junctions between the stiff arms and the connecting beams in the wide-column model, as well as at the column story levels. The rigidity of the wide-column arms can be simulated by assigning very high numerical values of axial areas and flexural rigidities to the members concerned. In practice, a value of 10^3–10^4 times the corresponding values for the flexible connecting beams has been found to provide results of sufficient accuracy without causing numerical problems in the solution. This procedure would effectively double the number of nodes and thus the number of degrees of freedom, or unknowns in the analysis, and would unnecessarily increase the amount of computation involved.

When forming the wide-column model, an extended length may again be used for the connecting beam to allow for the effect of the flexibility of the wall–beam junction.

It is frequently assumed that the axial deformations of the horizontal beams are considered negligible in comparison with the bending deformations, particularly as a result of the high inplane stiffness of the associated floor slabs. In that case, the horizontal displacements of all nodes will be the same at each floor level. Consequently, at three of the four nodes at each floor level, only two degrees of freedom (vertical and rotational displacement) will be required, while the remaining datum node will have the standard three degrees of freedom (vertical, horizontal, and rotational displacements). If the facility exists in the program used, the internodal constraint or the rigid floor option may be employed to ensure that all joint horizontal deflections are equal at each floor level.

Use of Stiffness Matrix for Rigid-Ended Beam Element.

Because of the rigid connecting segments, simple relationships exist between the actions at a column node and those at the adjacent wall–beam junction node, and it is possible to derive a composite stiffness matrix for the complete beam segment between column nodes that incorporates the influence of the stiff end segments.

The required stiffness matrix for the line element with rigid arms shown in Fig. 10.20 may be derived either by transforming the effects at the wall–beam junctions i and j to the nodes at the wall centroidal axes 1 and 2 by a transformation matrix, or by calculating the stiffness coefficients directly from first principles. In the latter case, unit vertical or rotational displacements may be imposed at nodes 1 or 2 and the resulting resisting moments and shears established from ordinary beam theory, giving directly the required unit force–displacement relationships.

Explicit forms of the resulting stiffness matrix for the wide column beam element have been published in the literature [e.g., 10.6].

If the analysis program includes a wide-column beam stiffness matrix, this may be used directly to represent the connecting beams in the analogous plane frame model of Fig. 10.19b.

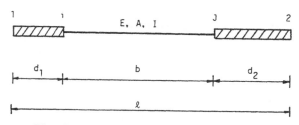

Fig. 10.20 Line element with rigid end arms.

10.4 COMPUTER ANALYSIS BY FRAME ANALOGY

It must be noted that the composite matrix analysis yields the beam moments at the wall centroidal nodes A and B on the rigid segment. In the real structure, the maximum moments in the connecting beams occur at their junctions with the walls, at points C and D (Fig. 10.21). In the absence of any lateral loads on the beams, the bending moment distribution varies linearly between nodes A and B. If the calculated beam moments at A and B are M_A and M_B, respectively, the true end moments on the beams will then be given by (Fig. 10.21)

$$M_C = M_A - \frac{d_1}{l}(M_A - M_B) \qquad (10.59)$$

$$M_D = M_B + \frac{d_2}{l}(M_A - M_B) \qquad (10.60)$$

Alternatively, if a point of contraflexure is again assumed to occur at the midspan position,

$$M_C = M_D = Q\frac{b}{2} \qquad (10.61)$$

where Q is the beam shear.

In some cases the connecting beams may be relatively deep, and this will increase considerably the stiffness of the wall-beam joint. The effect may readily be modeled in the equivalent frame by incorporating stiff vertical arms in the column element over the finite depth of the connecting beam, as shown in Fig. 10.22. The stiffness matrix for the resulting column element will then be of the same form as that for a rigid-ended beam. However, the analogy is less exact than that for the coupling beam, and care must be exercised when adding such stiff segments and in interpreting the ensuing results.

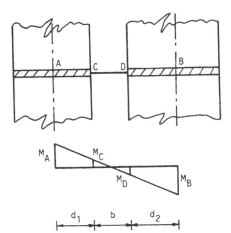

Fig. 10.21 Bending moment diagram for connecting beam.

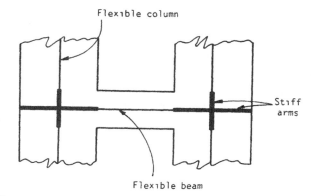

Fig. 10.22 Representation of coupled shear walls with deep connecting beams by equivalent frame.

Use of Haunched Member Facility. If a wide-column beam option is not available in the program, but a haunched member option is, the latter may be adopted to represent the rigid arms if specific large stiffness values are given to the cross-sectional area and flexural rigidity of the haunched ends. These values must be sufficiently large for the resulting deformations to be negligible, but not sufficiently large to cause computational problems from ill-conditioned equations. The stiffness of the end segments depends on the length as well as the cross-sectional properties, and the choice of the rigidities EA and EI for the stiff segments must reflect the effect of the ratio of the length of the arm to the span of the flexible connecting beam. End values of the order of 10^4 times the connecting beam values will generally be found acceptable.

Use of Equivalent Uniform Connecting Beams. In a symmetrical coupled wall structure, in which axial deformations of the connecting beams are assumed negligible, the rotations of the walls at any level will be equal. The rotations of the stiff-ended beams are also equal, and, consequently, it is possible to replace the stiff-ended beam by an equivalent uniform beam with an effective second moment of area I_e, thereby treating the wide-column frame as a normal plane frame of beams and columns.

Since the walls do not undergo relative horizontal deflections or rotations, only the vertical translational stiffness has to be reproduced correctly in the equivalent model.

The required second moment of area I_e of the equivalent beam may be related to the value I_b of the real flexible beam by equating the relative vertical displacements δ of the real and equivalent beams subjected to the same vertical shear P, as shown in Fig. 10.23.

Then,

$$\frac{P}{\delta} = \frac{12EI_b}{b^3} = \frac{12EI_e}{l^3} \tag{10.62}$$

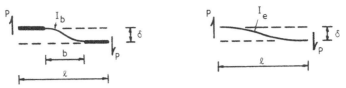

Fig. 10.23 Replacement of rigid-ended connecting beam by equivalent uniform beam.

or

$$I_e = \left(\frac{l}{b}\right)^3 I_b = \rho I_b \qquad (10.63)$$

where ρ is the stiffness amplification factor, equal to the cube of the ratio of the length between the wall centroidal nodes to the true flexible length of the connecting beam.

The coupled shear wall structure may then be represented by a frame having uniform beams of length l and flexural rigidity EI_e.

The required amplification factor ρ depends on the term $(l/b)^3$. In practice, the widths of the walls are generally much greater than the span of the connecting beams; a value of l/b is typically of the order of 3 to 5, resulting in corresponding values of ρ of 27 to 125. These figures illustrate clearly the large increase in effective stiffness of the connecting beams which can result from the wide-column effect of the walls.

If the connecting beams are relatively deep, so that the effects of shearing deformations may not be insignificant, the effective second moment of area to be assigned may be further modified to include this effect. The value of I_e must then be replaced by I_e', where

$$I_e' = \frac{I_e}{1+r} = \frac{1}{1+r}\left(\frac{l}{b}\right)^3 I_b \qquad (10.64)$$

where, again, $r = (12EI_e)/(GAb^2)\lambda$

Alternatively, in an analysis program that requires a shear area to be specified, a shear area $A_{se} = (l/b)A\lambda$ should be used for the equivalent beam, if A is the true cross-sectional area.

The wall actions, the shears in the connecting beams and the horizontal deflections are all given directly by the results from the analysis. However, the actual beams are again shorter than the equivalent beam, and the calculated beam end moments at A and B must be reduced by the factor b/l to give the maximum moments at the ends C and D of the real beams (Fig. 10.21).

Eqns. (10.63) and (10.64) do not strictly apply if the structure is not symmetrical. However, the walls in an unsymmetrical coupled shear wall structure of normal proportions rotate virtually identically at each floor level, and the connecting beams may safely be replaced by equivalent beams of effective second

252 COUPLED SHEAR WALL STRUCTURE

moment of area I_e or I_e' given by Eqs. (10.63) and (10.64). This allows a considerable simplification of the analysis, and yields results which are generally sufficiently accurate for practical purposes.

It is only if one wall is of much higher stiffness than the other, for example, if a wall is connected to a column, that the results should be used with caution. In that case, the more accurate analysis is recommended.

10.5 COMPUTER ANALYSIS USING MEMBRANE FINITE ELEMENTS

For most practical coupled shear wall structures, the equivalent frame technique will be the most versatile and accurate analytical method. In certain cases, however, notably with very irregular openings, such as those shown in Fig. 10.24, or with complex support conditions, it may prove difficult to model the structure with any degree of confidence using a frame of beams and columns. In that case, the use of membrane finite elements is the only feasible alternative.

In this technique, the surface concerned is divided into a series of elements, generally rectangular, triangular, or quadrilateral in shape, connected at a discrete set of nodes on their boundaries. Explicit or implicit forms of the corresponding stiffness matrices for different element shapes are presented in the literature, enabling the structure stiffness matrix to be set up and solved to give all nodal displacements and associated forces. The technique has the advantage that a refined mesh may be employed in regions of high stress gradient or particularly complex geometry, and a much coarser mesh in regions of low or uniform stress. It is also possible to model a structure by a combination of membrane finite elements in a complex region and an equivalent frame in the remaining more uniform region, provided that care is taken to establish the required conditions at the interface. As discussed in Chapter 5, transitions may be achieved by either triangular or quadrilateral elements; the latter approach, being more gradual, tends to yield more accurate results. Generally, when constructing the mesh for a finite element analysis, rectangular elements should be as square as possible, triangular elements

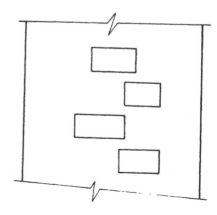

Fig. 10.24 Shear wall with irregular openings.

should be equilateral, and quadrilateral elements should be parallelograms with equal sides, to achieve most accurate results.

The method is now well established and documented [e.g., 10.7, 10.8], and may be used for practical structural analysis through general purpose programs that are widely available.

Particular difficulties arise when using the technique for structures such as coupled walls where relatively slender components such as coupling beams are connected to relatively massive components such as shear walls. Although it is perfectly acceptable to model the walls by rectangular membrane finite elements with two degrees of freedom at each node, it is inappropriate to use such elements for the connecting beams. The latter would require the use of high aspect ratio (length:depth) elements, which might lead to computational errors; in addition, a minimum of three elements would be required to model the double curvature form of bending in the connecting beams, which would increase considerably the size of the structure stiffness matrix and cost of solution. It is sufficiently accurate, and much more convenient, to model a beam by a standard line element, but in that case the node at the wall–beam junction would have to have three degrees of freedom associated with it (two translations and one rotation). It would not then be possible to ensure compatibility with the adjoining node of a plane stress element with only two degrees of freedom (two translations). Some other device is then required to achieve proper compatibility between beams and walls, and this may be achieved in different ways.

For example, it is possible to use special elements with an additional rotational degree of freedom at each node. Such special elements are still rarely available in general purpose programs, and they increase the number of degrees of freedom by 50%, although they avoid the necessity of horizontally long thin wall elements.

A simpler alternative is to add a fictitious, flexurally rigid, auxiliary beam to the edge wall element at the beam–wall junction. The fictitious beam must be connected to two adjacent wall nodes, either in the direction of, or normal to, the beam, as shown in Fig. 5.17. This allows the rotation of the wall, as defined by the relative transverse displacements of the ends of the auxiliary beam, and the moment, to be transferred to the beam. A similar device may be used to connect a column to a wall if the structure is modeled by a combination of a frame and plane stress finite elements.

If a large number of membrane elements are used to specify both beams and walls, as in the irregular structure of Fig. 10.24, the openings may always be modeled with reasonable accuracy by specifying very low values of the thickness of the elements in the openings to give stiffnesses that are negligible compared with those of the adjoining solid wall elements.

SUMMARY

The vertical interaction caused by the presence of connecting beams induces axial forces into a pair of coupled shear walls. The applied moments due to lateral forces are then resisted by a combination of moments in the walls and the couple arising from the axial forces in the walls. The greater the stiffness of the connecting beams, the greater is the proportion of the applied moments resisted by axial forces in the walls. The behavior of the structure then lies between that of two linked walls and a pair of dowelled walls acting as a monolithic composite cantilever. The presence of the moment-resisting connecting beams causes an increased lateral stiffness and a reduction in the maximum wall stresses.

An approximate theory is presented for uniform structures on the basis of a simplified model of the structure, in which the discrete set of connecting beams is replaced by an equivalent continuous connecting medium. A characteristic differential equation for either the wall axial force or the lateral deflection is derived, and general solutions obtained for a uniformly distributed lateral loading. A simplified design method is suggested, in which the wall stresses are obtained as a superposition of two simpler stress systems based on individual and fully composite cantilever actions of the structure. This technique is most suitable for uniform or quasiuniform structures under standard load systems.

For other than uniform wall systems, the most practical and versatile method involves a matrix stiffness analysis of an equivalent wide-column frame model. The coupled walls are represented by a plane frame of line columns along the centroidal axes of the walls and coupling beams, the finite width of the walls being represented by stiff arms connecting the wall axes to the ends of the beams. The analysis may be carried out by a general purpose plane-frame computer program using either special rigid-ended beam elements, haunched beam elements, or equivalent uniform elements for the connecting beams.

In particular cases of very complex geometries that are difficult to model by a grid of beams and columns, the alternative is to use a finite element analysis with the use of appropriate line or membrane elements for the structure considered.

REFERENCES

10.1 Timoshenko, S. P. and Gere, J. M. *Mechanics of Materials*, Chapter 6. Van Nostrand Reinhold, New York, 1973.

10.2 Coull, A. and Choudhury, J. R. "Stresses and Deflections in Coupled Shear Walls," *J. ACI* **64**, 1967, 65-72.

10.3 Coull, A. and Mukherjee, P. R. "Coupled Shear Walls with General Support Conditions." *Proc. of Conference on Tall Buildings*, Kuala Lumpur, 1974, pp. 4.24-4.30.

10.4 McGuire, W. and Gallagher, R. H. *Matrix Structural Analysis*. John Wiley, New York, 1979.

10.5 Coates, R. C., Coutie, M. G., and Kong, F. K. *Structural Analysis*, 3rd ed. Van Nostrand Reinhold, London, 1988.

10.6 MacLeod, I. A. "Lateral Stiffness of Shear Walls with Openings." In *Tall Buildings*, Pergamon Press, Oxford, 1967, pp. 223-244.

10.7 Gallagher, R. H. *Finite Element Analysis: Fundamentals*. Prentice-Hall, Englewood Cliffs, NJ, 1975.

10.8 Zienkiewicz, O. C. *The Finite Element Method*, 3rd ed. McGraw-Hill, New York, 1977.

CHAPTER 11

Wall-Frame Structures

A structure whose resistance to horizontal loading is provided by a combination of shear walls and rigid frames or, in the case of a steel structure, by braced bents and rigid frames, may be categorized as a wall-frame. The shear walls or braced bents are often parts of the elevator and service cores while the frames are arranged in plan, in conjunction with the walls, to support the floor system (Fig. 11.1).

When a wall-frame structure is loaded laterally, the different free deflected forms of the walls and the frames cause them to interact horizontally through the floor slabs; consequently, the individual distributions of lateral loading on the wall and the frame may be very different from the distribution of the external loading. The horizontal interaction can be effective in contributing to lateral stiffness to the extent that wall-frames of up to 50 stories or more are economical.

This chapter is concerned particularly with wall-frame structures that do not twist and, therefore, that can be analyzed as equivalent planar models. These are mainly plan-symmetric structures subjected to symmetric loading. Structures that are asymmetric about the axis of loading inevitably twist. Although the benefits from horizontal interaction between the walls and frames apply also to twisting structures, their consideration in a general way is extremely complex because the amount of interaction is highly dependent on the relative plan location of the bents.

Two examples of symmetric wall-frame arrangements are shown in plan in Fig. 11.2a and b, and an asymmetric structure is shown in Fig. 11.2c. In Fig. 11.2a the horizontal resistance is provided by walls and frames in parallel bents, which are constrained to deflect identically by the inplane rigidity of the floor slabs and, therefore, interact horizontally through shearing actions in the slabs. In Fig. 11.2b, each of the parallel bents consists of a wall and a frame in the same plane. In this case, the wall and frame in a planar bent interact horizontally through axial forces in the connecting beams or slabs.

The potential advantages of a wall-frame structure depend on the amount of horizontal interaction, which is governed by the relative stiffnesses of the walls and frames, and the height of the structure. The taller the building and, in typically proportioned structures, the stiffer the frames, the greater the interaction. It used to be common practice in the design of high-rise structures to assume that the shear walls or cores resisted all the lateral loading, and to design the frames for gravity loading only. Although this assumption would have incurred little error for buildings of less than 20 stories with flexible frames, it is possible that in many cases

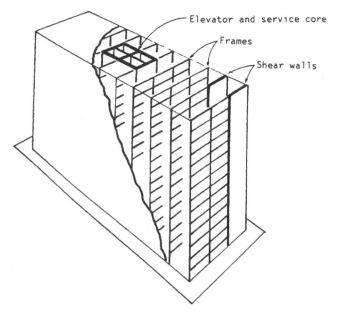

Fig. 11.1 Representative wall–frame structure.

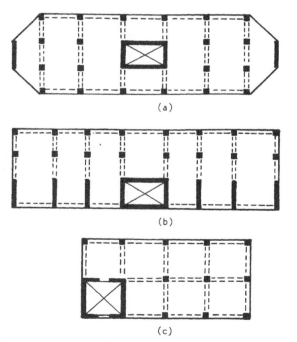

Fig. 11.2 (a) Plan-symmetric wall–frame structure: walls and frames in parallel bents; (b) plan symmetric wall–frame structure: walls and frames in same bents; (c) plan-asymmetric wall–frame structure.

where the frames were stiff and the buildings taller, opportunities were missed to design more rational and economical structures.

The principal advantages of accounting for the horizontal interaction in designing a wall-frame structure are as follows:

1. The estimated drift may be significantly less than if the walls alone were considered to resist the horizontal loading.
2. The estimated bending moments in the walls or cores will be less than if they were considered to act alone.
3. The columns of the frames may be designed as fully braced.
4. The estimated shear in the frames, in many cases, may be approximately uniform through the height; consequently, the floor framing may be designed and constructed on a repetitive basis, with obvious economy.

11.1 BEHAVIOR OF SYMMETRIC WALL-FRAMES

Considering the separate horizontal stiffnesses at the tops of a typical 10-story elevator core and a typical rigid frame of the same height, the core might be 10 or more times as stiff as the frame. If the same core and frame were extended to a height of 20 stories, the core would then be only approximately three times as stiff as the frame. At 50 stories the core would have reduced to being only half as stiff as the frame. This change in the relative top stiffness with the total height occurs because the top flexibility of the core, which behaves as a flexural cantilever, is proportional to the cube of the height, whereas the flexibility of the frame, which behaves as a shear cantilever, is directly proportional to its height. Consequently, height is a major factor in determining the influence of the frame on the lateral stiffness of the wall-frame.

A further understanding of the interaction between the wall and the frame in a wall-frame structure is given by the deflected shapes of a shear wall and a rigid frame subjected separately to horizontal loading, as shown in Fig. 11.3a and b. The wall deflects in a flexural mode with concavity downwind and a maximum slope at the top, while the frame deflects in a shear mode with concavity upwind and a maximum slope at the base. When the wall and frame are connected together by pin-ended links and subjected to horizontal loading, the deflected shape of the composite structure has a flexural profile in the lower part and a shear profile in the upper part (Fig. 11.3c). Axial forces in the connecting links cause the wall to restrain the frame near the base and the frames to restrain the wall at the top. Illustrations of the effects of wall-frame interaction are given by the curves for deflection, moments, and shears for a typical wall-frame structure, as shown in Fig. 11.4a, b, and c. The deflection curve (Fig. 11.4a) and the wall moment curve (Fig. 11.4b) indicate the reversal in curvature with a point of inflexion, above which the wall moment is opposite in sense to that in a free cantilever. Figure 11.4c shows the shear as approximately uniform over the height of the frame, except near the base where it reduces to a negligible amount. At the top, where

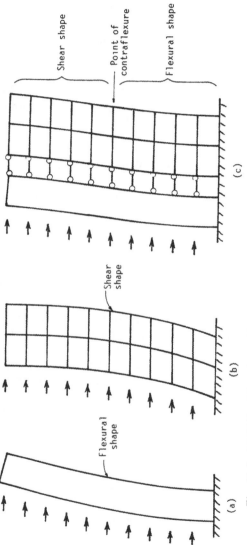

Fig. 11.3 (a) Wall subjected to uniformly distributed horizontal loading; (b) frame subjected to uniformly distributed horizontal loading; (c) wall–frame structure subjected to horizontal loading.

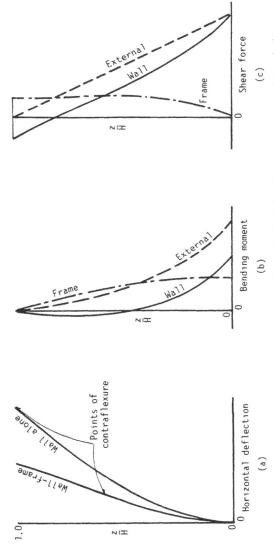

Fig. 11.4 (a) Typical deflection diagram of laterally loaded wall–frame structure; (b) typical moment diagrams for components of wall–frame structure; (c) typical shear diagrams for components of wall–frame structure.

the external shear is zero, the frame is subjected to a significant positive shear, which is balanced by an equal negative shear at the top of the wall, with a corresponding concentrated interaction force acting between the frame and the wall. Special consideration may have to be given in the design to transferring this interaction force through the top connecting slab or beam.

11.2 APPROXIMATE THEORY FOR WALL-FRAMES

Early analytical studies of wall-frame structures [11.1, 11.2] indicated the mode of interaction between the wall and frame and its potential for stiffening the structure.

The analytical treatment of nontwisting wall-frames, given here, allows a broad overview of their behavior and, simultaneously, gives a good qualitative and quantitative understanding of the relative influence of the wall and the frame. It also provides a rapid approximate hand method of analysis that is useful in the preliminary design of wall-frames [11.3, 11.4].

11.2.1 Derivation of the Governing Differential Equation

The planar wall-frame in Fig. 11.5a may be taken to represent either a structure with walls and frames interacting in the same plane, or one with walls and frames in parallel planes. Since, in a nontwisting structure, parallel walls and frames translate identically, they may be simulated by a planar linked model.

The analytical solution requires the structure to be represented by a uniform continuous model (Fig. 11.5b), with all components deflecting identically. The following assumptions are adopted to achieve this:

1. The properties of the wall and the frame members do not change over the height.
2. The wall may be represented by a flexural cantilever, that is, one which deforms in bending only.
3. The frame may be represented by a continuous shear cantilever, which deforms in shear only. This implies that the frame deflects only by reverse bending of the columns and girders, and that the columns are axially rigid.
4. The connecting members may be represented by a horizontally rigid connecting medium that transmits horizontal forces only and that causes the flexural and shear cantilevers to deflect identically.

Considering the wall and frame separately, as in Fig. 11.5c, w and q are, respectively, the distributed external loading and the distributed internal interactive force, whose intensities vary with height. Q_H is a horizontal concentrated force that, as will be demonstrated later, acts between the top of the wall and the frame.

The differential equation for shear in the flexural member is

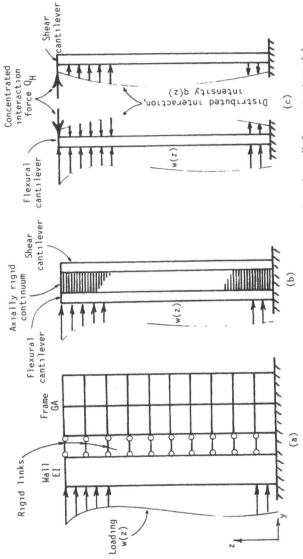

Fig. 11.5 (a) Planar wall–frame structure; (b) continuum analogy for wall–frame structure; (c) free body diagrams for wall and frame.

261

$$-EI \frac{d^3y}{dz^3} = \int_z^H [w(z) - q(z)] dz - Q_H \qquad (11.1)$$

and, for shear in the shear cantilever is

$$(GA) \frac{dy}{dz} = \int_z^H q(z) dz + Q_H \qquad (11.2)$$

in which the parameter (GA) represents the story-height averaged shear rigidity of the frame, as though it were a shear member with an effective shear area A and a shear modulus G. Note that G is not the shear modulus of the frame material nor is A the area of its members.

Differentiating and summing Eqs. (11.1) and (11.2) gives

$$EI \frac{d^4y}{dz^4} - (GA) \frac{d^2y}{dz^2} = w(z) \qquad (11.3)$$

or

$$\frac{d^4y}{dz^4} - \alpha^2 \frac{d^2y}{dz^2} = \frac{w(z)}{EI} \qquad (11.4)$$

in which

$$\alpha^2 = \frac{(GA)}{EI} \qquad (11.5)$$

Equation (11.4) is the characteristic differential equation for the deflection of a wall–frame.

11.2.2 Solution for Uniformly Distributed Loading

The solution of Eq. (11.4) for uniformly distributed external loading w can be written as

$$y(z) = C_1 + C_2 z + C_3 \cosh \alpha z + C_4 \sinh \alpha z - \frac{wz^2}{2EI\alpha^2} \qquad (11.6)$$

The boundary conditions for the solution of constants C_1 to C_4 are

1. fixity at the base

$$y(0) = \frac{dy}{dz}(0) = 0 \qquad (11.7)$$

11.2 APPROXIMATE THEORY FOR WALL-FRAMES

2. zero moment at the top of the flexural cantilever

$$M_b(H) = EI\frac{d^2y}{dz^2} = 0 \tag{11.8}$$

and

3. zero resultant shear at the top of the structure

$$EI\frac{d^3y}{dz^3}(H) - (GA)\frac{dy}{dz}(H) = 0 \tag{11.9}$$

Equations (11.7), (11.8), and (11.9) are used to determine C_1 to C_4 to give the deflection equation:

$$y(z) = \frac{wH^4}{EI}\left\{\frac{1}{(\alpha H)^4}\left[\frac{(\alpha H \sinh \alpha H + 1)}{\cosh \alpha H}(\cosh \alpha z - 1)\right.\right.$$
$$\left.\left. - \alpha H \sinh \alpha z + (\alpha H)^2\left[\frac{z}{H} - \frac{1}{2}\left(\frac{z}{H}\right)^2\right]\right]\right\} \tag{11.10}$$

In Eq. (11.10), the expression within the braces controls the shape of the deflection curve while the term wH^4/EI before the braces governs its magnitude. The deflected shape is a function of the dimensionless parameter αH, which represents the structural properties of the wall-frame where

$$\alpha H = H\sqrt{\frac{(GA)}{EI}} \tag{11.11}$$

αH characterizes the behavior of wall-frames so that, for similar distributions of applied loading, wall-frame structures with the same value of αH have similar deflection profiles and similar distributions of internal forces.

The first derivative of Eq. (11.10) is

$$\frac{dy}{dz}(z) = \frac{wH^3}{EI}\left\{\frac{1}{(\alpha H)^3}\left[\frac{(\alpha H \sinh \alpha H + 1)}{\cosh \alpha H}(\sinh \alpha z)\right.\right.$$
$$\left.\left. - \alpha H \cosh \alpha z + \alpha H\left(1 - \frac{z}{H}\right)\right]\right\} \tag{11.12}$$

Because the slope of a structure varies gradually with height, dy/dz may be taken to represent the story drift index, that is the story drift divided by the story height. The second and third derivatives of Eq. (11.10) are

$$\frac{d^2y}{dz^2}(z) = \frac{wH^2}{EI}\left\{\frac{1}{(\alpha H)^2}\left[\frac{(\alpha H \sinh \alpha H + 1)}{\cosh \alpha H}(\cosh \alpha z) - \alpha H \sinh \alpha z - 1\right]\right\}$$

(11.13)

and

$$\frac{d^3y}{dz^3}(z) = \frac{wH}{EI}\left\{\frac{1}{(\alpha H)}\left[\frac{(\alpha H \sinh \alpha H + 1)}{\cosh \alpha H}(\sinh \alpha z) - \alpha H \cosh \alpha z\right]\right\}$$

(11.14)

These will be used in determining the bending moment and shear distributions in the wall. Equations (11.12), (11.13), and (11.14) are similar to Eq. (11.10) in that each comprises an expression in terms of αH and z/H, controlling the distribution of the function, preceded by a term that governs its magnitude.

11.2.3 Forces in the Wall and Frame

Wall and Frame Moments. The wall behaves as a flexural cantilever; therefore, its bending moment is given by

$$M_b(z) = EI \frac{d^2y}{dz^2}(z) \qquad (11.15)$$

where suffix b refers to the wall. Substituting from Eq. (11.13) for the case of uniformly distributed loading, the wall moment is

$$M_b(z) = wH^2 \left\{\frac{1}{(\alpha H)^2}\left[\frac{(\alpha H \sinh \alpha H + 1)}{\cosh \alpha H}(\cosh \alpha z) - \alpha H \sinh \alpha z - 1\right]\right\}$$

(11.16)

The moment carried by the frame at any level is equal to the external moment minus the wall moment at that level; therefore, the moment carried by the frame is given by

$$M_s(z) = \frac{w(H-z)^2}{2} - M_b(z) \qquad (11.17)$$

where the suffix s refers to the frame and $M_b(z)$ is obtained from Eq. (11.16).

Wall and Frame Shears. The shear force in the wall is given by

$$Q_b(z) = -EI \frac{d^3y}{dz^3}(z) \qquad (11.18)$$

Substituting from Eq. (11.14) gives

$$Q_b(z) = -wH \left\{ \frac{1}{(\alpha H)} \left[\frac{(\alpha H \sinh \alpha H + 1)}{\cosh \alpha H} (\sinh \alpha z) - \alpha H \cosh \alpha z \right] \right\}$$
(11.19)

The shear carried by the frame at any level is equal to the external shear minus the wall shear at that level; therefore, the shear in the frame is given by

$$Q_s(z) = w(H - z) - Q_b(z) \quad (11.20)$$

where $Q_b(z)$ is obtained from Eq. (11.19).

Concentrated Interaction Force Q_H at the Top. The horizontal concentrated interaction force Q_H that acts between the wall and the frame at the top of the structure can be explained from the fact that the slope $dy/dz(H)$ at the top of the structure must have an associated shear at the top of the frame with a value

$$Q_s(H) = (GA) \frac{dy}{dz}(H) \quad (11.21)$$

But, because the total external shear at the top is zero, the shear in the frame can be equilibrated only by a reverse shear in the wall equal to

$$Q_b(H) = -EI \frac{d^3 y}{dz^3}(H) \quad (11.22)$$

This horizontal interaction between the top of the frame and the wall causing their respective shear forces is the force Q_H.

Equations (11.10) through (11.22) allow the derivation of the displacements, moments, and shears in the wall and frame of a uniform wall–frame structure subjected to uniformly distributed loading. The wall forces are adequate for the wall's design, whereas the frame forces, which are external to the frame, should be used in a separate analysis to determine the forces in the members of the frame. The frame member forces could be analyzed approximately by using the derived frame shears and the portal method or, for taller frames, by using the frame moments and the cantilever method (cf. Chapter 7).

The continuum representation of a multistory structure is acceptably accurate for structures of 10 stories or more. If, however, the frames are very tall and slender, axial deformations of the columns become significant and the method tends to underestimate the deflections.

Shear in the First Story of the Frame. The assumption in the analysis of zero inclination at the base of the structure, that is $dy/dz = 0$, implies that, at the base, the wall resists all the shear while the frame carries none. In an actual structure this is not realistic because the first floor is displaced, therefore the first story columns must be subjected to shear. Although the frame base shear is usually not large, it is useful to be able to estimate its magnitude. This can be determined from the first floor displacement $y(1)$, as obtained from Eq. (11.10) substituted into a rearrangement of the formulas relating the first story shear and displacement of a rigid frame, given in Chapter 7 [Eqs. (7.20) and (7.21)].

If the base connections are rigid, the shear in the first story of the rigid frame may be estimated by

$$Q_s(1) = \frac{12E\left(1 + \dfrac{C_1}{6G_1}\right)}{h_1^2 \left(\dfrac{1}{C_1} + \dfrac{2}{3G_1}\right)} y(1) \qquad (11.23)$$

or, if the base connections are hinged

$$Q_s(1) = \frac{12E}{h_1^2 \left(\dfrac{4}{C_1} + \dfrac{3}{2G_1}\right)} y(1) \qquad (11.24)$$

in which $C_1 = \Sigma(I_c/h_1)$ for the columns in the first story and $G_1 = \Sigma(I_g/L)$ for the girder spans in the first floor. I_c and I_g are the inertias of the columns and girders, respectively, h_1 is the first story height, and L is the girder span.

11.2.4 Solutions for Alternative Loadings

The above solutions relate to wall–frames subjected to uniformly distributed loading. Solutions for other types of loading distributions follow a similar procedure. The displacement equation and its derivatives for a top concentrated load and a triangularly distributed loading are summarized with those for the uniformly distributed loading in Appendix 2. These additional loading cases, and their combination with uniformly distributed loading, are useful for representing graduated wind loading and equivalent static earthquake loading.

11.2.5 Determination of Shear Rigidity (GA)

In the parameter αH, as defined by Eq. (11.11), the term (GA) represents the shear or racking rigidity of a rigid frame averaged over a story height. This is defined as the shear force to cause unit horizontal displacement per unit height. Referring to Fig. 11.6

$$(GA) = \frac{Qh}{\delta} \qquad (11.25)$$

11.2 APPROXIMATE THEORY FOR WALL-FRAMES

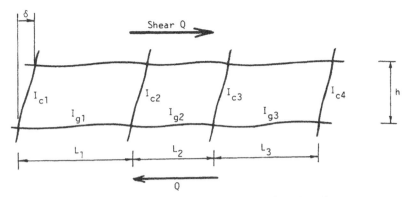

Fig. 11.6 Typical story of rigid frame subjected to shear.

which has been shown in Section 7.4.3 to be given for a typical story of a rigid frame by

$$(GA) = \frac{12E}{h\left(\dfrac{1}{G} + \dfrac{1}{C}\right)} \qquad (11.26)$$

where $G = \Sigma I_g/L$ for the girders across one floor level of a bent, and $C = \Sigma I_c/h$ for the columns in one story of the bent.

Rigid Frame and Wall Connected by Beams. It is common in reinforced concrete structures to connect a rigid frame in plane with a wall or a core by beams, as shown in Fig. 11.7a. The wide column effect of the wall may interact severely with the frame to cause a significant increase in the racking shear rigidity of the bent.

The shear rigidity of the composite bent can be estimated approximately by considering it in two parts, as in Fig. 11.7b, and summing their rigidities, $(GA)_I$ and $(GA)_{II}$.

$(GA)_I$ can be taken as half the value of (GA) for a coupled wall structure consisting of two similar walls, with centroidal axes spaced at $2(a + \alpha L_m)$ [11.5], that is,

$$(GA)_I = \frac{1}{2} \frac{12EI_g(2a + 2\alpha L_m)^2}{(2\alpha L_m)^3 h} \qquad (11.27)$$

in which α is given for the majority of practical structures by

$$\begin{aligned}
&\text{for } m = 1 &&\alpha = 0.566 + 0.024 \ln(\eta) + 0.0424\beta \\
&\text{and, for } m > 1 &&\alpha = 0.55 \ln(\beta) + 0.635 \\
&\text{where} &&\eta = a/L_m \quad \text{and} \quad \beta = EI_g/EI_c
\end{aligned} \qquad (11.28)$$

268   WALL–FRAME STRUCTURES

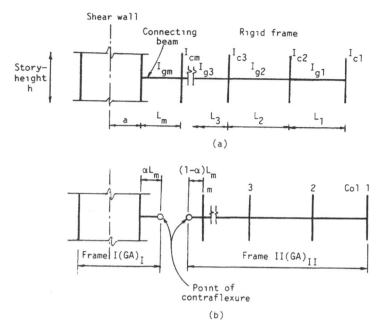

Fig. 11.7 (a) Story of planar wall–frame structure with connecting beams; (b) equivalent substructures of beam-connected wall–frame structure.

$(GA)_{\text{II}}$ is given for the frame by Eq. (11.26), where the term G includes a term to account for the mth girder of length $(1 - \alpha)L_m$.

11.3 ANALYSIS BY THE USE OF GRAPHS

Equations (11.10) for deflection, (11.12) for story drift index, (11.16) for wall moment, and (11.19) for wall shear must be evaluated for use in design. Each consists of a distribution expression that depends on the dimensionless parameters αH and z/H, and a magnitude term that accounts for the loading, the height and the flexural rigidity of the particular structure.

Equation (11.10) for deflection can be rewritten as

$$y(z) = \frac{wH^4}{8EI}\left\{\frac{8}{(\alpha H)^4}\left[\frac{\alpha H \sinh \alpha H + 1}{\cosh \alpha H}(\cosh \alpha z - 1)\right.\right.$$
$$\left.\left. - \alpha H \sinh \alpha z + (\alpha H)^2 \left(\frac{z}{H} - \frac{1}{2}\left(\frac{z}{H}\right)^2\right)\right]\right\} \quad (11.29)$$

or

$$y(z) = \frac{wH^4}{8EI} K_1(\alpha H, z/H) \quad (11.30)$$

where K_1, which represents the distribution expression in the braces, is a function of only αH and z/H, noting that αz is the product of the two.

As before, the expression within the braces controls the distribution of deflection for different values of the parameter αH of the wall-frame, while the term in front of the braces governs the magnitude of the distribution. Now, however, the term $wH^4/8EI$ represents the top deflection as though the wall alone resisted the loading, that is for the case of (GA) (and hence αH) equal to zero. The variation of K_1 over the height for $\alpha H = 0$, therefore, describes the deflection of a purely flexural cantilever, and has a maximum value at the top equal to unity. The difference in deflections between the curve for $\alpha H = 0$, and one for a wall-frame structure of a certain value of αH, represents the stiffening effect contributed by the frame and its interaction with the wall. The stiffer the frame, the greater the interaction and the greater the reduction in deflection.

The curves of K_1 for $\alpha H = 2.0$ and $\alpha H = 0$ are shown in Fig. 11.8a. The former represents the deflected shape for a well-proportioned, uniform wall-frame structure with an approximately uniform slope, and an approximately uniform shear in the frame, over most of the height. Similar normalized expressions for story drift index, wall moment, and wall shear can be developed by rewriting Eqs. (11.12), (11.16), and (11.19) and assigning coefficients K_2, K_3, and K_4, respectively, to represent the distribution expressions, thus

Story drift index $$\frac{dy}{dz}(z) = \frac{wH^3}{6EI} K_2(\alpha H, z/H) \quad (11.31)$$

Wall bending moment $$M_b(z) = \frac{wH^2}{2} K_3(\alpha H, z/H) \quad (11.32)$$

Wall shear $$Q_b(z) = wH K_4(\alpha H, z/H) \quad (11.33)$$

In each of Eqs. (11.31), (11.32), and (11.33), as in Eq. (11.30), the parameter K describes the distribution of the corresponding action over the height of the structure, as a function of αH, while the preceding term represents the maximum value of the action in the wall as if it were considered to act alone in carrying the loading. Consequently, the maximum values of K_2 to K_4 are all, as for K_1, unity. Representative curves of K_2, K_3, and K_4 for wall-frames with $\alpha H = 2.0$ and $\alpha H = 0$ are given in Fig. 11.8a and b. The K_3 diagram (Fig. 11.8b) for the wall moment in a structure with $\alpha H = 2$ indicates the point of contraflexure with a reverse moment in the upper region of the wall. The moment carried by the frame at a particular level may also be obtained from the graph of K_3, by subtracting K_3 for the considered structure from K_3 for $\alpha H = 0$, which represents the total external moment at the same level, and using the resulting difference of K_3 in Eq. (11.32). Similarly, the frame shear at a particular level may be obtained from Fig. 11.8b by subtracting K_4 for the considered structure from K_4 for $\alpha H = 0$, which represents the total external shear at that level, and using the resulting difference of K_4 in Eq. (11.33).

270 WALL–FRAME STRUCTURES

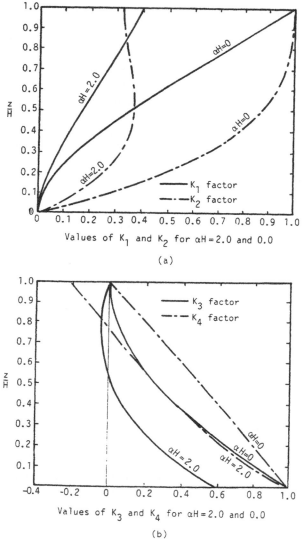

Fig. 11.8 (a) Representative curves for coefficients K_1 and K_2: (b) representative curves for coefficients K_3 and K_4.

Complete sets of curves for K_1 to K_4, and their formulas, are given for uniformly distributed, triangularly distributed, and concentrated top loading in Appendix 2.

These curves may also be used for the analysis of cores subjected to torsion, as explained in Chapter 13.

11.4 WORKED EXAMPLE TO ILLUSTRATE APPROXIMATE ANALYSIS

The theory and graphs developed in the preceding sections provide a practical method for the preliminary analysis of wall-frames. They should be used only for structures that do not twist and they are accurate only for structures with uniform properties over the height; however, they may serve additionally as a useful guide to the forces in nonuniform structures. The following outline procedure for an approximate analysis is illustrated by referring to an example of a nontwisting structure consisting of a central core and frames.

The plan of the structure in Fig. 11.9 is of a 35-story, 122.5 m-high, wall-frame structure. The horizontal resistance to wind acting on its long side is provided by six rigid frame bents and a central core.

It is required to determine deflections, maximum story drift, and forces in the core and frames for a wind loading of 1.5 kN/m², given the structural data in the figure.

The required procedures are first described and then illustrated numerically for the considered structure.

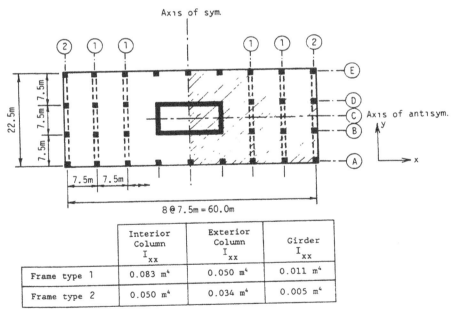

Fig. 11.9 Plan of 35-story wall-frame example structure.

272 WALL-FRAME STRUCTURES

Step 1. Determine Parameter αH

a. Add the flexural rigidities EI of all walls and cores to give the total $(EI)_t$. In this case there is only a core.

For the core $I = 313 \text{ m}^4$

Therefore $(EI)_t = 2.0 \times 10^7 \times 313 = 6.26 \times 10^9 \text{ kNm}^2$

b. Evaluate the shear rigidities (GA) of the rigid frame bents and any wall-frame bents, using Eqs. (11.26), (11.27), and (11.28), and sum them to give the total $(GA)_t$. In this case there are only the former.
The shear rigidities (GA) of the two types of frame in Fig. 11.9 are obtained by using the expression

$$(GA) = \frac{12E}{h\left(\dfrac{1}{G} + \dfrac{1}{C}\right)} \tag{11.26}$$

For frame Type 1:

$$(GA) = \frac{12 \times 2.0 \times 10^7}{3.5 \left[1 \bigg/ \dfrac{(3 \times 0.011)}{7.5} + 1 \bigg/ \dfrac{[2(0.083 + 0.050)]}{3.5} \right]}$$

$$= 2.85 \times 10^5 \text{ kN}$$

For frame Type 2:

$$(GA) = \frac{12 \times 2.0 \times 10^7}{3.5 \left[1 \bigg/ \dfrac{(3 \times 0.005)}{7.5} + 1 \bigg/ \dfrac{[2(0.050 + 0.034)]}{3.5} \right]}$$

$$= 1.32 \times 10^5 \text{ kN}$$

Total $(GA)_t = \Sigma(GA) = (4 \times 2.85 + 2 \times 1.32) \times 10^5$

Therefore $(GA)_t = 14.04 \times 10^5 \text{ kN}$

c. Use the values obtained in Items a and b to evaluate αH, using

$$\alpha H = H \sqrt{\frac{(GA)_t}{(EI)_t}} \tag{11.11}$$

For the given structure

$$\alpha H = 122.5 \sqrt{\frac{14.04 \times 10^5}{6.26 \times 10^9}}$$

11.4 WORKED EXAMPLE TO ILLUSTRATE APPROXIMATE ANALYSIS

Therefore

$$\alpha H = 1.83$$

An analysis for uniformly distributed loading is then made as follows:

Step 2. Determine Horizontal Displacements

The displacement at height z from the base is obtained by substituting αH and z/H in Eq. (11.10) or, alternatively, by taking the value of K_1 corresponding to the obtained values of αH and z/H from Fig. A2.1 and substituting it in the expression

$$y(z) = \frac{wH^4}{8(EI)_t} K_1(\alpha H, z/H) \qquad (11.30)$$

For the given structure, the wind loading per unit height

$$w = 1.5 \times 60 = 90 \text{ kN/m}$$

At the top

$$z/H = 1.0 \quad \text{and} \quad K_1 = 0.44$$

then, substituting these values in Eq. (11.30), the top displacement is obtained as

$$y(H) = \frac{90 \times 122.5^4 \times 0.44}{8 \times 6.26 \times 10^9}$$

$$= 0.178 \text{ m}$$

Displacement at other levels have been found similarly and are plotted in Fig. 11.10a.

Step 3. Determine Maximum Story Drift Index

The maximum story drift index is obtained by referring to Fig. A2.2, and scanning the appropriate αH curve to find the maximum value of K_2, which is then substituted in

$$\frac{dy}{dz}(\max) = \frac{wH^3}{6(EI)_t} K_2(\max) \qquad (11.31)$$

For the given structure, $K_2(\max) = 0.41$ at an approximate height $z/H = 0.55$. Therefore, the maximum story drift index is obtained as

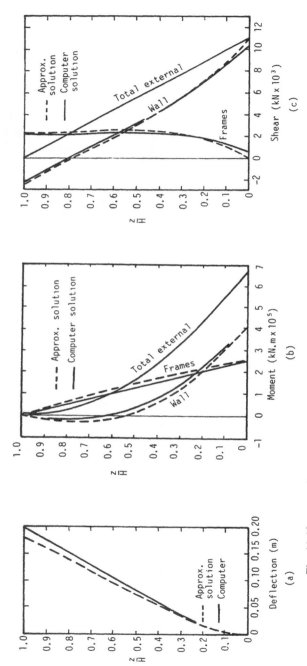

Fig. 11.10 (a) Deflections of example structure; (b) moments in example structure; (c) shears in example structure.

11.4 WORKED EXAMPLE TO ILLUSTRATE APPROXIMATE ANALYSIS

$$\frac{dy}{dz}(\max) = \frac{90 \times 122.5^3 \times 0.41}{6 \times 6.26 \times 10^9}$$

$$= 0.0018 \text{ or } 1/555$$

Step 4. Determine Bending Moments in the Wall and Frame

a. The total moment carried by the walls is obtained either by substituting αH and z/H in Eq. (11.16), or by taking the appropriate value of K_3 from Fig. A2.3 and substituting in

$$M_b(z) = \frac{wH^2}{2} K_3(\alpha H, z/H) \qquad (11.32)$$

For example, at the mid-ninth story level ($z = 29.75$ m, $z/H = 0.243$) of the structure considered, $K_3 = 0.25$. Therefore, the moment in the core is obtained as

$$M_b = \frac{90 \times 122.5^2 \times 0.25}{2}$$

$$= 1.69 \times 10^5 \text{ kNm}$$

For a structure consisting of multiple walls, the moment in any individual wall is then obtained by distributing the total wall moment between the walls in proportion to their flexural rigidities.

b. The total moment in the set of frames at a level z from the base, as expressed in Eq. (11.17), is equal to the difference between the total external moment and the total moment in the walls at that level.

$$M_s(z) = \frac{w(H-z)^2}{2} - M_b(z) \qquad (11.17)$$

At the same mid-ninth story level of the given structure, as in Item 4a, the moment carried by the frames is obtained as

$$M_s = \frac{90(122.5 - 29.75)^2}{2} - 1.69 \times 10^5 \text{ kNm}$$

$$= 2.18 \times 10^5 \text{ kNm}$$

The moment in the individual frames is obtained by distributing the total frame moment between the frames in proportion to their shearing rigidities. Therefore, the moment in frame Type 1

$$= \frac{2.85 \times 10^5}{14.04 \times 10^5} \times 2.18 \times 10^5 \text{ kNm}$$

$$= 4.43 \times 10^4 \text{ kNm}$$

and, the moment in frame Type 2

$$= \frac{1.32 \times 10^5}{14.04 \times 10^5} \times 2.18 \times 10^5 \text{ kNm}$$

$$= 2.05 \times 10^4 \text{ kNm}$$

The bending moments at other levels of the structure have been found similarly and are plotted in Fig. 11.10b.

Step 5. Determine Shear Forces in Wall and Frame

a. The total shear in the walls at a level z from the base may be obtained by substituting αH and z/H in Eq. (11.19) or by taking the value of K_4 from Fig. A2.4 and substituting in Eq. (11.33).

$$Q_b(z) = wHK_4(\alpha H, z/H) \qquad (11.33)$$

For example, at the mid-eighteenth story level ($z = 61.25$ m, $z/H = 0.5$) of the structure considered, $K_4 = 0.27$. Therefore, the shear in the core is obtained as

$$Q_b = 90 \times 122.5 \times 0.27$$

$$= 2.98 \times 10^3 \text{ kN}$$

For a structure consisting of multiple walls, the shear force in the individual walls is then obtained by distributing the total shear between the walls in proportion to their flexural rigidities.

b. The total shear in the frames at a height z is the difference between the external shear and the total wall shear at that level, as determined above:

$$Q_i(z) = w(H - z) - Q_b(z) \qquad (11.20)$$

At the same mid-eighteenth story level of the given structure, as in Item 5a, the shear carried by the frames is obtained as

$$Q_i = 90(122.5 - 61.25) - 2.98 \times 10^3$$

$$= 2.53 \times 10^3 \text{ kN}$$

The shear in the individual frames is given by distributing the total frame shear between the frames in proportion to their shearing rigidities. Therefore, the shear in frame Type 1

$$= \frac{2.85 \times 10^5}{14.04 \times 10^5} \times 2.53 \times 10^3 \text{ kNm}$$

$$= 514 \text{ kN}$$

and, the shear in frame Type 2

$$= \frac{1.32 \times 10^5}{14.04 \times 10^5} \times 2.53 \times 10^3 \text{ kNm}$$

$$= 238 \text{ kN}$$

The values of shear at other levels of the structure have been found similarly and are plotted in Fig. 11.10c.

11.5 COMPUTER ANALYSIS

The approximate method of analysis is valuable in providing an understanding of a wall–frame's behavior and in allowing the initial sizing of members as part of the preliminary design process. It does not allow, however, for changes of properties within the height of the structure or for the effects of axial deformations of the columns that in a tall slender frame could be significant. Therefore, a computer analysis, using one of the widely available structural analysis programs, should be used for the final design.

Modeling the wall–frame structure for a computer analysis will follow the principles outlined in Chapter 5. If the structure is symmetric on plan and subjected to symmetric loading, so that it does not twist, a planar model of only one-half of the structure subjected to one-half of the loading need be considered. Shear walls and shear-wall cores are represented by simple column cantilevers with corresponding moments of inertia, while the frames are represented by equivalent assemblies of beam elements. In the planar model the cantilever columns and frames are constrained at each floor level by the analysis program's nodal constraint option, if available, or connected by axially rigid links, to cause equal horizontal displacements of the bents, as imposed on the structure by the inplane rigidity of the floor slabs. The horizontal loads may be applied to the nodes of any convenient column or frame.

The wall–frame structure analyzed approximately in Section 11.4, and shown in plan in Fig. 11.9, can be modeled for a two-dimensional computer analysis as in Fig. 11.11a. One-half of the structure is used for the model and, because the Type 1 frames are identical, they are lumped into a single frame with members of twice the sectional properties. The double-symmetry of the structure's plan, about

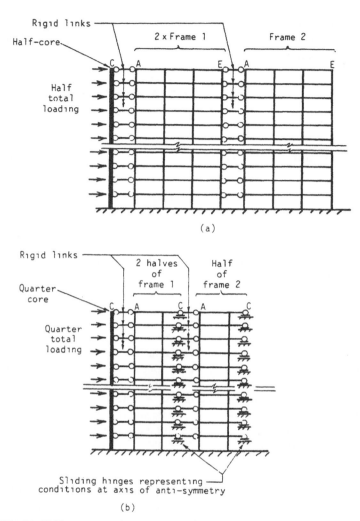

Fig. 11.11 (a) Half-structure planar model for computer analysis; (b) quarter-structure planar model for computer analysis.

axes perpendicular to and along the axis of loading, allow it to be analyzed as a quarter-plan structure. Constraints have to be applied to the cut ends of beams to represent the conditions on the axis of antisymmetrical behavior, as in Fig. 11.11b.

The results for the deflections, moments, and shears from the computer stiffness matrix analysis are compared with the results of the approximate analysis in Fig. 11.10a, b, and c. The discrepancy between the two methods for the deflections in the upper regions (Fig. 11.10a) is attributable to axial deformations in the columns of the frame, which are not considered in the approximate solution. The close comparison of the results of the shears and moments from these two methods is typical for uniform structures. A comparison of results from the two methods for a nonuniform structure, however, would probably show significant errors in those from the approximate method, especially in the forces around the change levels, where severe local interactions occur.

11.6 COMMENTS ON THE DESIGN OF WALL-FRAME STRUCTURES

A wall-frame high-rise structure typically consists of the walls and frames that evolve from the architectural plan of the building. The initial member sizes are usually determined from the gravity loading with some arbitrary increase to allow for the effects of horizontal loading. The benefit of accounting for the wall-frame interaction in the lateral load analysis, as opposed to assuming that the walls carry all the lateral loading, is that it recognizes the increased lateral stiffness due to the interaction, and allows the wall and frame members to be designed more correctly and economically.

11.6.1 Optimum Structure

Occasions may arise in very tall wall-frame structures when it is desirable to proportion the wall and frame components so as to optimize the desirable effect of the wall-frame interaction. Such an optimization would aim not only to achieve significant reductions in the deflections and the wall moments, but also to cause an approximately uniform distribution of shear over the height of the frame. This would then permit the repetitive design and construction of the floor framing.

To achieve such a well-proportioned uniform wall-frame structure, the combination of walls and cores should be sized in the preliminary stage of design so that when carrying their attributable gravity loading, together with two-thirds of the total horizontal loading, the tensile stresses in the walls and cores due to horizontal loading are suppressed.

The system of walls and cores should then be checked for drift. If, when subjected to the total horizontal loading, the maximum total drift or story drift of the walls and cores exceeds twice the allowable value, they should be stiffened to reduce the drift to that value. Adjustments of size in the lower region of the walls and cores are the most effective.

The resulting walls and cores should then be combined with the "gravity load designed" framing to assess the drift, and the wall and frame forces, of the whole structure subjected to the total horizontal loading. Any required additional stiffening should again be made by increasing the wall and core sizes in the lower region. A uniform wall-frame structure in which the set of walls and cores acting on its own deflects approximately twice as much as the combined wall-frame should carry an approximately uniform shear in the frame.

11.6.2 Curtailed or Interrupted Shear Walls

It is common in the design of practical wall-frame structures to reduce the size, or omit completely, the shear walls and cores in the upper parts of the building, where fewer elevator shafts are required. The question of how this curtailment will affect the stiffness of the building can be answered by considering the behavior of the wall–frame structure with full-height shear walls or cores.

An analysis of the "full-height" wall–frame structure shows that in the lower region of the structure the wall and frame both contribute to resisting the external moment and shear. In the upper region above the point of inflexion, where $d^2y/dz^2 = 0$, however, the moment in the wall is reversed to be of the same sense as the external moment; consequently, the moment in the frame exceeds the external moment. Further, in the uppermost region above the level where $d^3y/dz^3 = 0$, the shear in the wall is also reversed, and so the shear in the frame exceeds the external shear. Consequently, if the wall is reduced or eliminated above the point of contraflexure, the moment on the upper part of the frame is reduced, and if the wall is reduced or eliminated above the level where $d^3y/dz^3 = 0$, both the moment and the shear in the frame are reduced. In both cases the reduction or curtailment has little effect on the top deflection, and it may even lead to a slight reduction in deflection.

A practical approach to a curtailed-wall design, therefore, would be to first analyze the structure with the walls and cores that are to be reduced or curtailed included to the full height, without the proposed reduction or curtailment, and use the results to plot the deflection diagram and estimate the location of the point of inflexion. It would then be allowable to reduce or curtail the walls and cores at any level above the point of inflexion without causing a reduction in the lateral stiffness of the structure.

If it is intended to omit the walls or cores within one or more stories at an intermediate level, to provide for a "sky-lobby" or transition in the building, it may be deduced from the above explanation that this is permissible without any reduction in lateral stiffness, providing the omission occurs at a level above where the point of inflexion would be in the fully continuous structure. An analysis and design procedure similar to that suggested for the curtailed wall–frame structure would therefore be appropriate.

11.6.3 Increased Concentrated Interaction

A study of the wall–frame interaction, and especially of the concentrated interaction force Q_H at the top, leads to the notion that further stiffening of the structure could be achieved by increasing the magnitude of the top interaction force.

Referring to Eq. (11.21) gives the value of the force as

$$Q_H = (GA)\frac{dy}{dz}(H) \qquad (11.21)$$

that is, a function of the racking rigidity (GA) of the frame and the slope of the structure. The interaction force could be increased simply by increasing the racking rigidity of the story of the frame adjacent to the top of the wall, whether in "full-height" or "curtailed-wall" wall–frame structures. The increase in racking rigidity could be achieved in practice by increasing the inertias of the beams and columns in the story of the frame adjacent to the top of the wall, or by introducing a concrete diaphragm into the frame in that story to give a very large increase in the racking rigidity. Studies have shown that in wall-frame structures that are predominantly shear walls, reductions in the top deflections of up to 30% can be achieved. In such cases, particular attention must be given to designing the frames, and the members connecting the walls to the frames, for the locally high forces associated with the interaction.

SUMMARY

The horizontal interaction between the walls and frames in a wall-frame structure causes an increased lateral stiffness of the structure, reduced moments in the walls, and, in a uniform structure, an approximately uniform shear in the frame. The benefits of interaction increase with height so that wall-frames are economical for buildings of up to 50 stories or more.

The wall-frame horizontal interaction occurs because the different free-deflected shapes of the wall and the frame are made to conform to the same configuration by the axially stiff connecting girders and slabs.

An approximate theory is presented for nontwisting uniform wall-frames on the basis of a continuum model of the structure, with a flexural cantilever representing the walls, a shear cantilever representing the frames, and a horizontally stiff continuous linking medium representing the slabs and girders. A characteristic differential equation for deflection is written in terms of the two structural parameters of the wall-frame. This has been solved for three typical types of loading to obtain general formulas for the deflections, the story-drift, the shears, and moments on the walls and frames. Design curves are also developed that allow rapid estimates of the deflections and forces. The solutions by both formulas and graphs give close estimates of the deflections and forces in nontwisting uniform wall-frames of 10 stories or more, and approximate estimates of the forces in nontwisting, nonuniform wall-frames, which may be used as guidelines for their preliminary design. An accurate estimate of the deflections and forces in nonuniform or in twisting wall-frame structures requires a computer analysis.

The extent of the benefit that can be obtained from wall-frame interaction depends on the relative stiffnesses of the walls and frames and the height of the structure. In a very tall structure with a repetitive floor plan arrangement, an optimum proportioning of the wall and frame would be one that results in a practically uniform shear within each height region of the frame.

If the shear walls in a wall-frame structure are reduced or eliminated above a certain level, or the shear walls are omitted for one or two stories at an intermediate level, the lateral stiffness of the structure will be effectively not less than that of the corresponding fully continuous full-height structure, provided the changes are located above the point of inflection of the "fully continuous" structure.

REFERENCES

11.1 Rosenblueth, E. and Holtz, I. "Elastic Analysis of Shear Walls in Tall Buildings." *ACI J.* **56**(12), June 1960, 1209–1222.

11.2 Khan, F.R. and Sbarounis, J.A. "Interaction of Shear Walls and Frames." *J. Struct. Div.*, *Proc. ASCE* **90**, ST3, June 1964, 285–335.

11.3 Heidebrecht, A.C. and Stafford Smith, B. "Approximate Analysis of Tall Wall-Frame Structures." *J. Struct. Div.*, *Proc. ASCE* **99**, ST2, February 1973, 199–221.

11.4 Rosman, R. "Laterally Loaded Systems Consisting of Walls and Frames." *Proc. Symposium on Tall Buildings*, University of Southampton, England, 1966, pp. 273–289.

11.5 Nollet, M.-J. and Stafford Smith, B. "An Empirical Approach to the Evaluation of the Shear Rigidity of a Wall-Frame with Rigidly Jointed Link Beams." Structural Engineering Series Report No. 88-5, Department of Civil Engineering and Applied Mechanics, McGill University, November 1988.

CHAPTER 12

Tubular Structures

Chapter 4 discussed in broad terms the development of tubular structures for very tall buildings, involving a range of related structural forms: framed-tube, tube-in-tube, bundled-tube, braced-tube, and composite-tube systems. All have evolved from the traditional rigidly jointed structural frame. The basic design philosophy in all of these forms has been to place as much as possible of the load-carrying material around the external periphery of the building to maximize the flexural rigidity of the cross section.

The original development was the framed tube, which, under the action of wind loading, could suffer a considerable degree of shear lag in the normal-to-wind panels. The later more efficient bundled-tube and braced-tube systems were designed to produce a more uniform axial stress distribution in the columns of the "normal" panels. Some recent irregular "postmodern" buildings have involved a hybrid form of structure, in which only part of the periphery is of framed-tube construction while the remainder consists of a space-frame system.

This chapter is devoted to a discussion of the basic structural behavior of tubular structures, and to a description of the techniques used in the analysis of such structures under the action of lateral forces. The general analysis of three-dimensional tubular structures is considered briefly initially, and then the techniques that have been developed to reduce the amount of computation for symmetrical systems are described.

12.1 STRUCTURAL BEHAVIOR OF TUBULAR STRUCTURES

This section considers the structural behavior of the basic rectangular framed-tube structure when subjected to lateral forces, and the improvements that have been made in the subsequent bundled-tube and braced-tube developments. Some of the more important assumptions made in the modeling of these systems are discussed.

12.1.1 Framed-Tube Structures

The most basic framed-tube structure consists essentially of four orthogonal rigidly jointed frame panels forming a tube in plan, as shown in Fig. 12.1a. The frame panels are formed by closely spaced perimeter columns that are connected by deep

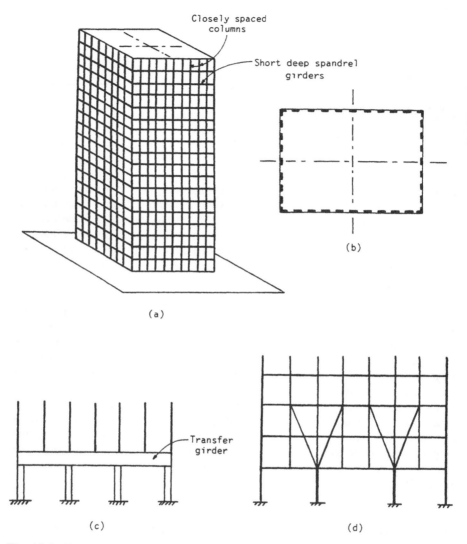

Fig. 12.1 Framed-tube structure. (a) Elevation; (b) structural plan; (c) transfer girder at base; (d) column collection at base.

spandrel beams at each floor level. In such structures, the "strong" bending direction of the columns is aligned along the face of the building (Fig. 12.1b), in contrast to the typical rigid frame bent structure where it is aligned perpendicular to the face. The basic requirement has been to place as much of the load-carrying material at the extreme edges of the building to maximize the inertia of the building's cross section. Consequently, in many structures of this form, the exterior tube is designed to resist the entire wind loading. The frames parallel to the wind act as the "webs" of the perforated tube cantilever, while the frames normal to the wind act as the "flanges." Vertical gravitational forces are resisted partly by the exterior frames and partly by some inner structure such as interior columns or an interior core, using the floor system that spans between the different vertical elements. Although tubular structures are most commonly of square or rectangular planform, they have also been employed in circular, triangular, and trapezoidal-shaped cross sections.

The essential uniformity of the system enables industrialized techniques to be used in the construction sequence. For steel structures, large elements of the facade frame may be prefabricated in a factory and transported to the site where they are hoisted into place and fixed. For concrete structures, the use of gang forms raised story by story enables very speedy construction rates to be achieved.

The closely spaced column configuration makes access difficult to the public lobby area at the base. In many buildings, larger openings at ground floor level have been achieved by using a large transfer girder to collect the vertical loads from the closely spaced columns and distribute them to a smaller number of larger more widely spaced columns at the base (Fig. 12.1c). Alternatively, several columns may be merged through an inclined column arrangement to allow fewer larger columns in the lowest stories, as shown in Fig. 12.1d.

In resisting the entire wind load by the peripheral frame, the tubular structure has the architectural advantage of allowing freedom in planning the interior. For example, a central core with long-span floors from the tube to the core provides the open spaces desired for office buildings, while distributed interior columns and walls, with shallow short-span floors, is very well suited for residential buildings.

Mode of Behavior. Although the structure has a tube-like form, its behavior is much more complex than that of a plain unperforated tube, and the stiffness may be considerably less. When subjected to bending under the action of lateral forces, the primary mode of action is that of a conventional vertical cantilevered tube, in which the columns on opposite sides of the neutral axis are subjected to tensile and compressive forces, as indicated by the broken lines in Fig. 12.2. In addition, the frames parallel to the direction of the lateral load (AD and BC of Fig. 12.2) are subjected to the usual inplane bending, and the shearing or racking action associated with an independent rigid frame. This primary action is complicated by the fact that the flexibility of the spandrel beams produces a shear lag that increases the stresses in the corner columns and reduces those in the inner columns of both the flange panels (AB and DC), and the web panels (AD and BC), as shown by the solid lines in Fig. 12.2.

This behavior may readily be appreciated by considering the basic mode of action involved in resisting lateral forces. The primary resistance comes from the side web panels, which deform so that the columns A and B are in tension and D and C are in compression (Fig. 12.2c). The principal interaction between the web

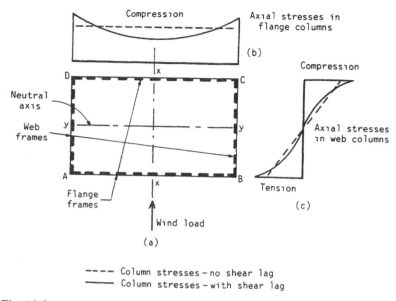

Fig. 12.2 Axial stress distribution in columns of laterally loaded framed tube.

and flange frames occurs through the vertical displacements of the corner columns. These displacements correspond to vertical shear in the girders of the flange frames, which mobilizes the axial forces in the flange columns. When column C, for example, suffers a compressive deformation, it will tend to compress the adjacent column C1 (Fig. 12.3) since the two are connected by the spandrel beams. The compressive deformations will not be identical since the flexible connecting spandrel beam will bend, and the axial deformation of the adjacent column will be less, by an amount depending on the stiffness of the connecting beam. (Pure tubular behavior would theoretically require connecting beams of infinite stiffness.) The deformation of column C1 will in turn induce compressive deformations of the next inner column C2, but the deformation will again be less. Thus each successive interior column will suffer a smaller deformation and hence a lower stress than the outer ones. Since the external applied moment must be resisted by the internal couple produced by the compressive and tensile forces on opposite sides of the neutral axis of the building, it follows that the stresses in the corner columns will be greater than those from pure tubular action, and those in the inner columns will be less.

The differences between the pure tubular stress distribution, as predicted by ordinary engineer's beam theory and the true situation is illustrated in Fig. 12.2. Because the column stresses are distributed less effectively than in a proper tube, the moment of resistance and the flexural rigidity are reduced. Thus, although an unbraced framed tube is a highly effective form of tall building construction, it

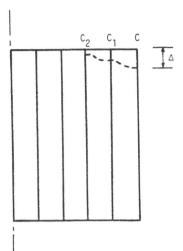

Fig. 12.3 Deformation of flange frame causing shear lag.

does not fully exploit the potential stiffness and strength of the structure because of the effects of shear lag in the perimeter frames.

The shear lag effect will produce bending of the floor slabs, since plane cross sections no longer remain plane; consequently, deformations of interior partitions and secondary structural components occur, which increase cumulatively throughout the height of the building. It is therefore of considerable importance to predict accurately the structural behavior of the system in order to produce an efficient and acceptable design.

Structural Analysis. For analytical purposes, it is usually assumed that the inplane stiffness of the floor system is so great that the floor slabs act as rigid diaphragms. Consequently, the cross-sectional shape is maintained at each story level, and cross sections at these positions undergo only rigid body movements in plan. All horizontal displacements may then be expressed in terms of two orthogonal translations and a rotation.

In addition, it is usually assumed that the out-of-plane stiffness of the floor slabs is so low that they do not resist bending or twisting. [Alternatively, their effective stiffness can be assessed as for flate plate or shear wall structures, (cf Chapters 7 and 10) and combined with that of the spandrel beams in assessing the stiffnesses of the frame panels.] The floor system is assumed unable to provide any coupling action between opposite normal-to-wind frames that act as flanges to the side frames. Both the side and normal frames are therefore subjected largely to inplane actions, and out-of-plane actions are generally negligible. When the building is subjected to lateral forces, the action of the floor system is then mainly to transmit the horizontal forces to the different vertical structural elements. Because the floor system does not participate otherwise in the lateral load resistance of the structure, then, provided the floor loadings are essentially constant throughout the height, a repetitive floor structure can be used with economy in the design and construction.

If, as is often the case, the framed-tube building is doubly symmetric in plan about both the major axes, any applied lateral load may be resolved into two orthogonal force components along the axes and a twisting moment about the vertical central axis. The structural behavior may then be obtained by a superposition of separate bending actions about axes XX and YY (Fig. 12.2) and a pure torsional action. The analysis of each individual action may then be considerably simplified by utilizing the double symmetry to allow the analysis of only one-half- or one-quarter-plan of the structure.

12.1.2 Bundled-Tube Structures

For very tall buildings, the shear lag experienced by conventional framed tubes may be greatly reduced by the addition of interior framed "web" panels across the entire width of the building to form a modular- or bundled-tube structure. (cf Fig. 4.13). When the building is subjected to bending under the action of lateral forces, the high-in-plane rigidity of the floor slabs constrains the interior web frames to deflect equally with the external web frames, and the shears carried by each will be proportional to their lateral stiffnesses. Since the end columns of the interior webs will be mobilized directly by the webs, they will be more highly stressed than in the single tube where they are mobilized indirectly by the exterior web, through the flange frame spandrels. Consequently, the presence of the interior webs reduces substantially the nonuniformity of column forces caused by shear lag, as shown in Fig. 12.4. The vertical stresses in the normal panels are more nearly uniform, and the structural behavior is much closer to that of a proper tube than the framed tube. Any interior transverse frame panels will act as flanges in a similar manner to the external normal frames.

The structure may be regarded as a set of modular tubes that are interconnected with common interior panels to form a perforated multicell tube, in which the frames in the wind direction resist the wind shears, while the flange frames carry most of the wind moments. The system is such that modules can be curtailed at different heights to reduce the cross section while structural integrity can still be maintained (Fig. 4.13). Any torsion resulting from the consequent unsymmetry is readily resisted by the closed-section form of the modules. The greater spacing of the columns, and shallower spandrels, permitted by the more efficient bundled-tube structure, provides the considerable advantage of larger window openings than are allowed in the single-tube structure.

Another possibility, which yields the same general form of structural behavior, is to use coupled shear walls to form the interior web of the framed tube, and thus create an alternative form of multicellular construction. The stress distribution in the flange frames will then be governed by the relative lateral stiffnesses of the frames and walls in the wind direction.

12 1 STRUCTURAL BEHAVIOR OF TUBULAR STRUCTURES   289

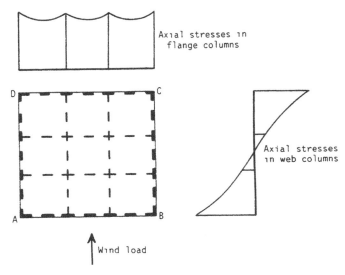

Fig. 12.4 Shear lag in bundled-tube structure.

12.1.3 Braced-Tube Structures

Although the framed tube is an efficient structure for resisting lateral forces, the potential stiffness of the tubular form is reduced by the web frames, racking due to the bending of the columns and beams, and by the shear lag in the flange frames that reduces their contribution to the overall moment of resistance, both effects in conjuction adding considerably to the lateral deflection.

A structural system that would simulate very closely the desired pure cantilever tube behavior could be achieved by eliminating all exterior columns and replacing them with diagonals in both directions (Fig. 12.5), spaced sufficiently closely to

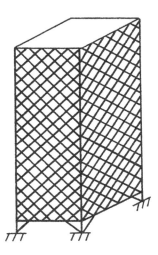

Fig. 12.5 Diagonal truss tube system.

represent bearing walls. By joining the diagonals where they cross, and at the four corners of the building, the structure will act effectively as a rigid tube in resisting horizontal forces. However, in addition to the large number of joints involved, and the awkward fenestration, structural disadvantages occur in the primary gravity loading response, because the vertical loads produce increased resolved forces in the inclined directions of the diagonal members, thus requiring a greater member cross-sectional area in comparison with a vertical column system.

Consequently, a more practical solution to increasing the efficiency of the framed-tube structure has been to add substantial diagonal bracing to the faces of the tube. The exterior columns may then be more widely spaced and the diagonals, generally inclined at about 45° to the vertical, serve to tie together the exterior columns and spandrel beams to form braced facade frames (Fig. 4.14). The bracing ensures that the exterior columns act together in resisting both gravity and horizontal wind forces. Consequently, a very rigid cantilever tube is produced, whose behavior under lateral load is very close to that of a pure rigid tube.

The exterior faces are generally provided with symmerically disposed double diagonal bracing, although, if the structure is rectangular in plan, the narrow faces may then have single diagonal bracing arranged in a zig-zag manner to allow the two sets of orthogonal diagonals to meet at the corners (Fig. 4.15).

The mode of behavior of a braced-tube structure subjected to either gravity or wind forces may readily be envisaged by considering the superposition of the effects of diagonal bracing on the behavior of the structure with vertical columns only.

Behavior under Gravity Loading. If the columns in a tube structure were of equal sectional area, the loading from their tributary floor areas would lead to the corner columns being less heavily stressed, and therefore shortening less, than the intermediate columns.

The mechanism by which the bracing contributes to the redistribution of the column loads can be envisaged readily by considering first the behavior of the structural components if the bracing members are not connected to the vertical columns, and then considering the interactive forces that would be mobilized if the two were subsequently connected together.

Consider initially a representative region of the facade frame (Fig. 12.6a), in which the diagonals are disconnected from the intermediate columns. Under the action of gravity loading, the connection points on the intermediate columns will displace downward by more than the corresponding points on the diagonals, whose displacements are now controlled by the vertical displacements of the less highly stressed corner columns. At this stage, the diagonal members must be in compression while the spandrel beams are in tension.

Now consider the forces that must be mobilized to provide vertical compatibility at the intersections when the intermediate columns and diagonals are connected together. Vertical forces must be provided that pull up on the columns and down on the diagonals, as shown in Fig. 12.6b. The initial compressive force in each intermediate column is now partially relieved by the upward force required at each

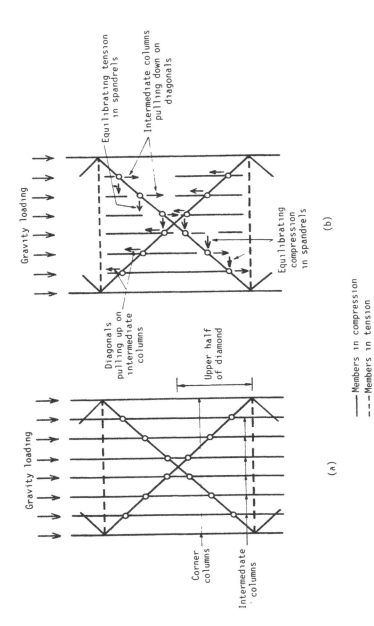

Fig. 12.6 Development of forces in braced tube due to gravity loading. (a) Diagonals disconnected from intermediate columns; (b) forces required to restore compatibility.

292 TUBULAR STRUCTURES

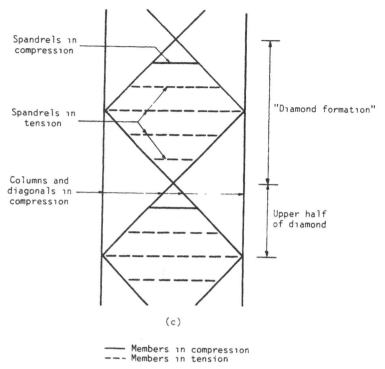

Fig. 12.6 (*Continued*) (c) Resulting forces in members.

of its intersections with a diagonal. The corresponding downward forces on each diagonal are carried at its ends by the corner columns, whose compressive forces are increased at each intersection with a diagonal. The net result tends to be an equalization of the stresses in the intermediate and corner columns.

The increments of force picked up by the diagonal result in a large compressive force at its lower end, which reduces in increments to a much lower compressive value at its upper end. At each intermediate intersection point in a diagonal, the horizontal thrust component must be balanced by an axial reaction in the intersecting spandrel, which will act as a strut in the upper half of each bracing "diamond" and as a tie in the lower half. Consequently, these actions reduce the initial tension in the spandrels in the upper halves of the bracing "diamonds," and increase the tension in the lower halves. The forces in both the intermediate and corner columns will change significantly at each diagonal intersection point. Over the vertical lengths between intersection points, changes will occur only by the increment of gravity load added at each floor level. The resulting force action in the facade panel is summarized qualitatively in Fig. 12.6c.

In narrow face, single zig-zag diagonally braced frames, the bracing is relatively ineffective in equalizing gravity load stresses in the columns since the diagonals are not provided with the very significant cross-tying or cross-strutting action of the spandrels which occurs in double-braced frames. As a consequence, the diagonals cannot provide the uplift required to transfer load from the intermediate to the corner columns [12.1].

Behavior under Lateral Loading. A similar procedure to that used for gravity loading may be used to determine the action of the braced tube in resisting wind loading.

Under the action of wind loading, the side frames act as the webs and the normal frames as the flanges. Consider for example the structural actions in the frame that acts as the tension flange. If the diagonals are initially disconnected from the intermediate columns, the columns and diagonals of the face will be in tension while the spandrels are in compression (Fig. 12.7a). Because of the shear lag effect, the intermediate columns will now be less highly stressed than the corner columns, and the connection points on the diagonals will be displaced upward by more than the corresponding points on the unconnected intermediate columns.

If the diagonals and intermediate columns are connected together, interactive vertical forces will be mobilized, which will pull up on the intermediate columns and down on the diagonals in order to establish compatibility at the connections (Fig. 12.7b). These upward forces cause an increase in tension in the intermediate columns, while the downward increments acting on the diagonals are transferred at their ends to the corner columns, thereby reducing the higher tensile forces that initially existed. In this way, the stresses in the corner and intermediate columns again tend to be equalized.

When superimposed on the original large tensile force in the diagonal, the increments of axial force acting down the diagonal produce a gradually reducing tension along the member, leading to a small net compression in the lowest one or two panels. As in the case of gravity loading, at each intersection point the thrust from the diagonal must be balanced by a horizontal reactive force in the spandrel. Spandrels in the upper halves of the bracing diamonds will now act as struts, while those in the lower halves act as ties.

A qualitative representation of the net forces in the windward face due to wind action is shown in Fig. 12.7c. The tensile forces in the intermediate columns increase down the structure by the increments applied at each intersection with a diagonal.

The forces in the columns, diagonals, and spandrels on the leeward face due to the lateral loading will be opposite in sense to those on the windward face.

The narrow-face web frames are subjected to bending and shearing actions as a result of wind loading. The typical distribution of axial forces in the web-frame columns of an unbraced framed tube is shown in Fig. 12.2c. Because of shear lag, the axial forces in the columns nearest to the corners have values that are higher than they would be in pure tubular action. An extension of the argument used earlier for the flange panels reveals that the action of the diagonals in a braced framed tube is again to reduce the high axial forces near the corners and bring them down to values closer to the pure linear tubular stress distribution.

As described in Chapter 6 for braced frames, the diagonals and spandrels of a diagonally braced framed tube serve as the web members in carrying the horizontal

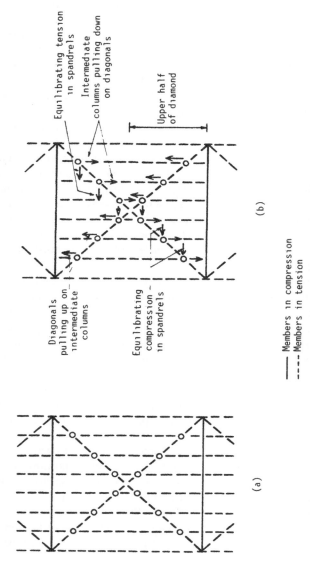

Fig. 12.7 Development of forces in braced tube due to wind loading. (a) Diagonals disconnected from intermediate columns; (b) forces required to restore compatibility.

12.1 STRUCTURAL BEHAVIOR OF TUBULAR STRUCTURES 295

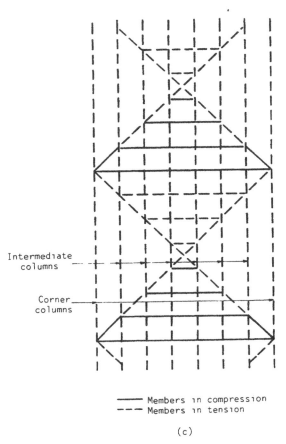

(c)

Fig. 12.7 (*Continued*) (c) Resulting forces in members.

shear, with the diagonals in either axial tension or compression, depending on their direction of inclination. Consequently, the shears, and hence the bending moments, which are carried by the vertical columns, are much reduced [12.1].

Combination of Gravity and Wind Loading. On superimposing the distributions of force in the flange frame panels due to gravity and wind loads, it is seen that all diagonals will tend to be in compression on the leeward face of the building, thus ensuring the availability of the full sectional areas of the vertical members in developing the tube's gross moment of inertia for resisting wind forces. The spandrels, however, develop net tension, an unavoidable penalty that must be paid in taking advantage of the braced-tube system.

On the windward side, the resulting forces will depend on the relative magnitudes of the compressive effects due to gravity loading and the tensile effects due to wind loading.

The members on the side faces will carry forces that are a combination of the compressive effects due to gravity loading and the compressive-tensile (leeward–windward) effects of the wind loading.

Differences in Structural Form between Traditional Framed Tubes and Diagonally Braced Framed Tubes. In traditional framed-tube structures, the columns are aligned so that their major stiffnesses lie in the plane of the perimeter frame, and the girder spans are kept short to produce stiff frames for the web panels and a low degree of shear lag in the flange panels. On the other hand, in a diagonally braced framed tube, the bracing effectively eliminates shear lag; the columns can therefore be turned with their narrow-face minor stiffness in the plane of the frame, the girder spans can be longer, the columns fewer, and, because the girder stiffnesses are no longer as important since they act mainly as horizontal ties, the spandrel girders can be shallower. This allows the windows to be both deeper and wider, a fact that finds approval with both the client and architect.

12.2 GENERAL THREE-DIMENSIONAL STRUCTURAL ANALYSIS

A tubular structure forms a highly indeterminate three-dimensional system consisting of a series of rigidly jointed frameworks connected together at the corners of the building, and to any interior web frames or interior core structure. The analysis of such a structure under any applied load system may be carried out conveniently by the stiffness method, using a general purpose frame analysis program. Appropriate modeling techniques have been described in Chapter 5.

A typical framed-tube structure will contain a large number of elements, and the number of degrees of freedom may run into thousands. Consequently, an analysis of the total structure will be costly and time consuming, although theoretically straightforward. The effective size of the structure stiffness matrix, and hence the amount of computation necessary, may be approximately halved or quartered if the structure is symmetric in plan about one or two central axes, respectively. The applied loads may be treated as combinations of symmetric or skew-symmetric systems, acting on either a half or a quarter of the structure. Appropriate boundary conditions are used at the lines of symmetry or skew symmetry, as described earlier in Chapter 5. However, if the structure is irregular in form, a complete three-dimensional analysis, or a full-structure two-dimensional analysis, as described in Section 5.6.3, will be necessary.

In the building, the high inplane stiffness of the floor slabs will have a considerable influence on the structural behavior, by ensuring that out-of-plane deformations of the frame panels will be effectively restrained at each floor level. The main actions will then be in the planes of the frame panels. As a further result of the inplane slab rigidity, it may be assumed that cross sections of the building will undergo only rigid body displacements, translation and rotation, in the horizontal plane at each floor level. To obtain an accurate estimate of structural behavior, it is essential to include this constraining action in the analysis of the three-dimensional framework.

The constraining action may be achieved in a number of different ways in a general frame program. If the available program has a dependent node, or master–slave, option, sets of nodes at the same level can be assigned as constrained to have related displacements. The option is useful in representing the inplane rigidity of the floor slabs by assigning one node at each floor level to be the master or datum one, and the remaining nodes as the slaves, so that their horizontal displacements and rotations are constrained to conform. This is possible since the horizontal displacements at all nodes in a horizontal plane may be expressed in terms of the two orthogonal translations and a rotation about a datum axis. The master–slave option produces a reduction of three degrees of freedom for each slave node at each floor level; consequently, the number of equations in the total stiffness matrix can be almost halved, for a relatively small penalty of an increase in the bandwidth of the stiffness matrix.

If the master–slave option is not available, the inplane rigidity of the slab can be represented at each floor by a peripheral frame of beams, assigned to be rigid in the horizontal plane, connecting the vertical elements.

Another less-satisfactory technique that has been used is to include fictitious axially rigid pin-ended horizontal diagonal bracing members connecting nodes on opposite corners at each floor level. The restrained corners then remain fixed relative to each other during any translation or rotation under applied loads. In addition, the axial stiffness of the beams at each story level can be assigned to be so large that any inplane axial deformations, or relative displacements between nodes, are negligible. The introduction of such diagonal bracing members has the disadvantage that it increases the bandwidth of the stiffness matrix and increases the solution time.

If the size of the problem is too large for the facilities available, further reductions in the amount of computation may be achieved by making use of the lumping techniques described in Chapter 5 to effectively reduce the number of stories of the structure that have to be treated.

12.3 SIMPLIFIED ANALYTICAL MODELS FOR SYMMETRICAL TUBULAR STRUCTURES

This section considers how, by recognizing the dominant structural actions involved, and neglecting the unimportant actions, it is possible to treat symmetrical three-dimensional tubular structures as simpler equivalent two-dimensional systems. The basic approach is initially considered in detail for the framed-tube system, and then the corresponding treatment of bundled tubes and braced tubes is described briefly.

12.3.1 Reduction of Three-Dimensional Framed Tube to an Equivalent Plane Frame

In this section is presented a simplified yet accurate approximate method for the analysis of symmetrical framed-tube structures subjected to bending produced by lateral forces. The method is intuitively appealing to the engineer since, by recognizing the dominant mode of behavior of the structure, it is possible to reduce the analysis to that of an equivalent plane frame, with a consequently large reduction in the amount of computation required for a conventional full three-dimensional analysis.

Consider initially the framed tube of Fig. 12.2 subjected to bending by lateral forces in the X direction. The lateral load is resisted primarily by the following actions.

1. The shearing actions in the web panels AD and BC parallel to the direction of the applied load.
2. The axial deformations of the normal frame panels AB and DC acting effectively as flanges to the web panels.

Due to the symmetry of the structure about the XX axis, and the very high inplane stiffness of the floor slabs, it may be assumed that out-of-plane actions of the web frames are negligible, and the frames are subjected only to planar actions. It is also assumed that the torsional rigidities of the girders are negligible.

The axial displacements of the corner columns in the web frames are restrained by the vertical rigidity of the two normal frames. Consequently, the interaction between the flange and web panels consists mainly of vertical interactive forces through the common corner columns, A, B, C, and D. As a result of these interactive forces, the flange panels AB and DC are subjected primarily to axial deformations, the uniformity of which across the panel will depend on the stiffnesses, that is, on the spans and flexural rigidities, of the connecting spandrel beams at each floor level.

Under the applied lateral loading, the shear forces will thus be resisted mainly by the web frames, while the bending moments will be resisted by the moments and axial forces in the columns of the web frames and the axial forces in the columns of the flange frames. By virtue of the large lever arm that exists between these flange panels, the wind moments will be resisted most effectively if the maximum amount of axial force can be induced in the columns of frames AB and CD.

All other torsional and out-of-plane actions may be considered to be secondary, apart from the out-of-plane bending of the columns in the flange frames whose horizontal deflections will be the same as those of the web frames. This action may be of significance in the lower levels of the building since bending then occurs about the weaker axis of the columns.

In analyzing the primary mode of behavior, the fundamental compatibility condition that must be established is that of equal vertical displacements at the corners where the orthogonal panels meet. In the analytical model, a mechanism is required that will allow vertical shear forces, but not horizontal forces or bending moments, to be transmitted from the web panels to the flange panels through the corner columns.

In addition, for the web frames, the joints must be free to rotate in the plane of the frame, to displace vertically, and to displace horizontally in unison in the plane of the frame at each floor level, due to the inplane rigidity of the floor slabs. For

12.3 SIMPLIFIED ANALYTICAL MODES FOR SYMMETRICAL TUBULAR STRUCTURES

the flange frames, the joints must be free to rotate in the plane of the frame, and to displace vertically. But flange joints on the line of symmetry must, and preferably all flange joints should, be constrained against horizontal displacement in the plane of the frame, as a result of the slab inplane rigidity.

For example, consider the simple framed tube shown in plan in Fig. 12.8a. Since the structure is symmetrical about both center lines XX and YY, only one-quarter, for example, EBH, need be considered in the analysis. The required boundary conditions at the lines of symmetry and skew symmetry are then introduced as described in Section 5.6.1. Because of symmetry about the XX axis, the shear force in the beams, and the slope in the Y direction, of panel AB must be zero at the line of symmetry (E). Conditions of skew symmetry about the YY axis require that the vertical deflection at the line of skew symmetry (H) must be zero. If, on the other hand, the web frame BC contains an even number of columns, so that no column is situated on the center line YY, the bending moment at the line of skew symmetry in the beam cut by the line of skew symmetry must also be zero (cf. Section 5.6.1). Appropriate support systems to simulate the required boundary conditions for the quadrant EBH of the structure of Fig. 12.8a are shown in Fig. 12.8b.

The equivalent planar system is obtained by "rotating" the normal half-panel EB through 90° into the plane of the web-half-panel BH. The inplane stiffness of the floor slabs constrains all members of the two web bents to have the same horizontal deflection in the X direction; therefore it can be assumed that one-quarter of the total lateral forces acting on the faces of the flange frames of the building can be applied in the plane of the half-web frame, as indicated in Fig. 12.8b. Since the beams are assumed axially rigid, the forces may be applied at any convenient nodes.

The desired vertical interaction between the web and flange panels may be achieved in various ways.

Most comprehensive modern general purpose structural analysis programs include an internodal constraint option. This allows the displacement relating to specified degrees of freedom at two or more nodes in a structure to be constrained to be identical. The appropriate nodes at the intersections of the web and flange frames may then be specified directly in the analysis to have equal vertical displacements. For conciseness, the technique is described briefly in Section 12.3.2 with reference to the more complex bundled-tube structure.

If this option is not available, some other device must be used to achieve the required vertical compatibility at the junctions.

One simple technique is to displace horizontally the intersection column of each flange frame by a small distance of, say, one-hundredth of the span of the adjacent beams, so that each common intersection joint is represented twice, once on each of the web and flange joints (B and B' in Fig. 12.8b). In numbering the nodes, the two nodes representing each intersection joint are numbered separately. The duplicate nodes are then joined by a fictitious stiff beam with a flexural rigidity of say 10,000 times that of the larger adjacent girder, and with one end assigned to be released for moment and axial force (Fig. 12.8c) [12.2]. By this device, vertical

300 TUBULAR STRUCTURES

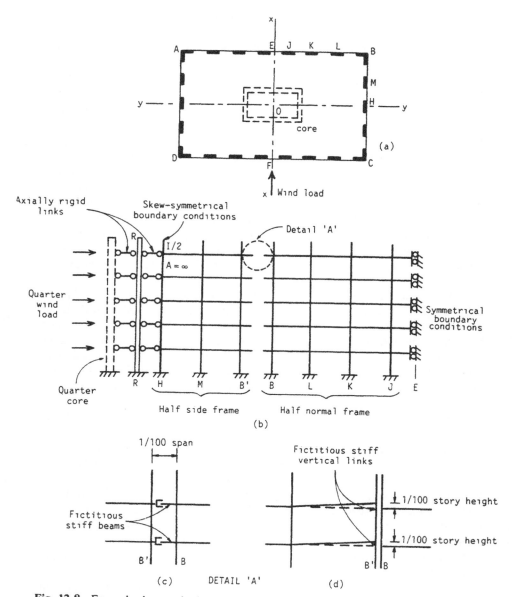

Fig. 12.8 Framed-tube or tube-in-tube structure. (a) Structural plan; (b) equivalent planar model; (c, d) detail of alternative joint connection in planar model.

compatibility is established and the required vertical shear transmitted between the web and flange frames, while decoupling the rotation and horizontal displacement. However, it has been found that the results may be sensitive to the stiffness assumed for the fictitious beams.

An alternative technique has been devised to improve the disconnection of the rotations and lateral displacements between the frame panels, and to reduce the sensitivity of the results to the stiffness of the connecting fictitious members [12.3]. The technique again involves the rotation of the flange frames into the plane of the web. The intersection line column of each frame is shown superimposed on the corresponding one of the web frame, but displaced vertically upward by a small distance, say less than one hundredth of the story height, as shown in Fig. 12.8d. Thus each intersection joint is duplicated, once on the web column, and, immediately above, on the flange column. These common nodes should again be numbered separately. Each pair of intersection joint nodes is then connected by a fictitious stiff vertical link with a large sectional area of, say, 10,000 times that of the intersection line column. The stiff links ensure vertical compatibility and transfer vertical shear between the web and flange frames while disconnecting rotations and vertical displacements.

In each model, the corner column is assigned its true inertias in the corresponding planes of the web and flange frames, but its area should be assigned wholly to the column B' in the web frame with a zero area assigned to the column B in the flange frame. Horizontal and rotational constraints are applied to the flange frame nodes on the vertical line of symmetry and preferably, as a means of reducing the total number of degrees of freedom, horizontal constraints are also applied to all other flange frame nodes.

If the structure has additional flange frames that connect part way along the web, as shown in Fig. 12.9a, they can be modeled simply in one of the previous ways described, with the frames overlaying but remaining separate from the web frame, as shown in Fig. 12.9b. In the graphic description of the model, it is usually clearer to show the web and flange frames separately, but to dimension them horizontally as though they were in the overlaid arrangements, as shown in Fig. 12.9c.

The resulting planar model may then be analyzed to give results similar to those from a full three-dimensional analysis.

The basic model does not include the out-of-plane bending of the columns in the normal frames, which may be of significance in the lower levels. These columns suffer the same horizontal deflections about their weaker axis as the columns in the side frames do about their stronger axis. The effect of the out-of-plane bending may be included in the basic model by adding an equivalent column (RR in Fig. 12.8b), whose flexural rigidity is equal to the sum of the out-of-plane flexural rigidities of one-quarter of the flange columns. The additional column is connected by pin-ended axially stiff links to the existing basic plane frame system. The links constrain the column to have the same horizontal deflection as the side panel members, and allow it to carry its share of the lateral forces. Once the total force and consequent moment has been determined for the effective column, it may be distributed to the individual columns in proportion to their flexural rigidities.

Hull-Core Structures. If the framed tube contains an inner core, it will bend with the same horizontal deflections as the outer tube, owing to the high inplane stiffness of the floor slab, and will carry a proportionate share of the lateral load. The core is frequently symmetric also, and so may conveniently be included by

302 TUBULAR STRUCTURES

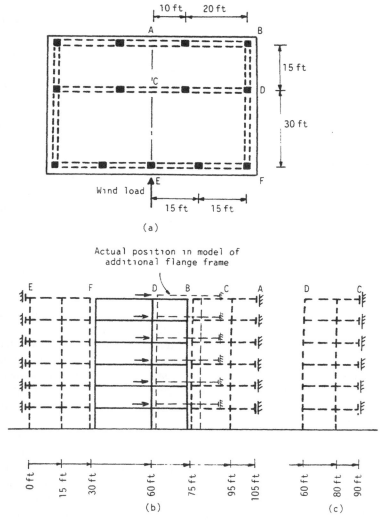

Fig. 12.9 Framed tube with additional interior flange frame. (a) Structural plan; (b, c) equivalent planar model.

adding one-quarter of it in the same planar model, connected by pin-ended axially rigid links to the web-frame system.

If the core acts as a simple cantilever, it may be modelled as a single equivalent column, as shown in Fig. 12.8b. If it is perforated, it may be treated as a wall with openings, as described in Chapter 10. Provided that the internal core can be modeled by an equivalent plane structure, it may always be linked to the outer framed-tube model to obtain the distribution of lateral forces on each component.

If the core cannot be treated as a plane element, or if the outer framed tube is

12.3 SIMPLIFIED ANALYTICAL MODES FOR SYMMETRICAL TUBULAR STRUCTURES

not symmetric, a three-dimensional analysis must again be performed. The nodes of the interior core must either be constrained by a "rigid floor" option to deflect horizontally with the nodes of the exterior frame, or be connected to them by a fictitious horizontal frame of axially stiff links. Either of these techniques will simulate the rigid-plane actions of the floor slabs, which span between the two components, in constraining the frame and core to translate and rotate in the horizontal plane as a single unit at each floor level. Since the inner core and outer frame will probably deform individually in predominantly different modes under the action of horizontal loads, a redistribution of horizontal shears will take place throughout the height of the building because they are constrained to deform in unison.

Torsion of Framed-Tube Structures. Plane frames are stiff in their own plane and relatively flexible out of plane and in torsion. Consequently, when a framed tube is subjected to twisting due to asymmetric lateral forces, the torsional moments are resisted primarily by couples resulting from horizontal shears in the planes of the peripheral frame panels. As for the web frames in the case of bending, the dominant action is planar shearing behavior, but in this case all the frame panels are subjected to similar shearing actions.

Under torsional loading, the actions of the two orthogonal frames will be coupled through the vertical displacements of the common corner columns. A direct plane-frame solution is not then feasible. However, by using the more sophisticated modeling technique described in Section 5.6.3, it is possible to derive an equivalent two-dimensional structure that can simulate directly the torsional behavior of a framed-tube structure.

12.3.2 Bundled-Tube Structures

In order to demonstrate the technique of reducing a non-twisting bundled-tube structure to an equivalent plane frame, consider the bundled tube of Fig. 12.10(a), which includes nine modular tubes. Since the structure is symmetrical about the YY axis, only one half, subjected to half the applied wind load, need be considered.

The two web frames ADGK and BEHL are first assembled in a single plane, in any order, with arbitrary spaces between them, as shown in Fig. 12.10b. The half-flange frames MLK, JHG, FED, and CBA are assembled in the same plane, also in any order, and arbitrarily spaced, and with the intersection line columns shown in both the web and flange frames, as in the framed-tube representation (Section 12.3.1). The planar system is dimensioned horizontally from the extreme left edge of the model to include the arbitrary spaces, and the nodes are numbered separately, as before.

Using the internodal constraint option, the nodes at each level of the set of web frames are specified as constrained to displace horizontally identical to, say, the extreme left web (datum) node at the level. This procedure simulates the horizontal

304 TUBULAR STRUCTURES

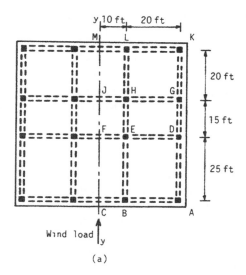

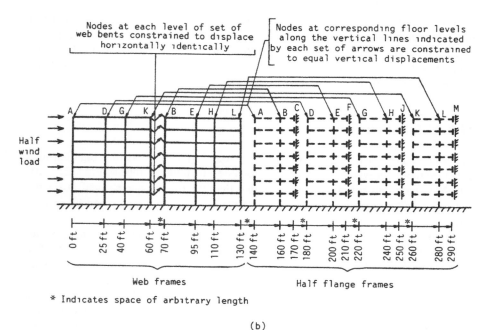

Fig. 12.10 Modelling of three-dimensional bundled tube by analogous plane frame structure. (a) Structural plan; (b) equivalent planar model.

constraint of the inplane rigid floor slabs. The vertical plane rotations and vertical displacements of the web nodes are then left free to behave independently.

The constraint option is then used to specify vertical constraint between the common nodes representing joints at the intersections of the web and flange frames. This procedure ensures that the required vertical compatibility between the web and flange frames is achieved.

All the flange nodes are left free to rotate, while those on the line of symmetry at M, J, F, and C (Fig. 12.10a), and preferably all other flange nodes also, are constrained against horizontal displacement to represent the effect of the high inplane rigidity of the floor slabs.

The horizontal loading may be applied to any vertical line of joints in one of the web frames (Fig. 12.10b). The equivalent two-dimensional model may then be analyzed by a plane frame program.

In addition to allowing the planar analysis of much more complex structures, the nodal constraint approach is to be preferred in all cases to the fictitious member technique because it achieves exact compatibility between the constrained nodes. It reduces the number of degrees of freedom and therefore the size of the computational problem, and it also avoids the possibility of numerical instability in the computation, which may arise from the very large differences in stiffness between the fictitious stiff members and the adjacent real ones in the equivalent structural model.

The two-dimensional analysis of the planar model gives results that match exactly those of a full three-dimensional analysis of the structure if the same assumptions are made in each case.

12.3.3 Diagonally Braced Framed-Tube Structures

In a nontwisting diagonally braced framed-tube structure, all four faces contribute significantly to the lateral resistance of the building. In resisting lateral forces, the side frames act as the webs and the normal frames act as the flanges of the tube. The main interactions between the web and flange frames again consist of vertical shears transmitted through the corner columns. Any symmetry of the structural plan about the line of the lateral load resultant again allows the analysis to be based on one-half of the structure, or one-quarter if the structure is doubly symmetrical.

The three-dimensional frame may thus be again replaced by a equivalent planar model using the same techniques described earlier for framed-tube structures. The required interactions and the vertical compatibility between web and flange frames at their intersections may again be achieved by an internodal constraint facility in the program, or, if this is not available, by introducing a set of fictitious auxiliary connecting members between web and flange nodes, as described in Section 12.3.1. If conditions of symmetry or skew symmetry are employed to reduce the size of the model, appropriate constraints must be included at the cut members on the lines of symmetry and skew symmetry, as discussed earlier in Chapter 5 (Section 5.6.1).

In devising the analytical model to be employed, the difference in construction between steel and concrete structures should be considered. Steel diagonal bracing will consist of additional inclined steel members attached to the columns and span-

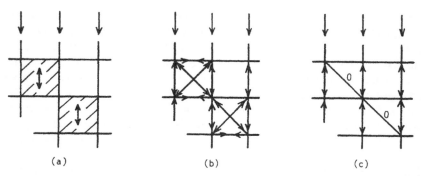

Fig. 12.11 Element of diagonally braced tube. (a) Infill panel; (b) equivalent double-diagonal bracing; (c) equivalent single-diagonal bracing.

drel girders. However, with concrete tubes, the diagonal bracing will generally be formed by infilling the window openings betwen the columns and spandrel girders along a diagonal line in the building perimeter frame. It then becomes important to model correctly the actions of the solid window panels in the analysis.

Under gravity loading, the loads in the columns will tend to compress the concrete bracing panels (Fig. 12.11a) and thus tend to suppress tensile stresses in the diagonal, when they arise, so that the panels can act as tension braces. In the analytical model, the bracing panels should be represented either by membrane finite elements (Fig. 12.11a), or as double-diagonal braced units in the frame (Fig. 12.11b). These are required to allow the model to be able to pick up the precompressive effects of gravity loads, which would not be possible if a single diagonal bracing member were used to model the infill panel (Fig. 12.11c). The stiffnesses of the double-diagonal bracing members can be estimated from a preliminary separate membrane finite element analysis of a typical individual panel subjected to a set of diagonal forces equivalent to resultant diagonal compressive loads. This will allow the effective area of the equivalent uniform strut to be determined.

When a framed-tube structure has deep spandrel girders with short spans, which makes the joint width and depth relatively large, it is advisable to include the wide-column deep-beam effect by the modeling techniques described in Section 5.6.5.

Because bending actions are less significant in braced-tube than in framed-tube structures, the member stiffnesses will not be reduced to the same extent in braced tubes by concrete cracking caused by tensile stresses.

SUMMARY

The basic philosophy in the design of tubular systems for very tall structures is to place as much as possible of the lateral load-resisting material at the exterior of the building in order to maximize the flexural rigidity of the system. The most efficient structure can only be obtained by tying together peripheral columns in such a way that they act as a rigid "box" or "tube" cantilevering out of the ground.

The original framed-tube structure consisted of closely spaced exterior columns connected at each floor level by deep spandrel girders to form a perforated tube. Although an effective system, the potential stiffness of the tubular form is reduced by the side web frames racking due to the bending of the columns and beams, and by shear lag in the normal flange frames, which reduces the moment of resistance of the structure's cross section. The deformations associated with shear lag can also cause distress to secondary non-load-bearing components in the building.

One innovation that reduces the degree of shear lag uses interior additional frames across the full width of the framed tube, in one or both directions, to produce a modular- or bundled-tube system. The additional web frames contribute not only their own shear and bending resistance to the building's stiffness, but also, in mobilizing directly more of the flange face columns, they reduce the shear lag in the flange frames, thereby increasing these frames' contribution to the structure's stiffness.

An even more efficient system has been developed by adding substantial widely spaced diagonal bracing members to the exterior faces to form a very rigid braced-tube or column-diagonal truss tube system. Under lateral loading, the structure behaves more like a braced frame with greatly reduced bending in the columns and girders of the frame, and a stress distribution that is very similar to that of a true rigid tubular structure. The bracing may be of double- or single-diagonal form, and may be provided over part of the periphery only to form a partial-tube system. More recent postmodern buildings have used tubular frameworks over a part of the perimeter only, and a space frame over the remainder.

The only suitable technique for the generalized analysis of such large complex framed structures is a full three-dimensional analysis. However, it is essential to include in the model the constraining effects of the high inplane stiffness of the floor slabs. It may be necessary to resort to lumping techniques to reduce the computation to a manageable size.

If the structure is symmetrical and no twisting occurs, it is possible, by recognizing the dominant modes of behavior of the structural components, to reduce the analysis to that of an equivalent plane frame, with a consequent large reduction in the amount of computation required for a full three-dimensional analysis, and with little or no loss of accuracy in the results.

REFERENCES

12.1 Grossman, J. S., Cruvellier, M., and Stafford Smith, B. "Behavior, Analysis and Construction of a Braced Tube Concrete Structure." *Concrete Int.*, **8**(9), September 1986, 32–42.

12.2 Rutenberg, A. V. "Analysis of Tube Structures Using Plane Frame Programs." *Proc. of Regional Conference on Tall Buildings*, Bangkok, Thailand, 1974, pp. 397–413.

12.3 Stafford Smith, B., Coull, A., and Cruvellier, M. "Planar Models for Analysis of Intersecting Bent Structures." *Computers and Structures*, **29**, 1988, 257–263.

CHAPTER 13

Core Structures

Elevator cores are primary components for resisting both horizontal and gravity loading in tall building structures. Reinforced concrete cores usually comprise an assembly of connected shear walls forming a box section with openings that may be partially closed by beams or floor slabs (Fig. 13.1a, b, and c). The moments of inertia of a reinforced concrete core are invariably large, so that it is often adequate in itself to carry the whole of the lateral loading. The horizontal load bending deflections and stresses of a core with a fully connected section are calculated conventionally, as for a vertical cantilever, on the basis of the core's moments of inertia about its principal axes.

If a building is also subjected to twist, as many are, the torsional stiffness of the core can be a significant part of the total torsional resistance of the building. The torsional behavior of the core and its analysis is a topic that is relatively unfamiliar to many engineers. The proportions of the height, length, and thickness of the walls of a typical building core classify it, in terms of its torsional behavior, as a thin-walled beam. Consequently, when the core twists, originally plane sections of the core warp (Fig. 13.2). Because the base section is prevented from warping by the foundation, the twisting induces vertical warping strains and stresses throughout the height of the core walls. In structures that are heavily dependent for their torsional resistance on the torsional stiffness of a core, the vertical warping stresses at the base of the core may be of the same order of magnitude as the bending stresses. In such cases warping stresses should not be neglected.

Partial closure of the core by beams or slabs across the openings restrains the core section from warping and thereby increases the core's torsional stiffness, while reducing its rotation and warping stresses. In providing the restraint, however, the connecting beams or slabs are subjected to shear and bending that may be of a sufficient magnitude to require consideration in their design.

The warping torsion action of the structural components of buildings has, in the past, been given relatively little attention; consequently, designers are generally not at ease with the concepts of warping behavior, nor with its methods of analysis. In the design of buildings that are structurally dependent on an elevator case, the designer should be able to appreciate whether a core is liable to twist and warp so that this may be taken into account in its analysis and design.

The aims of this chapter are first, to provide a simple introduction to the concept of restrained warping by explaining it from the principles of flexure; second, to

CORE STRUCTURES 309

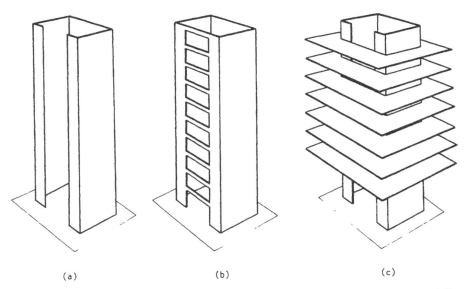

Fig. 13.1 (a) Open-section core; (b) core partially closed by beams; (c) core partially closed by floor slabs.

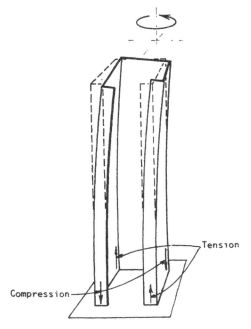

Fig. 13.2 Twisted core.

present a classical method of analysis for uniform cores that, through solutions obtained by the use of design curves, offers an understanding of the influence of certain structural parameters on warping; finally, to explain some methods of analysis that are more practical, in allowing the consideration of cores whose properties change throughout their height, and cores that interact with other structural assemblies. As a necessary adjunct to warping analysis, a section on the determination of the sectional and sectorial properties is also included.

13.1 CONCEPT OF WARPING BEHAVIOR

A simple example of restrained warping of a thin-walled core is an I section cantilever, fixed at its base, and subjected to torque at the top (Fig. 13.3a). The flanges in this case are unequal in size so that the section is singly-symmetric about its X axis. The web is assumed to be so slender as to contribute negligibly to the sectional properties.

Two points on the section (Fig. 13.3b) are particularly significant. The first is the center of area C, which is important in relation to vertical axial forces. If an axial force is applied through the center of area, only axial deformations and stresses will occur. If, however, an axial force is applied to the section through a point other than C, bending about the transverse axes, and possibly warping, can also occur. Neglecting the web, the position of the center of area is given by

$$\bar{x}_1 = \frac{A_2}{A_1 + A_2} L \quad \text{and} \quad \bar{x}_2 = \frac{A_1}{A_1 + A_2} L \qquad (13.1)$$

The second significant point on the section is the shear center D, which is important in relation to transverse forces on the core. If a transverse force acts through D, the member will only bend. If, however, a transverse force acts on the member elsewhere than through D, the member will twist and warp as well as bend. The shear center in this case is located along the X axis by

$$x_1 = \frac{I_2}{I_1 + I_2} L \quad \text{and} \quad x_2 = \frac{I_1}{I_1 + I_2} L \qquad (13.2)$$

An inspection of Eqs. (13.1) and (13.2) indicates that the center of area and the shear center generally will not coincide unless the section is doubly symmetric, in which case both points lie at the center of symmetry.

When a torque T about the Z axis is applied to the top of the member in Fig. 13.3a, it twists about the shear center axis with the flanges bending in their planes, about the X axis, and twisting about their vertical axes (Fig. 13.3c and d). The effect of the flanges bending is to cause the flange sections to rotate in opposite directions about their X axes so that initially plane sections through the member become nonplanar, or warped. Diagonally opposite corners b and e, in Fig. 13.3c, displace downward while a and f displace upward. At any level z up the height of

13.1 CONCEPT OF WARPING BEHAVIOR   311

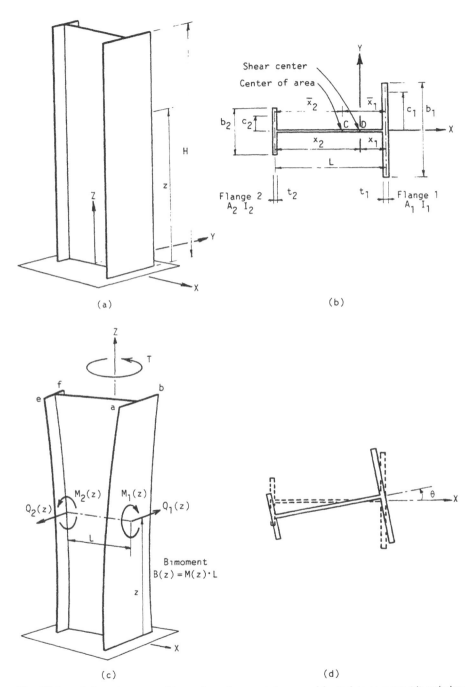

Fig. 13.3 (a) *I* section core; (b) section of core; (c) core subjected to torque; (d) twisting of flanges and web.

the core, the torque $T [= T(z)]$ is resisted internally by a couple $T_w(z)$ resulting from the shears in the flanges and associated with their inplane bending, and a couple $T_v(z)$ resulting from shear stresses circulating within the section and associated with the twisting of the flanges. Then

$$T_w(z) + T_v(z) = T(z) \tag{13.3}$$

The horizontal plane rotation of the member about its shear center axis at a height z from the base is $\theta(z)$, hence the horizontal displacement of flange #1 at that level is

$$y_1(z) = x_1 \theta(z) \tag{13.4}$$

and its derivatives are

$$\frac{dy_1}{dz}(z) = x_1 \frac{d\theta}{dz}(z) \tag{13.5}$$

$$\frac{d^2 y_1}{dz^2}(z) = x_1 \frac{d^2\theta}{dz^2}(z) \tag{13.6}$$

$$\frac{d^3 y_1}{dz^3}(z) = x_1 \frac{d^3\theta}{dz^3}(z) \tag{13.7}$$

Similar expressions hold for flange #2.

The shear associated with the bending in flanges #1 and #2 can be expressed by

$$Q_1(z) = -EI_1 \frac{d^3 y}{dz^3}(z) = -EI_1 x_1 \frac{d^3\theta}{dz^3}(z) \tag{13.8}$$

and

$$Q_2(z) = -EI_2 \frac{d^3 y}{dz^3}(z) = -EI_2 x_2 \frac{d^3\theta}{dz^3}(z) \tag{13.9}$$

Therefore, the torque contributed by these shear forces is

$$T_w(z) = Q_1 x_1 + Q_2 x_2 = -(EI_1 x_1^2 + EI_2 x_2^2) \frac{d^3\theta}{dz^3}(z) \tag{13.10}$$

or

$$T_w(z) = -EI_\omega \frac{d^3\theta}{dz^3}(z) \tag{13.11}$$

13.1 CONCEPT OF WARPING BEHAVIOR

where

$$I_\omega = I_1 x_1^2 + I_2 x_2^2 \qquad (13.12)$$

Incidentally, it may be deduced from horizontal equilibrium that $Q_1 = Q_2$.

I_ω is a geometric property of the section and is called the warping moment of inertia or warping constant. It expresses the capacity of the section to resist warping torsion. The torque resisted by the twisting of the flanges is

$$T_v(z) = GJ_1 \frac{d\theta}{dz}(z) \qquad (13.13)$$

where J_1 is the torsion constant of the section given by

$$J_1 = \frac{b_1 t_1^3}{3} + \frac{b_2 t_2^3}{3} \qquad (13.14)$$

in which b_1 and b_2 are the widths, and t_1 and t_2 are the thicknesses, of flanges #1 and #2, respectively.

Summing the two internal torques, (13.11) and (13.13), and equating them to the external torque as in Eq. (13.3).

$$-EI_\omega \frac{d^3\theta}{dz^3}(z) + GJ_1 \frac{d\theta}{dz}(z) = T \qquad (13.15)$$

Equation (13.15) is the fundamental equation for restrained warping torsion. It will be used and extended in the more direct presentation of warping theory, given later.

Considering the stresses in the walls due to bending, the compressive stress in flange #1 at c_1 from the X axis and z from the base is

$$\sigma_1(c_1, z) = \frac{M_1(z)}{I_1} c_1 \qquad (13.16)$$

The tensile stress in flange #2 at c_2 from the X axis is

$$\sigma_2(c_2, z) = \frac{M_2(z)}{I_2} c_2 \qquad (13.17)$$

Multiplying the right-hand side of Eq. (13.16) by the expression

$$\frac{L}{(x_1 + x_2)} \frac{x_1}{x_1}$$

which is equal to unity, and noting that since $Q_1 = Q_2$, and the flange moments $M_1 = M_2 = M$, gives

$$\sigma_1(c_1, z) = \frac{M(z) L x_1 c_1}{I_1 x_1^2 + I_1 x_1 x_2} \qquad (13.18)$$

and since, from Eq. 13.2

$$I_1 x_1 x_2 = I_2 x_2^2 \qquad (13.19)$$

Substituting Eq. (13.19) in (13.18)

$$\sigma_1(c_1, z) = \frac{M(z) L x_1 c_1}{I_1 x_1^2 + I_2 x_2^2} \qquad (13.20)$$

or

$$\sigma_1(c_1, z) = \frac{B(z) \omega(c_1)}{I_\omega} \qquad (13.21)$$

in which $B(z)$ [$= M(z) L$] is an action termed a *bimoment*, and $\omega(c_1)$ ($= x_1 c_1$), is a coordinate termed the sectorial area, or *principal sectorial coordinate*, for that point on the section. In its simplest form, as considered here, a bimoment consists of a pair of equal and opposite couples acting in parallel planes (Fig. 13.3c). Its magnitude is the product of the couple and the perpendicular distance between the planes.

The derivative of Eq. (13.15) is closely analogous to the Eq. (11.3) representing wall-frame behavior in Chapter 11.

$$EI \frac{d^4 y}{dz^4} - (GA) \frac{d^2 y}{dz^2} = w(z) \qquad (11.3)$$

in which y is the deflection of the structure, EI the flexural rigidity of the wall, (GA) the effective shear rigidity of the frame, and $w(z)$ the intensity of loading at level z.

Similarly Eq. (13.21) is analogous to the expression for stress in the wall of a wall-frame.

$$\sigma(c, z) = \frac{M(z) c}{I}$$

in which $\sigma(c, z)$ is the vertical stress in the wall at a distance c from the neutral axis.

These analogies indicate that the restrained torsion of a thin-walled member is the rotational counterpart of the horizontally loaded wall-frame. By reference to these analogies, a familiarity with the simpler wall-frame theory is useful in developing an understanding of warping theory.

The above elementary consideration of a twisting *I* section explains the concept of warping behavior and how the equations and parameters of restrained warping are related to the inplane flexure of the wall elements.

13.2 SECTORIAL PROPERTIES OF THIN-WALLED CORES SUBJECTED TO TORSION

The torsional resistance of an elevator core is provided by horizontal shear in the walls. Part of this, the warping shear, discussed in the previous section, is associated with the inplane bending of the walls. Additional torsion-resisting shear results from the plate twisting action, which causes shear stresses to circulate within the wall thickness (Fig. 13.4a) and, in a closed- or partly closed-section core, from further additional shear stresses that circulate unidirectionally around the core profile (Fig. 13.4b) [13.1].

In being the rotational counterpart of planar wall–frame behavior, restrained warping behavior involves a set of so-called sectorial parameters, each of which has a direct sectional counterpart in wall–frame theory. Since the sectorial parameters are generally unfamiliar to practicing engineers, their determination will be reviewed here. A worked example is given at the end of the section to illustrate the calculation of the parameters.

13.2.1 Sectorial Coordinate ω'

The sectorial coordinate at a point on the profile of a warping core is the parameter that expresses the axial response (i.e., displacement, strain, and stress) at that point, relative to the response at other points around the section. It is necessary in defining ω' to establish a system of axes for the core, and a sign convention. A right-handed axis system will be adopted with its origin at the base of the core and its Z axis vertically upward, as in Fig. 13.3a. In this system a positive rotation about the Z axis is clockwise when looking up the core from the base, or anticlockwise when looking down the core from the top. Because a building plan is invariably viewed from above, an anticlockwise rotation will be taken as positive.

A sectorial coordinate ω' is defined in relation to two points: a pole $0'$ at an arbitrary position in the plane of the section, and an origin P_0 at an arbitrary location on the profile of the section (Fig. 13.5a). The value of the sectorial coordinate at any point P on the profile is then given by

$$\omega'(s) = \int_0^s h \, ds \qquad (13.22)$$

where h is the perpendicular distance from the pole $0'$ to the tangent to the profile at P and s is the distance of P along the profile from P_0.

In effect, the sectorial coordinate ω' is equal to twice the area swept out by the

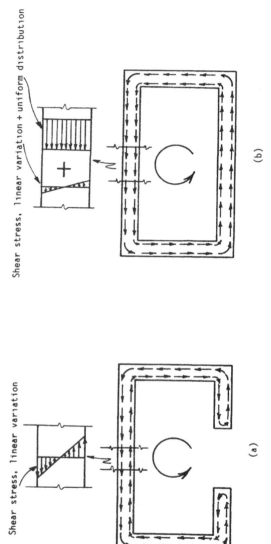

Fig. 13.4 (a) Twisting shear stresses in open section; (b) twisting shear stresses in closed section.

13.2 SECTORIAL PROPERTIES OF THIN-WALLED CORES SUBJECTED TO TORSION

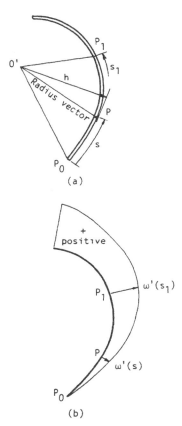

Fig. 13.5 (a) Profile of section; (b) sectorial coordinate ω' diagram.

radius vector $O'P$ in moving from P_0 to P. ω' increases positively for a radius vector sweeping anticlockwise and negatively for it sweeping clockwise. The sectorial coordinate diagram (Fig. 13.5b) indicates the values of ω' around the profile.

When the sectorial coordinates are related to the shear center as a pole, and to an origin of known zero warping displacement, Eq. (13.22) gives the principal sectorial coordinate values, ω, and their plot is the principal sectorial coordinate diagram. The principal sectorial coordinate of a section in warping theory is analogous to the distance c of a point from the neutral axis of a section in bending. The parameters ω and c are used in developing the corresponding warping and bending stiffness properties of the sections, and in determining the axial displacements and stresses.

13.2.2 Shear Center

The shear center of a core is a point in the plane of its section through which a load transverse to the core must pass to avoid causing torque and twist. It is also the point to which warping properties of a section are related, in the way that bending properties of a section are related to the neutral axis.

318 CORE STRUCTURES

Tall building cores are often singly or doubly symmetric in plan, which simplifies the location of the shear center. In doubly symmetric sections, the shear center lies at the center of symmetry while, in singly symmetric sections, it lies somewhere on the axis of symmetry. The location of the shear center along the axis of singly symmetric sections is considered here. For nonsymmetric sections the determination is more complex and the reader is referred for this to a more comprehensive text on the torsion of thin-walled members [13.1].

Considering the singly symmetric section in Fig. 13.6a, the location of the shear center may be determined as follows:

1. Construct the ω' diagram (Fig. 13.6b) by taking an arbitrary pole $0'$ on the line of symmetry, an origin D where the line of symmetry intersects the section, and by sweeping the ray $0'D$ around the profile.
2. Using the ω' and the y diagrams for the section, Figs. 13.6b and 13.6c, respectively, calculate the product of inertia of the ω' diagram about the X axis $I_{\omega'1}$ using

$$I_{\omega'1} = \int^A \omega' y \, dA \qquad (13.23)$$

in which $dA = t \, ds$, the area of a segment of the profile of thickness t and length ds. The integral in Eq. (13.23) may be evaluated simply for a straight-sided section by using the product integral table, Table 13.1, as shown for the worked example in Section 13.2.6.

3. Calculate I_{11}, the second moment of area of the section about the axis of symmetry.
4. Finally, calculate the distance α_1 of the shear center 0 from $0'$, along the axis of symmetry, using

$$\alpha_1 = \frac{I_{\omega'1}}{I_{11}} \qquad (13.24)$$

13.2.3 Principal Sectorial Coordinate (ω) Diagram

The ω diagram is related to the shear center 0 as its pole and a point of zero warping deflection as an origin. In a symmetrical section the intersection of the axis of symmetry with the profile at D defines a point of antisymmetrical behavior, and hence of zero warping deflection; therefore it may be used as the origin.

Values of ω can be found either from first principles, by sweeping the ray OD around the profile and taking twice the values of the swept areas, or by transforming the previously obtained values of ω', thus

$$\omega = \omega' - \alpha_1 y \qquad (13.25)$$

For the section of Fig. 13.6a, this gives the principal sectorial coordinate diagram in Fig. 13.6d.

13.2 SECTORIAL PROPERTIES OF THIN-WALLED CORES SUBJECTED TO TORSION

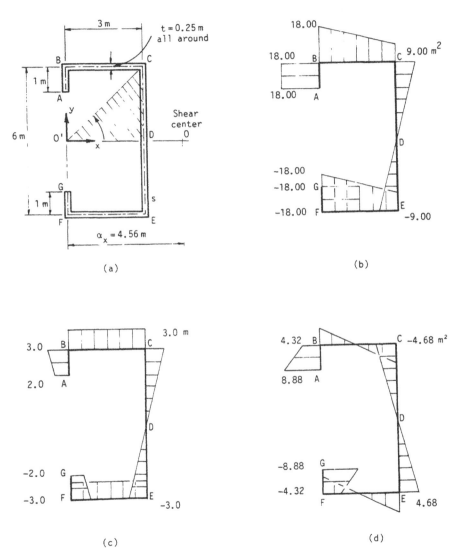

Fig. 13.6 (a) Singly symmetric core section; (b) ω' diagram; (c) y diagram; (d) principal sectorial ω diagram.

TABLE 13.1 Product Integrals

PRODUCT INTEGRAL TABLE $\int_0^L F_1(x)F_2(x)dx$

$F_1(x)$ \ $F_2(x)$	rectangle b, L	right triangle b, L	right triangle (reverse) b, L	trapezoid b_1, b_2, L
rectangle a, L	abL	$\frac{1}{2}abL$	$\frac{1}{2}abL$	$\frac{aL}{2}(b_1+b_2)$
trapezoid a_1, a_2, L	$\frac{bL}{2}(a_1+a_2)$	$\frac{bL}{6}(a_1+2a_2)$	$\frac{bL}{6}(2a_1+a_2)$	$\frac{L}{6}(2a_1b_1+a_1b_2 + a_2b_1+2a_2b_2)$
triangle a, L	$\frac{1}{2}abL$	$\frac{1}{3}abL$	$\frac{1}{6}abL$	$\frac{1}{6}aL(b_1+2b_2)$
parabola a, L	$\frac{2}{3}abL$	$\frac{1}{3}abL$	$\frac{1}{3}abL$	$\frac{1}{3}aL(b_1+b_2)$
parabolic tangent a, L	$\frac{1}{3}abl$	$\frac{1}{4}abL$	$\frac{1}{12}abL$	$\frac{aL}{12}(b_1+3b_2)$
tangent parabolic a, L	$\frac{2}{3}abL$	$\frac{5}{12}abL$	$\frac{1}{4}abL$	$\frac{aL}{12}(3b_1+5b_2)$
stepped c, d, e, L	$\frac{Lb}{6}(c+4d+e)$	$\frac{Lb}{6}(2d+e)$	$\frac{Lb}{6}(c+2d)$	$\frac{L}{6}[b_1(c+2d) + b_2(2d+e)]$

13.2.4 Sectorial Moment of Inertia I_ω

This geometric parameter expresses the warping torsional resistance of the core's sectional shape. It is analogous to the moment of inertia in bending.

The sectorial moment of inertia is derived from the principal sectorial coordinate distribution using

$$I_\omega = \int^A \omega^2 \, dA \qquad (13.26)$$

As the warping of a section is associated with axial strains and, therefore, with the elastic modulus E, the parameter EI_ω is the warping rigidity of the core.

13.2 SECTORIAL PROPERTIES OF THIN-WALLED CORES SUBJECTED TO TORSION

A worked example illustrating the calculation of I_ω, and using the ω diagram of Fig. 13.6d, is given in Section 13.2.6.

13.2.5 Shear Torsion Constant J

When an open-section core is subjected to torque (Fig. 13.4a) each wall twists and shear stresses circulate within the thickness of the wall. The stresses are distributed linearly across the thickness of the wall, acting in opposite directions on opposite sides of the wall's middle line. As the effective lever arm of these stresses is equal to only two-thirds of the wall thickness, the torsional resistance of these stresses is low. The torsion constant for this plate twisting action is

$$J_1 = \tfrac{1}{3} \sum_{}^{n} bt^3 \tag{13.27}$$

in which b is the width and t the thickness of a wall. The summation includes the n walls that comprise the section. The plate twisting rigidity of an open section core is given by GJ_1.

When a closed section core is subjected to torque (Fig. 13.4b), a torque resistance, additional to the walls' twisting resistance, is given by shear stresses that circulate around the profile and that are uniform across the walls' thickness. Because the lever arm of these stresses is approximately the breadth of the core profile, the torque resistance of the uniform circulating stresses is very much larger than that of the plate twisting stresses. The torsion constant for the uniform stress action is

$$J_2 = \frac{\Omega^2}{\oint ds/t} \tag{13.28}$$

in which Ω is twice the area enclosed by the profile of the section. The integral $\oint$ is taken completely around the profile.

The total shear torsion constant for the combined plate twisting and uniform circulating shears is, therefore

$$J = J_1 + J_2 = \frac{1}{3} \sum_{}^{n} bt^3 + \frac{\Omega^2}{\oint ds/t} \tag{13.29}$$

and the corresponding shear torsional rigidity is GJ.

13.2.6 Calculation of Sectorial Properties: Worked Example

Determine, for the open-section shown in Fig. 13.6a,

1. the location of the shear center,

2. the principal sectorial coordinate diagram,
3. the sectorial moment of inertia, I_ω, and
4. the St. Venant torsion constant J.

1. **Location of Shear Center.** Following the steps outlined in Section 13.2.2, the ω' diagram is constructed as in Fig. 13.6b. Then, using Eq. (13.23) together with the ω' and y diagrams (Fig. 13.6b and 13.6c)

$$I_{\omega'_1} = \int^A \omega' y \, dA = \int^s \omega' y \, t \, ds$$

From the product integral table, Table 13.1:

	ω'	y	$\int \omega' y \, t \, ds$
For DC			$3 \times 9 \times 3 \times 0.25/3 = 6.750 \text{ m}^5$
CB			$3(18 + 9) \times 3 \times 0.25/2 = 30.375 \text{ m}^5$
BA			$18(2 + 3) \times 1 \times 0.25/2 = 11.250 \text{ m}^5$
			For half of section = 48.375 m^5
			For whole section $I_{\omega'_1} = 2 \times 48.375 = 96.750 \text{ m}^5$

The moment of inertia of the section about the X axis

$$I_{11} = \tfrac{1}{12}(3.25 \times 6.25^3 - 2.75 \times 5.75^3 - 0.25 \times 4.0^3) = 21.22 \text{ m}^4$$

From Eq. (13.24) the distance of shear center from the pole $0'$ is

$$\alpha_1 = \frac{I_{\omega'_1}}{I_{11}} = \frac{96.75 \text{ m}^5}{21.22 \text{ m}^4} = 4.56 \text{ m}$$

2. **Principal Sectorial Coordinate Diagram.** This may now be constructed by using the shear center 0 as a pole and sweeping the ray $0D$ around the profile, or by calculating the ordinates using Eq. (13.25).

$$\omega = \omega' - \alpha_1 y \qquad (13.25)$$

At A = $18 - 4.56 \times 2 = 8.88 \text{ m}^2$
At B = $18 - 4.56 \times 3 = 4.32 \text{ m}^2$
At C = $9 - 4.56 \times 3 = -4.68 \text{ m}^2$

13.3 RESTRAINED WARPING OF UNIFORM CORES SUBJECTED TO TORSION

From antisymmetry the respective values at G, F, and E have the same magnitudes but are of opposite signs. The resulting principal sectorial coordinate diagram is given in Fig. 13.6d.

3. Sectorial Moment of Inertia I_ω. From Eq. (13.26)

$$I_\omega = \int^A \omega^2 \, dA = \int^s \omega^2 \, t \, ds \qquad (13.26)$$

Using the ω diagram (Fig. 13.6d) and the product integral table (Table 13.1):

Variation of ω	$\int \omega^2 \, t \, ds$
For DC	$\tfrac{1}{3} \times (4.68)^2 \times 0.25 = 5.48 \text{ m}^6$
CB	$\tfrac{1}{6} \times 2[(4.68)^2 + (-4.68)(4.32) + (4.32)^2] \times 0.25 = 5.09 \text{ m}^6$
BA	$\tfrac{1}{6} \times 2[(8.88)^2 + (4.32)(8.88) + (4.32)^2] \times 0.25 = 11.32 \text{ m}^6$
	For the half section $\Sigma \int \omega^2 \, t \, ds = 21.89 \text{ m}^6$
	$\therefore I_\omega$ for the whole section $= 2 \times 21.89 = 43.8 \text{ m}^6$

4. Torsion Constant J. For the open section core, using Eq. (13.27)

$$J = J_1 = \tfrac{1}{3} \sum^n bt^3 = \tfrac{1}{3}(1 + 3 + 6 + 3 + 1) \times 0.25^3 = 0.073 \text{ m}^4$$

13.3 THEORY FOR RESTRAINED WARPING OF UNIFORM CORES SUBJECTED TO TORSION

The theory for the warping torsion of thin-walled members is relatively recent compared with the theories for other modes of action [13.2, 13.3]. One of the most significant contributions was made by Vlasov [13.4], who is credited with the sectorial coordinate and bimoment concepts. The close analogy between warping torsion theory in the twisting mode and the wall–frame theory in a planar mode has emerged from more recent research on tall building structures [13.5, 13.6].

13.3.1 Governing Differential Equation

Consider the core in Fig. 13.7, fixed at the base, free at the top, and subjected to a distributed torque of intensity $m(z)$ at a height z from the base. It is assumed for the analysis that the core has an undeformable cross section with uniform dimensions and properties throughout the height.

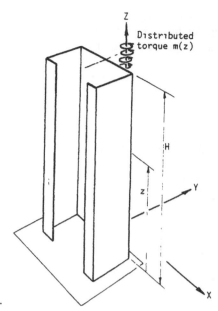

Fig. 13.7 Core subjected to distributed torque.

As described in Section 13.1, the external torque $T(z)$ at z is carried partly by a torque $T_w(z)$ associated with in-plane flexure of the core walls, that is with warping, and partly by a torque $T_v(z)$ corresponding to shears circulating within the walls and around the profile.

$$T_w(z) + T_v(z) = T(z) \qquad (13.3)$$

The torque associated with warping was shown in Eq. (13.11) to be

$$T_w(z) = -EI_\omega \frac{d^3\theta}{dz^3}(z) \qquad (13.11)$$

while that associated with the circulating shears is

$$T_v(z) = GJ \frac{d\theta}{dz}(z) \qquad (13.30)$$

Substituting Eqs. (13.11) and (13.30) in Eq. (13.3) gives the fundamental differential equation

$$-EI_\omega \frac{d^3\theta}{dz^3}(z) + GJ \frac{d\theta}{dz}(z) = T(z) = \int_z^H m\,dz \qquad (13.31)$$

Differentiating Eq. (13.31) and dividing by EI_ω, and noting that $dT/dz = -m(z)$, gives

$$\frac{d^4\theta}{dz^4}(z) - \alpha^2 \frac{d^2\theta}{dz^2}(z) = \frac{m(z)}{EI_\omega} \qquad (13.32)$$

in which

$$\alpha^2 = \frac{GJ}{EI_\omega} \qquad (13.33)$$

Equation (13.32) is the characteristic differential equation representing the warping torsion of a core.

13.3.2 Solution for Uniformly Distributed Torque

Following Eq. (11.6) in the analogous wall-frame theory, the solution for the rotation of a core subjected to a uniformly distributed torque m is

$$\theta(z) = C_1 + C_2 z + C_3 \cosh \alpha z + C_4 \sinh \alpha z - \frac{mz^2}{2EI_\omega \alpha^2} \qquad (13.34)$$

The boundary conditions for the determination of constants C_1 to C_4 are

1. fixity at the base

$$\theta(0) = \frac{d\theta}{dz}(0) = 0 \qquad (13.35)$$

2. zero rate of change of twist at the top

$$\frac{d^2\theta}{dz^2}(H) = 0 \qquad (13.36)$$

3. zero resultant torque at the top

$$-EI_\omega \frac{d^3\theta}{dz^3}(H) + GJ \frac{d\theta}{dz}(H) = 0 \qquad (13.37)$$

Equations (13.35), (13.36), and (13.37) lead to the solution of C_1 to C_4, which on substitution in Eq. (13.34) give

$$\theta(z) = \frac{mH^4}{EI_\omega} \left\{ \frac{1}{(\alpha H)^4} \left[\frac{(\alpha H \sinh \alpha H + 1)}{\cosh \alpha H} (\cosh \alpha z - 1) \right. \right.$$
$$\left. \left. - \alpha H \sinh \alpha z + (\alpha H)^2 \left[\frac{z}{H} - \frac{1}{2}\left(\frac{z}{H}\right)^2 \right] \right] \right\} \qquad (13.38)$$

326 CORE STRUCTURES

Equation (13.38) for the core rotation has two distinct parts. The expression within the braces defines the distribution of rotation over the height. It is a function of the dimensionless structural parameters αH and z/H, noting that $\alpha z = \alpha H(z/H)$ where

$$\alpha H = H\sqrt{\frac{GJ}{EI_\omega}} \qquad (13.39)$$

The preceding term mH^4/EI_ω, which includes the loading, height, and magnitude of the core structural properties, defines the magnitude of the rotational distribution.

The structural parameter αH characterizes the behavior of the core; consequently, cores having the same value of αH will have similar distributions of rotations and actions under similar distributions of loading.

The expression for the rotation of a core [Eq. (13.38)] is similar to that for the lateral deflection of a wall–frame [Eq. (11.10)]. Consequently, a core and a wall–frame with the same values of αH, and subjected, respectively, to similar distributions of torque and horizontal loading, will have correspondingly similar distributions of rotation and deflection.

A typical distribution of rotation for a core subjected to a uniformly distributed torque is shown in Fig. 13.8a. The derivatives of Eq. (13.38) are

$$\frac{d\theta}{dz}(z) = \frac{mH^3}{EI_\omega}\left\{\frac{1}{(\alpha H)^3}\left[\frac{(\alpha H \sinh \alpha H + 1)}{\cosh \alpha H}(\sinh \alpha z) - \alpha H \cosh \alpha z\right.\right.$$
$$\left.\left. + \alpha H\left(1 - \frac{z}{H}\right)\right]\right\} \qquad (13.40)$$

$$\frac{d^2\theta}{dz^2}(z) = \frac{mH^2}{EI_\omega}\left\{\frac{1}{(\alpha H)^2}\left[\frac{(\alpha H \sinh \alpha H + 1)}{\cosh \alpha H}(\cosh \alpha z) - \alpha H \sinh \alpha z - 1\right]\right\} \qquad (13.41)$$

$$\frac{d^3\theta}{dz^3}(z) = \frac{mH}{EI_\omega}\left\{\frac{1}{(\alpha H)}\left[\frac{(\alpha H \sinh \alpha H + 1)}{\cosh \alpha H}(\sinh \alpha z) - \alpha H \cosh \alpha z\right]\right\} \qquad (13.42)$$

13.3.3 Warping Stresses

The warping effects that are of concern to a designer include the vertical stresses in the core walls and, in cases where an open-section core is partially closed by beams, the shear and moment in the beams.

13.3 RESTRAINED WARPING OF UNIFORM CORES SUBJECTED TO TORSION

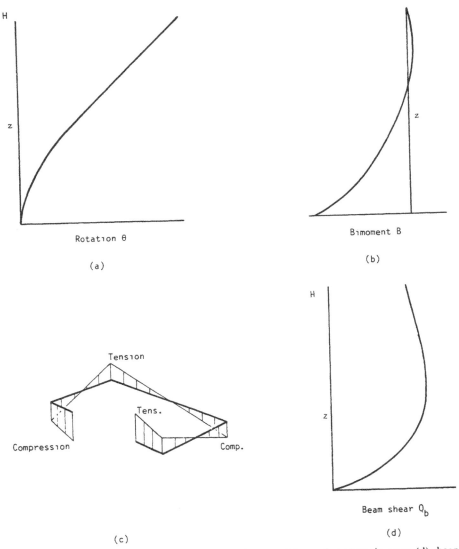

Fig. 13.8 (a) Rotation of core; (b) bimoment in core; (c) warping stress in core; (d) shear force in beams of partially closed core.

The principal warping action is the bimoment. The magnitude of the bimoment at a particular height in the core governs the magnitude of the vertical stress distribution at that level. The bimoment is given by

$$B(z) = -EI_\omega \frac{d^2\theta}{dz^2}(z) \tag{13.43}$$

Substituting from Eq. (13.41) for a core subjected to a uniformly distributed torque, the bimoment is given by

$$B(z) = -mH^2 \left\{ \frac{1}{(\alpha H)^2} \left[\frac{(\alpha H \sinh \alpha H + 1)}{\cosh \alpha H} (\cosh \alpha z) - \alpha H \sinh \alpha z - 1 \right] \right\}$$

(13.44)

Equations (13.43) and (13.44) are analogous to Eqs. (11.15) and (11.16) for the wall moment in a wall–frame structure.

The vertical displacement w at a point P, distance s from the origin on the section at height z, is given by

$$w(s, z) = -\omega(s) \frac{d\theta}{dz}(z)$$

(13.45)

Differentiating to obtain the vertical strain

$$\epsilon(s, z) = \frac{dw}{dz}(s, z) = -\omega(s) \frac{d^2\theta}{dz^2}(z)$$

(13.46)

from which the vertical stress at (s, z) is

$$\sigma(s, z) = E\epsilon(s, z) = -E\omega(s) \frac{d^2\theta}{dz^2}(z)$$

(13.47)

Substituting for $d^2\theta/dz^2$ from Eq. (13.43) gives

$$\sigma(s, z) = \frac{B(z)\,\omega(s)}{I_\omega}$$

(13.21)

which is the same as the expression derived for the I section in Section 13.1 on the basis of bending theory.

Equation (13.21) shows that, for a particular core, the magnitude of the stress distribution at a height z is governed by the bimoment at that level, while the distribution of axial stress over the section is defined by the principal sectorial coordinate (ω) diagram.

For a particular form of applied torque, the distribution of bimoment over the height of the core is governed by the characteristic parameter αH. For a core with a typical value of $\alpha H = 2.0$, subjected to a uniformly distributed torque, the bimoment distribution is as shown in Fig. 13.8b. A typical distribution of warping stress over a simple core section is given in Fig. 13.8c. The bimoment curve (Fig. 13.8b) shows that the warping stress distribution in the upper part of the core is opposite in sense to that in the lower part, while there is a level of "contrawarping", that is with a zero bimoment and zero warping stresses, at the transition.

13.3.4 Elevator Cores with a Partially Closed Section

Open-section elevator cores are often partially closed by beams or slabs at each floor level, as in Fig. 13.1b and c. When the core twists, the walls' edges on opposite sides of an opening undergo vertical displacements in opposite directions and vertical plane rotations in the same direction. These two types of relative displacement combine in subjecting the connecting beams or slabs to shear and bending. The vertical shear at the ends of the beams induces in the core walls complementary horizontal shears that circulate around the core. These circulatory shear stresses are similar to those that circulate in a closed section and result in a large increase in the effective J of the core, which causes a reduction in the core rotations and in the warping deformations and vertical stresses. The open-section theory for cores, as developed in the preceding sections, may be used to make an approximate analysis of cores with slender beams or slabs across the openings.

Assuming that in a partially closed core (Fig. 13.1b) the effect of the connecting beams on the circulating shear may be represented by an equivalent continuous shear diaphragm (Fig. 13.9a), the equivalent core would be a closed section consisting of the original walls, which behave in shear and flexure, and a diaphragm that behaves in shear only.

Accounting only for flexure of the beams, because they are assumed to be slender, the thickness of the equivalent shear diaphragm is shown later, in Section 13.6.1, to be

$$t_1 = \frac{12EI_b}{GL^2h} \tag{13.48}$$

in which I_b is the inertia of a connecting beam (or slab equivalent) about its horizontal axis, L is its length, and h is the vertical spacing of the beams or slabs.

The equivalent shear torsion constant of the section may now be obtained using

$$J = \frac{1}{3}\sum bt^3 + \frac{\Omega^2}{\oint ds/t} \tag{13.29}$$

in which the first term includes only the real walls of the core, while the second term includes the whole profile of the equivalent closed section. For practical-sized connecting beams or slabs, the thickness t_1 of the equivalent diaphragm is very small relative to the wall thickness of the core. Consequently the very large contribution of the diaphragm to the denominator of the second term in Eq. (13.29) means that the contribution of the real walls may be neglected, then

$$J = \frac{1}{3}\sum bt^3 + \frac{\Omega^2}{L/t_1} \tag{13.49}$$

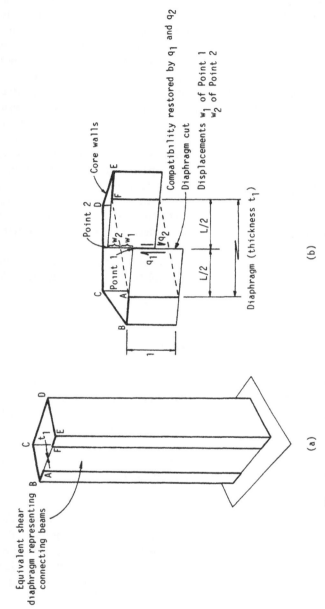

Fig. 13.9 (a) Beams represented by diaphragm; (b) diaphragm cut down middle.

13.3 RESTRAINED WARPING OF UNIFORM CORES SUBJECTED TO TORSION

and, substituting for t_1 from Eq. (13.48)

$$J = \frac{1}{3} \sum bt^3 + \frac{12EI_b\Omega^2}{GhL^3} \qquad (13.50)$$

The resulting value of J for the partially closed section is substituted in Eq. (13.39) to determine the effective αH for the core. The rotation, bimoments, and warping stresses in the core are then determined as described previously for an open-section core.

If the opening in a core is connected across at each floor level by a flat slab (Fig. 13.1c) the slabs should be transformed into equivalent beams that may then be treated as above. Provided there is no slab within the profile of the core, the size of an equivalent beam may be taken as given in Appendix 1 for slabs connecting shear walls.

13.3.5 Forces in Connecting Beams

The shears and moments in the connecting beams, caused by warping of the core, depend on the resulting relative displacements of the beam ends where they connect to the core. The vertical displacement at a point on the section of a core at height z is given by Eq. (13.45)

$$w(s, z) = -\omega(s)\frac{d\theta}{dz}(z) \qquad (13.45)$$

in which $d\theta/dz$ may be obtained from Eq. (13.40).

If the equivalent diaphragm-connected core (Fig. 13.9a) is cut vertically down the middle of the shear diaphragm AF, as in Fig. 13.9b, the sectorial coordinates ω_1 and ω_2 of points 1 and 2 at the edges of the cut can be evaluated for the resulting open-section core. The relative vertical displacement of 1 and 2, caused by the warping $d\theta/dz$, is then

$$w_2 - w_1 = (\omega_1 - \omega_2)\frac{d\theta}{dz} \qquad (13.51)$$

The shear force per unit height required to restore compatibility of the two halves, A1 and 2F, at the cut is then

$$q_1 = -q_2 = \frac{Gt_1(w_2 - w_1)}{L} \qquad (13.52)$$

and substituting for t_1 from Eq. (13.48) and for $w_2 - w_1$ from Eq. (13.51)

$$q_1 = -q_2 = \frac{12EI_b(\omega_1 - \omega_2)}{L^3 h}\frac{d\theta}{dz} \qquad (13.53)$$

Summing the distributed shear over a story height h to obtain the beam shear gives approximately

$$Q_b = \frac{12EI_b(\omega_1 - \omega_2)}{L^3}\frac{d\theta}{dz} \qquad (13.54)$$

Since $\omega_1 - \omega_2 = $ twice the enclosed area of the section $= \Omega$, Eq. (13.54) becomes

$$Q_b = \frac{12EI_b}{L^3}\Omega\frac{d\theta}{dz} \qquad (13.55)$$

Substituting for $d\theta/dz$ from Eq. (13.40), the shear in a beam at height z in a core subjected to a uniformly distributed torque m is

$$Q_b(z) = \frac{12I_b\Omega}{L^3}\frac{mH^3}{I_\omega}\left\{\frac{1}{(\alpha H)^3}\left[\frac{(\alpha H \sinh \alpha H + 1)}{\cosh \alpha H}(\sinh \alpha z)\right.\right.$$
$$\left.\left. - \alpha H \cosh \alpha z + \alpha H\left(1 - \frac{z}{H}\right)\right]\right\} \qquad (13.56)$$

The maximum bending moment in the connecting beam is then

$$M_b(z) = Q_b(z)\frac{L}{2} \qquad (13.57)$$

A typical distribution of shear in the connecting beams over the height of a uniform core is shown in Fig. 13.8d. The above method of analysis for partially closed cores is valid for beams across any opening in the planar wall of a core, whether symmetrically or asymmetrically located, provided the walls to which the beams connect are in the plane of the beam and that the inplane bending stiffnesses of the connected walls are significantly greater than those of the beams.

13.3.6 Solutions for Alternative Loadings

Although the above solutions refer to a uniformly distributed torque, solutions for other torque distributions follow similar procedures. The cases of a triangularly distributed torque and a concentrated torque at the top, and their superpositions with a uniformly distributed torque, are useful in representing graduated wind loading and static equivalent earthquake loading. Solutions for these additional two cases are given in Appendix 2.

13.4 ANALYSIS BY THE USE OF DESIGN CURVES

Similarly to the solutions for displacements and actions in wall–frame analysis, the solutions for the rotations and actions in a core can be expressed in the form of design curves. This provides a rapid method of hand analysis, while the curves give a useful guide to variations in a core's behavior for changes in the structural parameters.

For example, Eq. (13.38), expressing the rotation of a core subjected to a uniformly distributed torque, may be written in the form

$$\theta(z) = \frac{mH^4}{8EI_\omega} K_1(\alpha H, z/H) \qquad (13.58)$$

in which

$$K_1 = \frac{8}{(\alpha H)^4} \left\{ \frac{(\alpha H \sinh \alpha H + 1)}{\cosh \alpha H} (\cosh \alpha z - 1) \right.$$

$$\left. - \alpha H \sinh \alpha z + (\alpha H)^2 \left[\frac{z}{H} - \frac{1}{2}\left(\frac{z}{H}\right)^2 \right] \right\} \qquad (13.59)$$

K_1, which defines the distribution of θ over the height, is a function of αH and z/H. The term $mH^4/8EI_\omega$ defines the magnitude of the rotation distribution. The factor 8 is introduced into the two terms so that when $\alpha H = 0$, that is, when the torsion coefficient J is zero and the core resists torsion only by restrained warping, $K_1(H) = 1.0$, and $mH^4/8EI_\omega$ defines the rotation at the top. Curves of K_1 are plotted in Fig. A2.1, which allows a rapid hand solution for the core rotations.

The connecting beam shear and the bimoment, as represented by Eq. (13.56) and (13.44), respectively, can also be written concisely as

$$Q_b(z) = \frac{2mH^3}{I_\omega} \frac{I_b \Omega}{L^3} K_2(\alpha H, z/H) \qquad (13.60)$$

and

$$B(z) = -\frac{mH^2}{2} K_3(\alpha H, z/H) \qquad (13.61)$$

K_2 and K_3 are also functions of αH and z/H. They represent the distributions of the respective actions over the height of the core, while the functions preceding them govern the magnitudes of the distributions. Curves of K_2 and K_3 are given in Figs. A2.2 and A2.3 for rapid hand solutions.

The case of a triangularly distributed torque, and a concentrated torque at the top, may be solved similarly using the additional design curves given in Appendix 2. Their results may be superposed with those for uniformly distributed loading for more complex loading cases.

13.5 WORKED EXAMPLE TO ANALYZE A CORE USING FORMULAS AND DESIGN CURVES

The closed solution for the core, as described above, may be used to determine the rotations and stresses in uniform structures whose torque resistance depends entirely on the core. These include cores combined with unbraced frames, and cores with hanging or cantilevered floors. Although the method is accurate only for cores with a uniform section throughout the height, it may be used to obtain an approximate estimate of the actions in nonuniform cores for the purpose of a preliminary design.

To demonstrate the analysis of a core structure, a 70-m-high building, consisting of 20 3.5-m-high stories, with the plan arrangement shown in Fig. 13.10, is considered. The resistance to forces in the Y direction and to torque is provided by a core whose dimensions and properties are shown in Fig. 13.6a. The building is subjected to a uniformly distributed wind pressure of 0.5 kN/m².

It is required to determine the maximum deflection at the top of the structure and the vertical direct stresses at the base due to bending and twisting, for two cases. In the first case the core has an open section and, in the second, the core opening is partially closed by 0.25-m-wide by 0.6-m-deep beams at 3.5 m centers up the height of the structure. For the second case determine also the maximum value of the shear and the bending moment in the connecting beams.

An elastic modulus $E = 2.0 \times 10^7$ kN/m² and the shear modulus $G = 0.9 \times 10^7$ kN/m² are assumed for the concrete properties. The procedures of the analysis are first described and then illustrated numerically for the considered structure.

Step 1. Determine the Sectorial Properties. As explained in Section 13.2 for open-section cores, and in Sec. 13.3.4 for cores partially closed by beams, the sectorial properties can be obtained. For the given structure, the location of the shear center, the principal sectorial coordinate (ω) diagram, the moment of inertia I_{xx}, and the values of I_ω and J_1 are obtained from the results of the worked example in Section 13.2.6.

The distance of the shear center is 4.56 m from the pole 0' (Fig. 13.6a) and the parameters of the core are $I_{xx} = 21.22$ m⁴, $I_\omega = 43.8$ m⁶, and $J = 0.073$ m⁴.

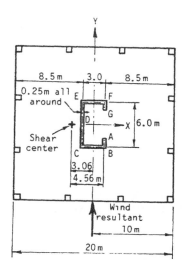

Fig. 13.10 Example core structure.

13.5 WORKED EXAMPLE TO ANALYZE A CORE

Step 2. Determine the Line of Action of the Horizontal Loading Resultant and Its Eccentricity e from the Shear Center. The resultant wind force per unit height of the building is equal to $20 \times 0.5 = 10$ kN/m and it acts 3.06 m to the right of the shear center, that is, the eccentricity, e, from the shear center is 3.06 m. Since the external torque is the product of the horizontal loading and its eccentricity, the torque due to the wind is 30.6 kN m/m height, anticlockwise.

To determine the maximum deflection at the top of the structure and the stresses at the base, a bending analysis has to be performed. Deflection at the top due to bending

$$y_b(H) = \frac{wH^4}{8EI_{11}} = \frac{10 \times 70^4}{8 \times 2.0 \times 10^7 \times 21.22} = 0.071 \text{ m}$$

Bending moment at the base

$$M_b(0) = \frac{wH^2}{2} = \frac{10 \times 70^2}{2} = 24{,}500 \text{ kNm}$$

Bending stress at C, at the base of the core (Fig. 13.10), is given by

$$\frac{M_b c}{I_{11}} = \frac{24{,}500 \times 3.125}{21.22} = 3608 \text{ kN/m}^2$$

Stresses at other points are obtained similarly. The bending stress diagram is given in Fig. 13.11a.

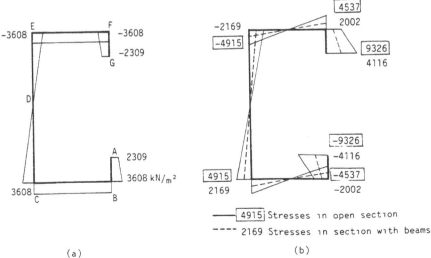

Fig. 13.11 (a) Bending stresses at base of example core structure; (b) warping stresses at base of example core structure.

The following procedures, from Step 3 onward, refer to the torsional analysis.

Step 3. Determine the parameter αH using Eq. (13.39)

$$\alpha H = H\sqrt{\frac{GJ}{EI_\omega}}$$

a. For the case of the open-section:

$$\alpha H = 70\sqrt{\frac{0.9 \times 10^7 \times 0.073}{2.0 \times 10^7 \times 43.8}} = 1.9$$

b. For the case of the partially closed section: Consider the core with its opening partially closed by beams at 3.5 m centers. The beams are 0.25 m wide by 0.6 m deep. The moment of inertia of a beam is

$$I_h = 0.25 \times 0.6^3/12 = 0.0045 \text{ m}^4$$

The modified torsion constant J is given by Eq. (13.50)

$$J = \frac{1}{3}\sum bt^3 + \frac{12EI_h\Omega^2}{GhL^3}$$

in which Ω is twice the area enclosed by the profile, h is the story height, and L is the span of the beam

$$J = 0.073 + \frac{12 \times 2.0 \times 10^7 \times 0.0045 \times 36^2}{0.9 \times 10^7 \times 3.5 \times 4^3} = 0.767 \text{ m}^4$$

Therefore,

$$\alpha H = H\sqrt{\frac{GJ}{EI_\omega}} = 70\sqrt{\frac{0.9 \times 10^7 \times 0.767}{2.0 \times 10^7 \times 43.8}} = 6.2$$

Step 4. Determine Rotations and Total Deflections. The rotation at any level of the core for a uniformly distributed torque m may be obtained either from Eq. (13.38) or by taking the appropriate value of K_1 from Fig. A2.1 and using Eq. (13.58).

a. For the case of the open-section: The rotation is given by Eq. (13.38)

$$\theta(z) = \frac{mH^4}{EI_\omega}\left\{\frac{1}{(\alpha H)^4}\left[\frac{(\alpha H \sinh \alpha H + 1)}{\cosh \alpha H}(\cosh \alpha z - 1)\right.\right.$$
$$\left.\left. - \alpha H \sinh \alpha z + (\alpha H)^2\left|\frac{z}{H} - \frac{1}{2}\left(\frac{z}{H}\right)^2\right|\right]\right\} \quad (13.38)$$

13 5 WORKED EXAMPLE TO ANALYZE A CORE

At the top, $z = H = 70$ m, and $\alpha z = \alpha H = 1.9$, and $z/H = 1.0$

$$\theta(H) = \frac{30.6 \times 70^4}{2.0 \times 10^7 \times 43.8} \left\{ \frac{1}{(1.9)^4} \left[\frac{(1.9 \sinh 1.9 + 1)(\cosh 1.9 - 1)}{\cosh 1.9} \right. \right.$$

$$\left. \left. - 1.9 \sinh 1.9 + (1.9)^2 (1 - 0.5) \right] \right\}$$

Thus, $\theta(H) = 0.0448$ rads anticlockwise.

Therefore, the additional deflection of the right-hand side of the building due to twist

$$y_t(H) = \theta(H) \times \text{distance of right-hand side from shear center}$$

$$= 0.0448 \times 13.06 = 0.585 \text{ m}$$

Note that the displacement due to torsion is much larger than that due to bending.

The total deflection at the top floor right-hand side due to bending and twist

$$y_b(H) + y_t(H) = 0.071 + 0.585 = 0.656 \text{ m}$$

This deflection represents $\frac{1}{107}$ of the total building height, an unacceptably large value.

b. For the case of the partially closed section: The rotation at the top is given by

$$\theta(H) = \frac{30.6 \times 70^4}{2.0 \times 10^7 \times 43.8} \left\{ \frac{1}{(6.2)^4} \left[\frac{(6.2 \sinh 6.2 + 1)}{\cosh 6.2} (\cosh 6.2 - 1) \right. \right.$$

$$\left. \left. - 6.2 \sinh 6.2 + (6.2)^2 (1 - 0.5) \right] \right\}$$

Thus $\theta(H) = 0.00796$ rads anticlockwise.

Then the total deflection at the top floor right-hand side due to bending and twist is

$$y = 0.071 + 0.00796 \times 13.06 = 0.175 \text{ m}$$

This deflection is $\frac{1}{400}$ of the height and is therefore generally acceptable.

Note the stiffening effect of the beams in reducing the displacement due to torsion.

Step 5. Determine Bimoments and Warping Stresses. The warping stresses at a height z are determined from the bimoment B at that level. The bimoment is obtained for a uniformly distributed loading either from Eq. (13.44) or by taking K_3 from Fig. A2.3 and using Eq. (13.61).

Then, at a point on the section where the principal sectorial coordinate is $\omega(s)$, the vertical warping stress is obtained from Eq. (13.21).

The resulting warping stresses must be combined with the bending stresses obtained by considering the core as a flexural cantilever about its major axes, to give the total axial stresses due to horizontal loading.

a. For the case of the open-section: The vertical stresses at the base due to warping are determined from the bimoment at the base: this is given by Eq. (13.44).

$$B(z) = -mH^2 \left\{ \frac{1}{(\alpha H)^2} \left| \frac{(\alpha H \sinh \alpha H + 1)}{\cosh \alpha H} (\cosh \alpha z) \right. \right.$$
$$\left. \left. - \alpha H \sinh \alpha z - 1 \right] \right\} \qquad (13.44)$$

At the base $z = 0$ and $\alpha z = 0$

$$B(0) = -30.6 \times 70^2 \left\{ \frac{1}{(1.9)^2} \left| \frac{(1.9 \sinh 1.9 + 1)}{\cosh 1.9} (\cosh 0) \right. \right.$$
$$\left. \left. - 1.9 \sinh 0 - 1 \right] \right\}$$

Thus $B(0) = -4.6 \times 10^4$ kNm^2

The warping stress

$$\sigma(s, z) = \frac{B(z) \omega(s)}{I_\omega} \qquad (13.21)$$

Values at ω for points on the section are given in Fig. 13.6d. Then the vertical warping stress at G, for example, is

$$\sigma_\omega(G) = \frac{-4.6 \times 10^4 \times (-8.88)}{43.8} = 9326 \text{ kN/m}^2, \text{ tension}$$

Warping stresses at other points are obtained similarly. The distribution of warping stress over the section is given in Fig. 13.11b. In this case, the maximum warping stresses (± 9326 kN/m^2) are significantly larger than the maximum bending stresses (± 3608 kN/m^2).

b. For the case of the partially closed section: To determine the warping stresses at the base due to torsion, the bimoment at $z = 0$ is given by

13 5 WORKED EXAMPLE TO ANALYZE A CORE

$$B(0) = -30.6 \times 70^2 \left\{ \frac{1}{(6.2)^2} \left[\frac{(6.2 \sinh 6.2 + 1)}{\cosh 6.2} (\cosh 0) \right. \right.$$

$$\left. \left. - 6.2 \sinh 0 - 1 \right] \right\}$$

Thus, $B(0) = -2.03 \times 10^4$ kNm^2 from which the vertical warping stress at G is

$$\sigma_\omega(G) = \frac{-2.03 \times 10^4 \times (-8.88)}{43.8} = 4116 \text{ kN/m}^2 \text{ tension}$$

This is less than half the value of the warping stress in the core without beams. The distribution of warping stresses around the base of the core is compared with that of the open section in Fig. 13.11b.

Step 6. Determine Shear and Moment in the Connecting Beams. This step is only for the case of the partially closed section with connecting beams. The shear Q_b may be obtained either from Eq. (13.56), or by taking K_2 from Fig. A2.2 and using Eq. (13.60).

The maximum value of the beam shear may be obtained by scanning the appropriate curve of K_2 to find the maximum value and using it in Eq. (13.60) for shear.

To determine the shear force and bending moment in the connecting beams, the shear in a beam at height z from the base is given by Eq. (13.56).

$$Q_b(z) = \frac{12 I_b \Omega}{L^3} \frac{m H^3}{I_\omega} \left\{ \frac{1}{(\alpha H)^3} \left[\frac{(\alpha H \sinh \alpha H + 1)}{\cosh \alpha H} (\sinh \alpha z) \right. \right.$$

$$\left. \left. - \alpha H \cosh \alpha z + \alpha H \left(1 - \frac{z}{H}\right) \right] \right\} \quad (13.56)$$

For the shear in the beam at four stories above the base,

$$Q_b = \frac{12 \times 0.0045 \times 36 \times 30.6 \times 70^3}{4^3 \times 43.8} \left\{ \frac{1}{(6.2)^3} \left[\frac{(6.2 \sinh 6.2 + 1)}{\cosh 6.2} \right. \right.$$

$$\left. \left. \cdot (\sinh 1.24) - 6.2 \cosh 1.24 + 6.2(1 - 0.2) \right] \right\}$$

Therefore, $Q_b = 96.88$ kN. The maximum moment, at the ends of the beam, is then

CORE STRUCTURES

$$M_b(z) = Q_b(z) \frac{L}{2}$$

$$M_b = 96.88 \times 2 = 193.76 \text{ kNm} \tag{13.57}$$

This example illustrates that when an open or partially closed-section core is the primary torque-resisting component of a building, the vertical stresses due to warping may be very significant. Also, beams connecting across the opening of a core can significantly stiffen it against torsion, thereby reducing the twist and the warping stresses.

The above results were calculated from formulas. They could have been obtained more rapidly by design curve solutions. This procedure is illustrated by repeating the analysis of the core with beams.

The rotation at the top is obtained by using Eq. (13.58)

$$\theta(z) = \frac{mH^4}{8EI_\omega} K_1(\alpha H, z/H) \tag{13.58}$$

where K_1 for $\alpha H = 6.2$ and $z/H = 1.0$ is read from Fig. A2.1 as 0.077 then

$$\theta(H) = \frac{30.6 \times 70^4 \times 0.077}{8 \times 2 \times 10^7 \times 43.8}$$

$$= 0.00807 \text{ rads anticlockwise}$$

The bimoment at the base is obtained by using Eq. (13.61).

$$B(z) = -\frac{mH^2}{2} K_3(\alpha H, z/H) \tag{13.61}$$

where K_3 for $\alpha H = 6.2$ and $z/H = 0$ is read from Fig. A2.3 as 0.27; then

$$B(0) = \frac{-30.6 \times 70^2 \times 0.27}{2}$$

$$= -2.02 \times 10^4 \text{ kNm}^2$$

The shear in the beams is obtained by using Eq (13.60).

$$Q_b(z) = \frac{2mH^3}{I_\omega} \frac{I_b \Omega}{L^3} K_2(\alpha H, z/H) \tag{13.60}$$

where K_2 is obtained for $\alpha H = 6.2$ and the appropriate value of z/H from Fig. A2.2. The level of the beam carrying the maximum shear is identified by scanning the curve corresponding to $\alpha H = 6.2$ and finding the maximum value of K_2.

K_2 (max) for $\alpha H = 6.2$ is read as 0.083 at $z/H = 0.3$. Then

$$Q_h(\max) = \frac{2 \times 30.6 \times 70^3 \times 0.0045 \times 36 \times 0.083}{43.8 \times 4^3}$$

$$= 100.7 \text{ kN}$$

Although not as exact as the results from the detailed formulas, the results from the design curves are accurate enough for practical purposes.

13.6 COMPUTER ANALYSES OF CORE STRUCTURES

The classical theory for cores, described in the previous sections, is useful for the analysis of single core structures with uniform properties. It is useful also for obtaining an understanding of warping behavior and the response of a core to variations in the structural parameters. In the majority of building structures, however, cores act in combination with other types of assembly, such as rigid frames and shear walls, and their properties vary with height. The classical theory cannot be used to analyze such complex assemblies and it is necessary to revert to a stiffness matrix computer analysis, which can consider the total structure including the core and its discrete variations.

Three categories of computer model are considered: the first includes the finite element and equivalent frame models, the second is a two-column model, and the third is a single-column model. Although they are referred to in order of increasingly concise representation and, hence, increasing efficiency in use of computer time, they are unfortunately in the reverse order in the simplicity of their application and in the ease of interpreting their results. Each, however, has virtues sufficient to merit its consideration here.

13.6.1 Membrane Finite Element Model Analysis

A significant advantage of a membrane element model analysis of a core is that it does not require any knowledge of warping theory, nor does it require the calculation of the warping sectorial properties.

Assuming the cross section of a building core to be fixed in shape due to the horizontal rigidity of the floor slabs, the principal interaction between the walls is vertical shear along their connecting edges. Consequently, an elevator core can be modeled for analysis by plane stress rectangular or quadrilateral membrane elements, in conjunction with auxiliary beams as described in Chapter 5. The auxiliary beams are added around the profile at each mesh level (Fig. 13.12) and are assigned to be flexurally rigid in the horizontal plane and to be rigidly connected at the corners. The auxiliary beams serve to maintain the cross-sectional shape of the core when it is subjected to loading and, by stiffening the auxiliary beams at the wall-edge element against vertical bending, they allow beams to connect across

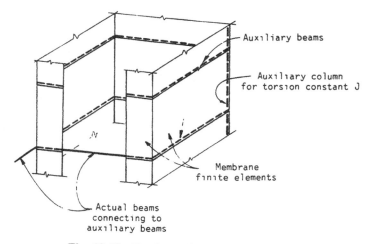

Fig. 13.12 Membrane finite element model.

openings in the core, or beams to connect to the core from other parts of the total structure. The behavior of an open-section core acting alone can be sensitive to its torsion constant J_1, where $J_1 = \Sigma \frac{1}{3}bt^3$. As this property cannot be represented by a plane stress membrane element model, an auxiliary column should be added down one of the wall edges with the value of J_1 of the core assigned to the column.

In most situations, full wall-width elements provide an adequate representation of a core's behavior and its interaction with other connecting structures. The simpler types of membrane elements give a linear distribution of vertical stress in each wall. In certain locations shear lag may allow significantly higher axial stresses to develop in the walls, especially at the free vertical edges of the walls in open-section cores. It may be desirable, therefore, at critical levels such as near the base, or in regions where the walls change thickness, to refine the mesh by using quadrilateral elements, as described in Chapter 5.

It has been found in modeling cores for analysis that incompatible mode plane stress elements, which allow for inplane bending deformations, give consistently more accurate results than compatible mode elements. Incompatible mode elements may also be used in much greater height-to-width ratios, up to 6:1 or more, while maintaining an acceptable accuracy.

An alternative way of modeling beams that connect across an opening in a membrane element model of a core is to use story-height equivalent plane stress membrane elements with a vertical shearing stiffness equal to the transverse bending stiffness of the beams (Fig. 13.13). Referring to Fig. 13.14a and b, the thickness of the equivalent membrane element representing a beam can be found by considering the vertical shear deformation of the membrane element. The vertical displacement

$$\delta_{vm} = \frac{VL}{Ght_1} \tag{13.62}$$

13.6 COMPUTER ANALYSES OF CORE STRUCTURES

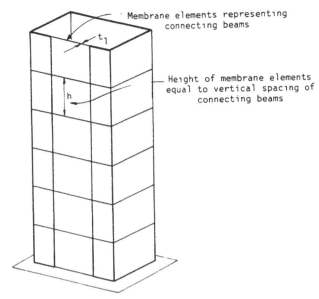

Fig. 13.13 Membrane elements representing connecting beams.

must be equal to the double-bending and shear deflection of the replaced beam

$$\delta_{vb} = \frac{VL^3}{12EI_b} + \frac{VL}{GA_v} \qquad (13.63)$$

Equating (13.62) to (13.63), assuming that the shear area of the beam $A_v = A_b/1.2$, and solving for the thickness of the equivalent membrane element gives

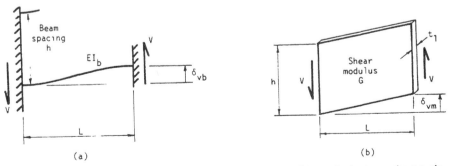

Fig. 13.14 (a) Transversely displaced connecting beam; (b) equivalent membrane element.

$$t_1 = \cfrac{1}{h\left(\cfrac{L^2}{12I_\text{b}}\cfrac{G}{E} + \cfrac{1.2}{A_\text{b}}\right)} \qquad (13.64)$$

in which A_b and I_b are, respectively, the sectional area and the inertia of the connecting beam.

The second term in Eq. (13.63) and the second term in the demoninator of Eq. (13.64) account for shear deformation of the connecting beam. If the length-to-depth ratio of the beam is 5 or more, the shear deflection of the beam will be relatively insignificant compared with its bending deflection, in which case the thickness of the equivalent membrane element will be

$$t_1 = \frac{12EI_\text{b}}{GL^2h} \qquad (13.65)$$

13.6.2 Analogous Frame Analysis

If a membrane finite-element option is not available as part of the structural analysis program, an analogous frame can provide a satisfactory alternative model for analyzing a core. Similarly to the membrane element model, an analogous frame model does not require any knowledge of warping theory or any calculation of the warping parameters.

The asymmetric braced frame analogy described in Chapter 9 has proved to give results for deflections and stresses comparable in accuracy with plane stress membrane element analyses. It is approximately as efficient in computer use as a membrane element analysis but, in the absence of specially written subroutines, it takes additional time in initially calculating the member sizes and finally transforming the frame forces into the stress results.

The frame analogy works well in representing a core by wall-width modules (Fig. 13.15), while for a more refined analysis, subwall-width modules should be

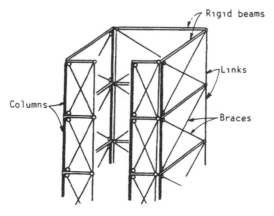

Fig. 13.15 Frame analogy.

13.6 COMPUTER ANALYSES OF CORE STRUCTURES

used. Beams across openings, or beams connecting parts of a surrounding structure to the core, are joined to the ends of the modules' rigid arms. In an open-section core, for which it is desirable to include the torsional constant $\Sigma \frac{1}{3} bt^3$ of the walls, the value of the constant for each wall segment can be assigned to the column of the corresponding frame module.

13.6.3 Two-Column Analogy

A relatively simple two-column model can be used for the computer analysis of either a separate core subjected to lateral loading or a core that is part of, and interacts horizontally with, a larger surrounding structure [13.7]. The analogous model represents the warping and St. Venant torsional modes of behavior in addition to the bending modes. It is, therefore, well suited for open or partially closed-section cores. It can represent cores that change section within the height, including changes in the directions of the principal bending axes and in the locations of the shear centers.

To simplify the explanation, a multisectional core, whose sections are monosymmetric about the X axis, will be considered. Using the same technique, models can be formed for multisectional cores with differently asymmetric sections and different principal axes of bending. The example core shown in Fig. 13.16a has a lower region L, and an upper region U, having different sections, both sections being monosymmetric about the X axis, but with differently located shear centers.

If the principal moments of inertia, the St. Venant torsion constant, and the warping moment of inertia of the lower region of the core are denoted by I_X^L, I_Y^L, J^L, and I_ω^L, the core in that region can be represented by two columns located on one of the principal bending axes at distances a from, and on opposite sides of, the shear center. The properties of the core are shared between the columns so that, for each column

$$I_{c1}^L = \tfrac{1}{2} I_1^L \qquad (13.66)$$

$$I_{c1}^L = \tfrac{1}{2} I_1^L \qquad (13.67)$$

$$J_{c2}^L = \tfrac{1}{2} J_2^L \qquad (13.68)$$

while the warping moment of inertia is developed by locating the columns at distances from the shear center, such that for opposite directions of bending in parallel planes (Fig. 13.16b).

$$2 I_{c1}^L (a^L)^2 = I_\omega^L \qquad (13.69)$$

to give

$$a^L = \sqrt{\frac{I_\omega^L}{2 I_{c1}^L}} \qquad (13.70)$$

346 CORE STRUCTURES

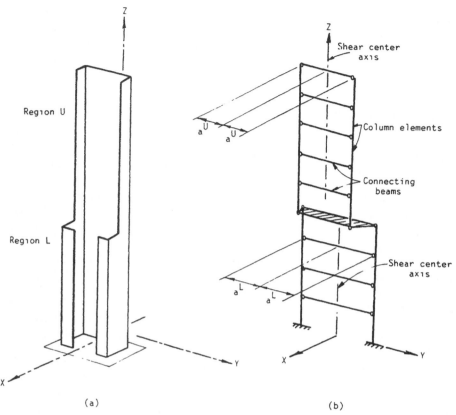

Fig. 13.16 (a) Multisection core; (b) two-column analogy.

or, substituting from Eq. (13.67)

$$a^L = \sqrt{\frac{I_\omega^L}{I_1^L}} \qquad (13.71)$$

The upper region U can also be represented by two columns with properties determined from the properties of the upper core section, with the columns located on one of the upper core's principal bending axes at distances a^U from its shear center.

The remaining problem in forming the two-sectional model is to connect the upper columns to the lower columns in such a way as to represent properly the behavior of the core at the transition level. The pairs of columns in the two regions are located with respect to different shear centers, at different distances apart along different principal axes. Their connections must establish compatibility at the transition level with respect to all the modes of behavior. These requirements are, for bending

$$\left(\frac{dx}{dz}\right)^L = \left(\frac{dx}{dz}\right)^U \tag{13.72}$$

and

$$\left(\frac{dy}{dz}\right)^L = \left(\frac{dy}{dz}\right)^U \tag{13.73}$$

for St. Venant torsion

$$\theta_z^L = \theta_z^U \tag{13.74}$$

and for warping torsion

$$\left(\frac{d\theta}{dz}\right)^L = \left(\frac{d\theta}{dz}\right)^U \tag{13.75}$$

The compatibility requirements for bending and St. Venant torsion could be achieved simply by using a direct constraint option to make the vertical plane and horizontal plane rotations of the upper and lower columns conform. This, however, would not satisfy warping compatibility, which requires that $d\theta/dz$ for the upper and lower columns conform at the transition level. For this, it is shown in Fig. 13.17a that the vertical plane angles of rotation due to warping must be related as follows:

$$\beta^U = a^U \frac{d\theta}{dz} \quad \text{and} \quad \beta^L = a^L \frac{d\theta}{dz} \tag{13.76}$$

$$\therefore \beta^L = \beta^U \frac{a^L}{a^U} \tag{13.77}$$

where β^L, β^U, a^L, a^U, and θ are as shown in Fig. 13.17a.

A mechanism to connect the upper and lower columns, which satisfies all the compatibility conditions simultaneously, and which is suitable for computer analysis, is shown in Fig. 13.17b. Using the rigid floor option, which is available in the more comprehensive structural analysis programs, a horizontally rigid diaphragm is placed at the transition level, and a second "auxiliary" diagram is located a small distance, say one-tenth of a story height, below the transition level. The upper columns, which terminate at the transition level, have rigid stub column extensions down to the auxiliary diaphragm level. Nodes are assigned on the upper column lines at the two diaphragm levels. Similarly, the lower columns are extended upwards from the auxiliary diaphragm level, by rigid stub columns, to the transition diaphragm level, with nodes assigned on the lower column lines at the two levels.

348 CORE STRUCTURES

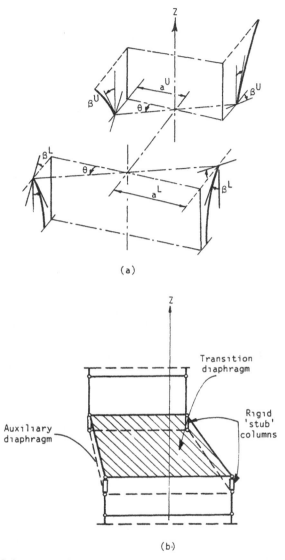

Fig. 13.17 (a) Column warping displacements at transition level; (b) mechanism to represent transition level.

The horizontal plane relative rotations of the transition level and the auxiliary diaphragm apply inclinations through the rigid stub columns to the upper and lower columns, which conform with Eq. (13.77). These cause the upper pair of columns to have the same $d\theta/dz$ as the lower pair. If, simultaneously, the auxiliary diaphragm and the transition plane translate relatively in the X or Y directions, corresponding to bending in these directions, the proposed mechanism applies the same change of inclination to the upper and lower columns. The model is, therefore, capable of simultaneously achieving compatibility at the transition level of the X and Y inclinations for bending, the rotation θ for the St. Venant action, and $d\theta/dz$ for the warping action.

Comparisons of results from the two-column analogy models with those from the more detailed and accurate shell element models have shown the deflections to be generally within 10% and the stresses within 20%; this can be considered as reasonable for such a simple model of a complex structure. Although the two-column model has the great advantage of simplicity, its main disadvantage is in requiring a knowledge of warping theory, and in having to first calculate the sectional and sectorial properties of the core. The ultimate value of the model will be achieved, therefore, only when used in combination with a computer program to determine the sectional and sectorial properties of thin-walled member sections.

13.6.4 Single Warping-Column Model

In the stiffness matrix analysis of a structure, a core could be represented in all its aspects of behavior, except warping, by a stack of simple column elements, with six degrees of freedom per node, located on the shear center axis. Its assigned properties would include inertias I_X and I_Y, to represent its resistance to bending about its principal axes, and J to represent its St. Venant resistance to torsion. If the section were completely closed, so that warping was negligible, or if the structure were not likely to twist, such a column model would allow an acceptably accurate analysis. If, however, the section were open or only partially closed, and the structure was subjected to a significant torque, then the warping torsional resistance of the core would be a factor in the behavior of the structure and the simple column model would not be an adequate representation of the core.

A single column model, with an extra, seventh, degree of freedom per node to represent warping, has been developed to include all the modes of behavior of a core [13.5]. It is particularly suitable for cores that are uniform over the height, but can also be used for cores that change properties with height provided the shear center axes of the different regions lie on approximately the same vertical line.

The seventh, warping, degree of freedom is taken to be $d\theta/dz$ which, as used for the two-column model in the previous section, expresses the magnitude of warping and may be used as the warping degree of freedom, while B, the bimoment, is the corresponding action.

With seven degrees of freedom per node, the warping column element (Fig. 13.18) has a 14 × 14 stiffness matrix. The warping column element is located on the shear center axis of the core with its stiffness coefficients referred to that axis. A complete core would be represented by a vertical stack of story-height elements with nodes in the floor levels (Fig. 13.19). The terms of the stiffness matrix corresponding to twist and to warping are given in Table 13.2. The remaining terms of the total 14 × 14 stiffness matrix, as referred to the shear center, are given in Table 13.3 [13.8]. The principal advantage of the single warping-column model

350 CORE STRUCTURES

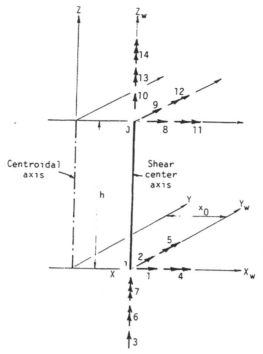

Fig. 13.18 Single warping-column element.

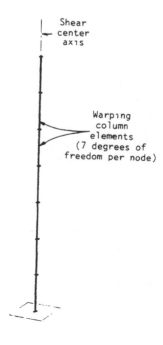

Fig. 13.19 Single-column model of core.

TABLE 13.2 Single Warping-Column Model: Stiffness Matrix for Twisting and Warping

$$K_{\theta\theta} = \left(\dfrac{GJ}{2 + \alpha H \sinh \alpha H - 2 \cosh \alpha H}\right)$$

	6	7	13	14
6	$\alpha \sinh \alpha H$	$(1 - \cosh \alpha H)$	$-\alpha \sinh \alpha H$	$(1 - \cosh \alpha H)$
7	$(1 - \cosh \alpha H)$	$\dfrac{1}{\alpha}(\alpha H \cosh \alpha H - \sinh \alpha H)$	$-(1 - \cosh \alpha H)$	$\dfrac{1}{\alpha}(\sinh \alpha H - \alpha H)$
13	$-\alpha \sinh \alpha H$	$-(1 - \cosh \alpha H)$	$\alpha \sinh \alpha H$	$-(1 - \cosh \alpha H)$
14	$(1 - \cosh \alpha H)$	$\dfrac{1}{\alpha}(\sinh \alpha H - \alpha H)$	$-(1 - \cosh \alpha H)$	$\dfrac{1}{\alpha}(\alpha H \cosh \alpha H - \sinh \alpha H)$

TABLE 13.3 Single Warping-Column Model: Total Stiffness Matrix

	1	2	3	4	5	6	7	8	9	10	11	12	13	14
1	$\dfrac{12EI_1}{h^3}$													
2	0	$\dfrac{12EI_1}{h^3}$												
3	0	0	$\dfrac{AE}{h}$											
4	0	$-\dfrac{6EI_1}{h^2}$	0	$\dfrac{4EI_1}{h}$										
5	$\dfrac{6EI_1}{h^2}$	0	$x_0\dfrac{AE}{h}$	0	$\dfrac{4EI_1}{h}+\dfrac{x_0^2 AE}{h}$									
6	0	0	0	0	0	a								
7	0	0	0	0	0	a	a							
8	$-\dfrac{12EI_1}{h^3}$	0	0	0	$-\dfrac{6EI_1}{h^2}$	0	0	$\dfrac{12EI_1}{h^3}$						
9	0	$-\dfrac{12EI_1}{h^3}$	0	$\dfrac{6EI_1}{h^2}$	0	0	0	0	$\dfrac{12EI_1}{h^3}$					
10	0	0	$-\dfrac{AE}{h}$	0	$-\dfrac{x_0 AE}{h}$	0	0	0	0	$\dfrac{AE}{h}$				
11	0	$-\dfrac{6EI_1}{h^2}$	0	$\dfrac{2EI_1}{h}$	0	0	0	0	$\dfrac{6EI_1}{h^2}$	0	$\dfrac{4EI_1}{h}$			
12	$\dfrac{6EI_1}{h^2}$	0	$-x_0\dfrac{AE}{h}$	0	$\dfrac{2EI_1}{h}-\dfrac{x_0^2 AE}{h}$	0	0	$-\dfrac{6EI_1}{h^2}$	0	$\dfrac{x_0 AE}{h}$	0	$\dfrac{4EI_1}{h}+\dfrac{x_0^2 AE}{h}$		
13	0	0	0	0	0	a	a	0	0	0	0	0	a	
14	0	0	0	0	0	a	a	0	0	0	0	0	a	a

Symmetrical

[a] Terms with rows or columns 6, 7, 13, 14 are as given in Table 13.2.

is its extremely concise form of representing warping. Its disadvantages, however, are that it cannot be used to represent a multisectional core with shifts in the shear center, that it requires a knowledge of warping theory to evaluate the sectorial properties of the core and to interpret the results, and that it requires a specially written subroutine to form the 14 × 14 matrix in the structural analysis program.

SUMMARY

A shear wall elevator core can provide a major part of the bending and torsional resistance in a building structure. The bending actions may be analyzed conventionally, as for a vertical cantilever located on the shear center axis. Because of a core's proportional similarity to a thin-walled beam, however, any significant twisting action should be analyzed to include both warping and shear torsion. The warping effects include vertical stresses in the walls, and shear and bending in beams or slabs that connect across openings in the core section.

The concept of warping torsion can be developed by considering an I-shaped section subjected to twist. The external torque is resisted partly by the inplane bending of the flanges and partly by their twisting resistance. A characteristic differential equation for warping torsion is developed through the use of simple bending theory.

A classical analysis of warping torsion requires the prior evaluation of the shear center location, the principal sectorial coordinate diagram, the warping moment of inertia, and the torsion constant.

A uniform-section core subjected to torque can be analyzed by formulating and solving the characteristic differential equation for rotation of the equivalent continuum model. Useful solutions for simple distributions of applied torque can be written in terms of nondimensional structural parameters. The parameters are functions of the relative shear to flexural torque stiffnesses and the height of the core. Solutions are available for the rotation and its derivatives, which lead to expressions for the warping stresses in the walls, and for the forces in the connection beams. Design curve solutions of the characteristic equation provide a rapid means of estimating the rotations and actions in a uniform core, as well as allowing a more general appreciation of the factors that influence a core's torsional behavior. The closed solution of the warping problem demonstrates clearly the analogy between the torsional behavior of a core and the lateral load behavior of a wall–frame.

A torsion analysis of a core whose properties vary with height, or a core that interacts with other assemblies as part of a total building structure, requires a method that allows the core to be considered as an assembly of discrete elements. A finite element analysis using a standard structural analysis program is the most appropriate. The model may consist of plane stress membrane elements with an auxiliary system of beams. Alternatively, if only a frame analysis program is available, an analogous frame model can be used with negligible loss in accuracy.

A simple model for the approximate computer analysis of a core consists of two columns placed on one of the core's principal bending axes and located on opposite sides of the shear center. By the use of a transition mechanism, multisection cores with changing locations of shear center can be analyzed. The model is useful for representing simply a complex core that is part of a larger structure.

The most concise model for the computer analysis of a core is the single warping column model. This consists of a vertical stack of column elements, each of whose usual six degrees of freedom per node has been augmented by a seventh, warping, degree of freedom. The warping column model, however, cannot be analyzed by a standard frame analysis program unless a special subroutine, to introduce the additional warping degree of freedom, is incorporated. Also, the warping column model cannot be used to represent multisectional cores because there is no mechanism to represent the changes of section, as is available for the two-column model.

REFERENCES

13.1 Kollbrunner, C.F. and Basler, K. *Torsion in Structures, An Engineering Approach.* Springer-Verlag. New York, 1969.

13.2 Umanskii, A.A. "Torsion and Bending of Thin-Walled Aircraft Structures" (in Russian). *Oborongiz* 1939.

13.3 Timoshenko, S.P. "Theory of Bending, Torsion and Buckling of Thin-Walled Members of Open Cross-Section." *J. Franklin Inst.* **239,** March, April, May 1945, 201–219, 249–268, 343–361, respectively.

13.4 Vlasov, V.Z. *Thin-Walled Elastic Beams,* 2nd ed. (translated from the Russian by the Israel Program for Scientific Translations for the N.S.F. and Department of Commerce, U.S.A.). Office of Technical Services, Washington, D.C., 1961.

13.5 Stafford Smith, B. and Taranath, B.S. "Analysis of Tall Core-Supported Structures Subject to Torsion." *Proc. Inst. Civ. Engineers.* **53,** September 1972, 173–187.

13.6 Heidebrecht, A.C. and Stafford Smith, B. "Approximate Analysis of Open-Section Shear Walls Subject to Torsional Loading." *Proc. ASCE* ST12, December 1973, 2355–2373.

13.7 Stafford Smith, B. and Jesien, W. "Two-Column Model for Static Analysis of Mono-Symmetrical Thin-Walled Beams." *Structural Engineering Report No. 88-3.* Department of Civil Engineering and Applied Mechanics, McGill University, May 1988.

13.8 Weaver, W.J., Brandow, G.E., and Manning, T.A. "Tier Buildings with Shear Cores, Bracing and Setbacks." *Computers Structures* **1,** 1971, 57–83.

CHAPTER 14

Outrigger-Braced Structures

An outrigger-braced high-rise structure consists of a reinforced-concrete or braced-steel frame main core connected to the exterior columns by flexurally stiff horizontal cantilevers. The core may be located between the column lines with outriggers extending on both sides (Fig. 14.1a) or it may be located on one side of the building with cantilevers connecting to columns on the other side (Fig. 14.1b).

When horizontal loading acts on the building, the column-restrained outriggers resist the rotation of the core, causing the lateral deflections and moments in the core to be smaller than if the free-standing core alone resisted the loading (Fig. 14.2a, b, and c). The result is to increase the effective depth of the structure when it flexes as a vertical cantilever, by inducing tension in the windward columns and compression in the leeward columns.

In addition to those columns located at the ends of the outriggers, it is usual to also mobilize other peripheral columns to assist in restraining the outriggers. This is achieved by including a deep spandrel girder, or "belt," around the structure at the levels of the outriggers, as in Fig. 14.3; hence, the occasionally used name, "belt-braced structure."

To make the outriggers and belt girder adequately stiff in flexure and shear, they are made at least one, and often two, stories deep. Consequently, to minimize the obstruction they cause, they are usually located at plant levels.

A building can be stiffened effectively by a single level of outriggers at the top of the structure, in which case it is sometimes referred to as a "top-hat" structure. Each additional level of outriggers increases the lateral stiffness, but by a smaller amount than the previous additional level. Up to four outrigger levels may be used in very tall buildings.

While the outrigger system is very effective in increasing the structure's flexural stiffness, it does not increase its resistance to shear, which has to be carried mainly by the core.

In this chapter an approximate method of analysis [14.1, 14.2] is presented for uniform outrigger structures, that is, structures with a uniform core, uniform columns, and similar-sized outriggers at each level. Although practical structures differ significantly from the uniform structure, in having vertical members that reduce in size up the height, the method of analysis is useful in allowing a rough estimate of the deflections and forces for use in a preliminary design. This approximate method of analysis is also useful in providing an estimate of the optimum levels

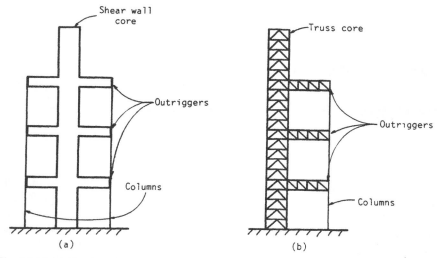

Fig. 14.1 (a) Outrigger structure with central core; (b) outrigger structure with offset core.

of the outriggers for minimizing the drift. The information provides guidance for the general arrangement of the structure.

14.1 METHOD OF ANALYSIS

Approximate methods of analysis for other types of high-rise structures, such as coupled walls and wall–frames, have been able to take advantage of the repetitive story-to-story arrangement in adopting a continuum approach. The outrigger-braced structure, with at most four outriggers, is not strictly amenable to a continuum analysis and has to be considered in its discrete arrangement. A compatibility method is chosen, in which the rotations of the core at the outrigger levels are matched with the rotations of the corresponding outriggers.

14.1.1 Assumptions for Analysis

The method of analysis is based on the following assumptions:

1. The structure is linearly elastic.
2. Only axial forces are induced in the columns.
3. The outriggers are rigidly attached to the core and the core is rigidly attached to the foundation.
4. The sectional properties of the core, columns, and outriggers are uniform throughout their height.

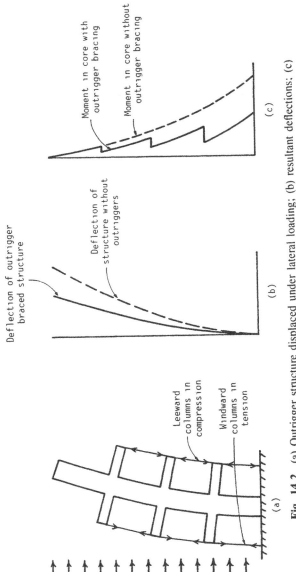

Fig. 14.2 (a) Outrigger structure displaced under lateral loading; (b) resultant deflections; (c) resultant core moments.

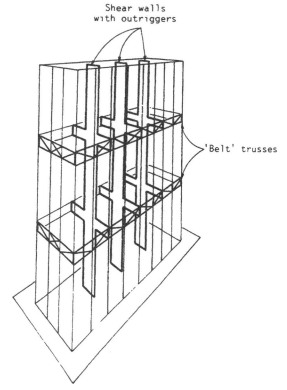

Fig. 14.3 Outrigger structure with belt girders.

Assumption 4 is less restrictive than at first it might appear. Although in a practical structure there is a reduction up the height in the inertia of the core and the sectional area of the columns, the factors of concern in a preliminary analysis (i.e., the drift at the top, the moment at the base of the core, and axial forces in the columns) are predominantly influenced by the properties of the structure in the lowest region. Consequently, an analysis of a uniform structure, with the "lowest region" properties of the actual structure, will give results of sufficient accuracy for a preliminary design.

14.1.2 Compatibility Analysis of a Two-Outrigger Structure

A two-outrigger structure will be used to demonstrate the method of analysis because it includes all the necessary steps in their simplest form. The analysis of structures with more or fewer than two outriggers can then easily be performed on the basis of the method for the two-outrigger case.

Starting from the statically determinate free-standing core, a one-outrigger structure is once redundant, a two-outrigger structure twice redundant, and so on.

The number of compatibility equations necessary for a solution corresponds to the degree of redundancy. The compatibility equations state, for each outrigger level, the equivalence of the rotation of the core to the rotation of the outrigger. The rotation of the core is expressed in terms of its bending deformation, and that of the outrigger in terms of the axial deformations of the columns and the bending of the outrigger.

The model for analysis is the two-outrigger structure shown in Fig. 14.4a, subjected to a uniformly distributed horizontal load. Two other loading cases have been studied, and some of their results will be summarized at the end of the chapter.

The bending moment diagram for the core (Fig. 14.4e) consists of the external-load moment diagram (Fig. 14.4b) reduced by the outrigger restraining moments that, for each outrigger, are introduced at the outrigger level and extend uniformly down to the base (Fig. 14.4c and d).

From the moment-area method, the core rotations at levels 1 and 2 (Fig. 14.4a) are, respectively,

$$\theta_1 = \frac{1}{EI} \int_{x_1}^{x_2} \left(\frac{wx^2}{2} - M_1 \right) dx + \frac{1}{EI} \int_{x_2}^{H} \left(\frac{wx^2}{2} - M_1 - M_2 \right) dx \quad (14.1)$$

and

$$\theta_2 = \frac{1}{EI} \int_{x_2}^{H} \left(\frac{wx^2}{2} - M_1 - M_2 \right) dx \quad (14.2)$$

in which EI and H are the flexural rigidity and total height of the core, w is the intensity of horizontal loading, x_1 and x_2 are the respective heights of outriggers 1 and 2 from the top of the core, and M_1 and M_2 are their respective restraining moments on the core.

Expressions for the rotations of the outriggers at the points where they are connected to the core (i.e., at the "inboard" ends) will now be developed. Each rotation consists of two components: one allowed by the differential axial deformations of the columns, and the other by the outriggers bending under the action of the column forces at their "outboard" ends.

The rotation of the "inboard" ends of the outrigger at level 1 is

$$\theta_1 = \frac{2M_1(H - x_1)}{d^2(EA)_c} + \frac{2M_2(H - x_2)}{d^2(EA)_c} + \frac{M_1 d}{12(EI)_0} \quad (14.3)$$

and for the outrigger at level 2

$$\theta_2 = \frac{2(M_1 + M_2)(H - x_2)}{d^2(EA)_c} + \frac{M_2 d}{12(EI)_0} \quad (14.4)$$

360 OUTRIGGER-BRACED STRUCTURES

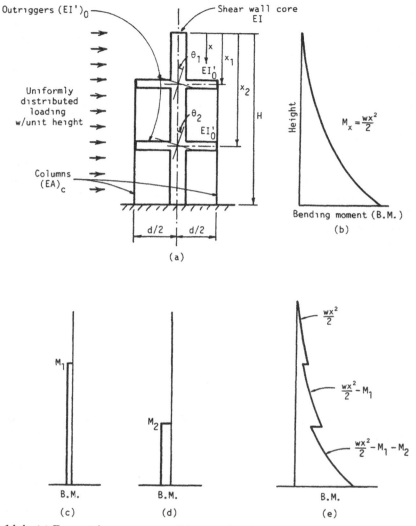

Fig. 14.4 (a) Two-outrigger structure; (b) external moment diagram; (c) M_1 diagram; (d) M_2 diagram; (e) core resultant moment diagram.

in which $(EA)_c$ is the axial rigidity of the column and $d/2$ is its horizontal distance from the centroid of the core. $(EI)_0$ is the effective flexural rigidity of the outrigger (Fig. 14.5b), allowing for the wide-column effect of the core, which can be obtained from the actual flexural rigidity of the outrigger $(EI')_0$ (Fig. 14.5a) as

$$(EI)_0 = \left(1 + \frac{a}{b}\right)^3 (EI')_0 \qquad (14.5)$$

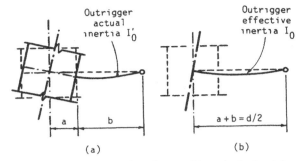

Fig. 14.5 (a) Outrigger attached to edge of core; (b) equivalent outrigger beam attached to centroid of core.

Equating the rotations of the core and outrigger at level 1, Eqs. (14.1) and (14.3), respectively

$$\frac{2M_1(H-x_1)}{d^2(EA)_c} + \frac{2M_2(H-x_2)}{d^2(EA)_c} + \frac{M_1 d}{12(EI)_0}$$
$$= \frac{1}{EI}\left[\int_{x_1}^{x_2}\left(\frac{wx^2}{2}-M_1\right)dx + \int_{x_2}^{H}\left(\frac{wx^2}{2}-M_1-M_2\right)dx\right] \quad (14.6)$$

and similarly for the rotations at level 2, equating Eqs. (14.2) and (14.4),

$$\frac{2(M_1+M_2)(H-x_2)}{d^2(EA)_c} + \frac{M_2 d}{12(EI)_0} = \frac{1}{EI}\int_{x_2}^{H}\left(\frac{wx^2}{2}-M_1-M_2\right)dx \quad (14.7)$$

Rewriting Eqs. (14.6) and (14.7) gives

$$M_1[S_1 + S(H-x_1)] + M_2 S(H-x_2) = \frac{w}{6EI}(H^3 - x_1^3) \quad (14.8)$$

and

$$M_1 S(H-x_2) + M_2[S_1 + S(H-x_2)] = \frac{w}{6EI}(H^3 - x_2^3) \quad (14.9)$$

in which S and S_1 are

$$S = \frac{1}{EI} + \frac{2}{d^2(EA)_c} \quad (14.10)$$

$$S_1 = \frac{d}{12(EI)_0} \quad (14.11)$$

362 OUTRIGGER-BRACED STRUCTURES

14.1.3 Analysis of Forces

The simultaneous solution of Eqs. (14.8) and (14.9) gives the restraining moment applied to the core by the outrigger at level 1

$$M_1 = \frac{w}{6EI} \left[\frac{S_1(H^3 - x_1^3) + S(H - x_2)(x_2^3 - x_1^3)}{S_1^2 + S_1 S(2H - x_1 - x_2) + S^2(H - x_2)(x_2 - x_1)} \right] \quad (14.12)$$

and, for the moment applied to the core by the outrigger at level 2.

$$M_2 = \frac{w}{6EI} \left[\frac{S_1(H^3 - x_2^3) + S[(H - x_1)(H^3 - x_2^3) - (H - x_2)(H^3 - x_1^3)]}{S_1^2 + S_1 S(2H - x_1 - x_2) + S^2(H - x_2)(x_2 - x_1)} \right]$$
$$(14.13)$$

Having solved the outrigger restraining moments M_1 and M_2, the resulting moment in the core, which is shown in Fig. 14.4e, can be expressed generally by

$$M_x = \frac{wx^2}{2} - M_1 - M_2 \quad (14.14)$$

in which M_1 is included only for $x > x_1$, and M_2 is included only for $x > x_2$.

The forces in the columns due to the outrigger action are $\pm M_1/d$ for $x_1 < x < x_2$ and $(M_1 + M_2)/d$ for $x \geq x_2$.

The maximum moment in the outriggers is then $M_1 \cdot b/d$ for level 1 and $M_2 \cdot b/d$ for level 2, where b is the net length of the outrigger (Fig. 14.4a).

14.1.4 Analysis of Horizontal Deflections

The horizontal deflections of the structure may be determined from the resulting bending moment diagram for the core by using the moment-area method. A general expression for deflections throughout the height could be written; it would, however, be very complicated. Concentrating, therefore, on the top drift only, this may be expressed as

$$\Delta_0 = \frac{wH^4}{8EI} - \frac{1}{2EI} [M_1(H^2 - x_1^2) + M_2(H^2 - x_2^2)] \quad (14.15)$$

in which the first term on the right-hand side represents the top drift of the core acting as a free vertical cantilever subjected to the full external loading, while the two parts of the second term represent the reductions in the top drift due to the outrigger restraining moments M_1 and M_2.

14.2 GENERALIZED SOLUTIONS OF FORCES AND DEFLECTIONS

So far, the method of analysis of only a two-outrigger structure has been presented. The same method of analysis applied to structures with more, or fewer, than two outriggers leads to expressions for restraining moments and top drift having the same form as Eqs. (14.14) and (14.15), respectively. By induction from these, generalized expressions may be developed, as follows.

14.2.1 Restraining Moments

The restraining moments for a uniform structure subjected to uniformly distributed loading may be expressed concisely in matrix form for simultaneous solution by computer:

$$\begin{bmatrix} M_1 \\ M_2 \\ \vdots \\ M_i \\ \vdots \\ M_n \end{bmatrix} = \frac{w}{6EI} \begin{bmatrix} S_1 + S(X - X_1) & S(H - X_2) & \cdots & S(H - X_i) & \cdots & S(H - X_n) \\ S(H - X_2) & S_1 + S(H - X_2) & \cdots & S(H - X_i) & \cdots & S(H - X_n) \\ \vdots & \vdots & \vdots & \vdots & \vdots & \vdots \\ S(H - X_i) & S(H - X_i) & \cdots & S_1 + S(H - X_i) & \cdots & S(H - X_n) \\ \vdots & \vdots & \vdots & \vdots & \vdots & \vdots \\ S(H - X_n) & S(H - X_n) & \cdots & S(H - X_n) & \cdots & S_1 + S(H - X_n) \end{bmatrix}^{-1} \begin{bmatrix} H^3 - X_1^3 \\ H^3 - X_2^3 \\ \vdots \\ H^3 - X_i^3 \\ \vdots \\ H^3 - X_n^3 \end{bmatrix} \quad (14.16)$$

in which n is the number of levels of outriggers. Equation (14.16) requires that the properties of the structure, the levels of the outriggers, and the magnitude of loading be specified.

A general expression for the moment in the core between outriggers j and $j + 1$ is then

$$M_x = \frac{wx^2}{2} - \sum_{i=1}^{j} M_i \qquad (14.17)$$

In the region between the top of the structure and the first outrigger from the top, the second term on the right-hand side of Eq. (14.17) will be zero.

14.2.2 Horizontal Deflections

Substitution of restraining moments M_1 to M_n, from the solution of Eq. (14.16) into Eq. (14.18) gives the resultant deflection at the top of the structure.

$$\Delta_0 = \frac{wH^4}{8EI} - \frac{1}{2EI} \sum_{i=1}^{n} M_i (H^2 - x_i^2) \qquad (14.18)$$

14.3 OPTIMUM LOCATIONS OF OUTRIGGERS

The preceding analysis is useful in not only providing estimates of the core moments and horizontal deflections, but also allowing an assessment of the optimum levels of the outriggers to minimize the horizontal top deflection. This is achieved by maximizing the drift reduction [i.e., the second term on the right-hand side of Eq. (14.18)].

Illustrating the procedure by continuing to consider the two-outrigger structure, the second term of its deflection equation [Eq. (14.15)] is maximized by differentiating with respect first to x_1, then to x_2, thus

$$(H^2 - x_1^2) \frac{dM_1}{dx_1} + (H^2 - x_2^2) \frac{dM_2}{dx_1} - 2x_1 M_1 = 0 \qquad (14.19)$$

and

$$(H^2 - x_1^2) \frac{dM_1}{dx_2} + (H^2 - x_2^2) \frac{dM_2}{dx_2} - 2x_2 M_2 = 0 \qquad (14.20)$$

in which dM_1/dx_1, dM_2/dx_1, dM_1/dx_2, and dM_2/dx_2 are the derivatives of M_1 and M_2 from Eqs. (14.12) and (14.13), respectively, with respect to x_1 and x_2. Substituting for M_1 and M_2 and their derivatives, Eqs. (14.19) and (14.20) can be solved simultaneously for the values of x_1 and x_2 that define the optimum levels of the outriggers. The solution of Eqs. (14.19) and (14.20) is, obviously, complex; therefore, a numerical method of solution by computer is necessary.

In the complete expressions for Eqs. (14.19) and (14.20), the structural properties were expressed initially by parameters S and S_1, as defined by Eqs. (14.10) and (14.11). Equations (14.19) and (14.20) can be rewritten in terms of more meaningful nondimensional parameters α and β, which represent the core-to-column and core-to-outrigger rigidities, respectively, as follow

$$\alpha = \frac{EI}{(EA)_c (d^2/2)} \quad (14.21)$$

and

$$\beta = \frac{EI}{(EI)_0} \frac{d}{H} \quad (14.22)$$

It is then possible to simplify Eqs. (14.19) and (14.20) further by combining α and β into a single parameter ω, as defined by

$$\omega = \frac{\beta}{12(1 + \alpha)} \quad (14.23)$$

The parameter ω, which is nondimensional, is the characteristic structural parameter for a uniform structure with flexible outriggers. It is useful in that it allows various aspects of the behavior of outrigger structures to be expressed in a very concise form.

It may be deduced from Eqs. (14.21) to (14.23) that, with all other properties remaining constant, there is a reduction in ω as the outriggers' flexural stiffnesses are increased, and that ω increases as the axial stiffnesses of the columns increase.

Equations (14.19) and (14.20) may be solved to find the optimum outrigger levels, 1 and 2, for a range of values of ω, to give the results plotted graphically in Fig. 14.6b. Thus, for any uniform two-outrigger structure having specified member properties, ω may be calculated and Fig. 14.6b used to determine the optimum locations for the outriggers in order to achieve minimum drift.

The described procedure leading to the results plotted in Fig. 14.6b for a two-outrigger structure has been repeated for one-, three-, and four-outrigger structures. The graphs showing the optimum outrigger locations are given in Fig. 14.6a, c, and d, respectively.

14.4 PERFORMANCE OF OUTRIGGER STRUCTURES

Although the method of analysis for uniform structures presented earlier is useful in estimating the forces and drift for preliminary design, it is of further value in providing general information about the most efficient arrangement of the structure. This guidance relates particularly to the appropriate number and location of the outriggers, as follows.

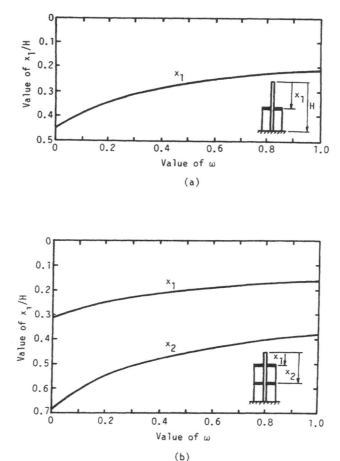

Fig. 14.6 (a) Optimum outrigger location for one-outrigger structure; (b) optimum outrigger locations for two-outrigger structure.

14.4.1 Optimum Locations of Outriggers

Considering, for purposes of comparison, hypothetical structures in which the outriggers are flexurally rigid (i.e., with ω equal to zero) the curves in Fig. 14.6a–d may be interpreted to yield simple approximate guidelines for the location of the outriggers to minimize the deflection: the outrigger in a one-outrigger structure should be at approximately half-height; the outriggers in a two-outrigger structure should be at approximately one-third and two-thirds heights; in a three outrigger structure they should be at approximately the one-quarter, one-half, and three-quarters heights, and so on. Generally, therefore, for the optimum performance of an n-outrigger structure, the outriggers should be placed at the $1/(n+1)$, $2/(n+1)$, up to the $n/(n+1)$ height locations.

14.4 PERFORMANCE OF OUTRIGGER STRUCTURES   367

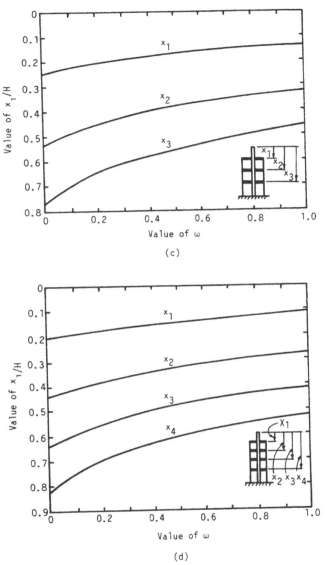

Fig.14.6 (*Continued*)   (c) Optimum outrigger locations for three-outrigger structure; (d) optimum outrigger locations for four-outrigger structures.

In any outrigger system it is, perhaps unexpectedly, structurally inefficient to locate an outrigger at the top, and it should only be done when other reasons (e.g., a plant floor located at the top) prevail.

Indeed, it has been shown [14.1] that a structure with n optimally located outriggers is almost as effective in its lateral resistance as the same structure with an additional outrigger at the top. In a uniform structure, the lowest outrigger always induces the maximum restraining moment, with the outriggers above carrying successively less. In an optimally arranged structure, the moment carried by an outrigger ranges from one-half to two-thirds that carried by the outrigger below. However, in an optimally arranged structure, but with an additional outrigger at the top, the top outrigger carries approximately one-sixth of the moment of the outrigger below. This illustrates clearly the inefficiency of including a top outrigger.

14.4.2 Effects of Outrigger Flexibility

In real structures, the flexibility of the outriggers makes it necessary to modify the above guidelines. The curves in Fig. 14.6a–d show that the larger the value of ω (i.e., the more flexible the outriggers) the further up the structure the set of outriggers must be located in order to minimize the drift. The relative intervals between the outriggers and the top, however, remain approximately the same.

14.4.3 "Efficiency" of Outrigger Structures

A useful measure of the effectiveness of an outrigger system in reducing the free-standing core's lateral deflection and base moment is to express the resulting reductions as percentages of the corresponding reductions that would occur if the core and columns behaved fully compositely. Fully composite action implies that, in overall flexure of the structure, the stresses in the vertical components are proportional to their distances from their common centroidal axis, with the structure having an overall flexural rigidity equal to

$$(EI)_t = \frac{(EA)_c d^2}{2} + EI \qquad (14.24)$$

It has been shown [14.2] that the percentage efficiencies for both drift and base moment reductions can be expressed in terms of ω. These are plotted for one- to four-outrigger structures in Fig. 14.7a and b, respectively.

Considering the one-outrigger structure with a flexurally rigid outrigger, the maximum efficiency in drift reduction is 87.5%, and the corresponding efficiency in core base moment reduction is 58.3%. For two-, three-, and four-outrigger structures, the respective efficiencies are 95.5%, 70.3%; 97.8%, 77.1%; and 98.5%, 81.3%. Evidently, for structures with very stiff outriggers (i.e., with low values of ω) there is little to be gained in drift control by exceeding four outriggers, hence the reason for the graphs being plotted for up to only four outriggers.

The optimum outrigger arrangement for the maximum reduction in drift does not simultaneously cause the maximum reduction in the core base moment. For this, the outriggers would have to be lowered toward the base.

In cases where drift control is not critical, the design emphasis could change to controlling the core moments by lowering the levels of the outriggers until the drift limitations are just satisfied. Such a refined consideration of the structure is rarely

14.4 PERFORMANCE OF OUTRIGGER STRUCTURES

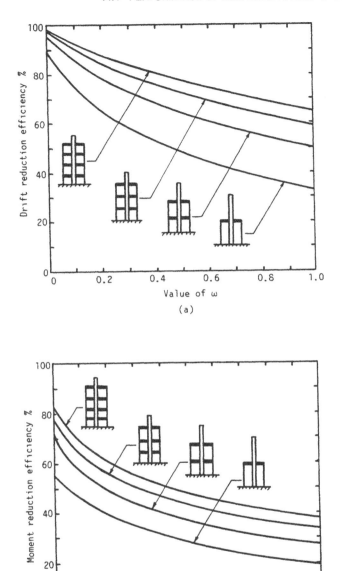

Fig. 14.7 (a) Efficiency in drift reduction; (b) efficiency in moment reduction.

possible in practice. However, the designer's general appreciation of how various factors influence the performance of an outrigger structure should lead to his making a better choice when alternative outrigger arrangements are possible.

14.4.4 Alternative Loading Conditions

Analysis of outrigger braced structures subjected to a triangular load distribution, with its apex at the base, and a concentrated top load have also been made. These are useful for considering graduated wind loading and static equivalent earthquake loading, [14.3].

It was deduced that the "minimum top drift" optimum levels of the outriggers for the triangular loading were only slightly higher than those deduced for uniformly distributed loading. The optimum levels for the concentrated top load were slightly higher than those for the triangularly distributed loading.

Considering that wind loading distributions are usually in the form of a trapezium, consisting of superposed uniform and triangular loading distributions, and the static equivalent earthquake loading is usually given as a triangular distribution with a relatively small concentrated load at the top, it may be concluded from the above observations that the "minimum top drift" optimum levels may reasonably be taken as approximately those obtained for uniformly distributed loading.

SUMMARY

A compatibility method of analysis for the forces and deflections in uniform outrigger-braced structures subjected to uniformly distributed lateral loading is presented. Taking as an example a two-outrigger structure, the rotation of the core at each outrigger level (derived from the distribution of moments in the core) is matched with the rotation of the "inboard" end of the corresponding outrigger (derived from the axial deformation of the columns and the flexure of the outrigger). The resulting compatibility equations are solved to give the outrigger restraining moments. From these, the core moment diagram, the outrigger moments, and the column axial forces may be determined. The horizontal deflections of the structure are then obtained from the core moment diagram by, for example, the moment-area method.

The method of analysis presented for a two-outrigger structure can be extended to structures with other numbers of outriggers, leading by induction to generalized equations for the restraining moments and horizontal displacements.

Optimum levels of the outriggers for minimum top drift are obtained by maximizing the drift reduction caused by the outriggers. These are plotted as a function of a nondimensional relative stiffness parameter ω, for structures with up to four outriggers and with variable stiffnesses of the core and outrigger system. The efficiencies of uniform-with-height outrigger structures are assessed with regard to drift and core base moment reductions. These are plotted as functions of ω for structures with one to four outriggers, and for outrigger systems of varying stiffness properties.

The optimum levels of outriggers for the minimum drift of structures subjected to a triangularly distributed load, with its apex at the base, are slightly higher than for structures subjected to a uniformly distributed load, while the optimum levels in structures subjected to a concentrated top load are slightly higher again.

REFERENCES

14.1 Stafford Smith, B. and Nwaka, I. O. "Behavior of Multioutrigger Braced Tall Buildings." *ACI Special Publication SP-63*, 1980, 515-541.

14.2 Stafford Smith, B. and Salim, I. "Parameter Study of Outrigger-Braced Tall Building Structures." *Proc. ASCE* **107,** ST10, October 1981, 2001-2014.

14.3 Coull, A. and Lau, W. H. O. "Outrigger-Braced Structures Subjected to Equivalent Static Seismic Loading." *Proc. 4th Int. Conf. on Tall Buildings*, Hong Kong and Shanghai, Y. K. Cheung and P. K. K. Lee (Eds.), Vol. 1, Hong Kong, 1988, pp. 395-401.

CHAPTER 15

Generalized Theory

Reviewing previous chapters on the various forms of high-rise structure, it can be stated in general that high-rise buildings resist horizontal loading by stiff-inplane bents such as shear walls, coupled walls, rigid frames, and vertical trusses. As the structure deflects horizontally, the column or wall "chord" members of each bent interact with the transversely connecting girder, slab, or brace "web" members of the bent. The chord and web components resist the external moment and shear, respectively.

It has been shown elsewhere [15.1, 15.2] that the various types of lateral load resisting bents described above may be considered to be members of a family of cantilever structures that, when subjected to transverse loading, deflect with different amounts of flexural and shear mode responses.

The flexural response consists of a combination of overall bending of the total assembly, due to axial deformations of the vertical members, with single-curvature bending of the vertical members. The shear response is a frame-racking action due to deformation of the web members. In braced frames the racking results from axial deformation of the braces and beams, in coupled walls it is due to double-curvature bending of the connecting beams, and in rigid frames it results from double-curvature bending of the columns and beams.

Consider three extreme examples of shear-flexure behavior: a tall braced frame adopts a primarily flexural shape (Fig. 15.1a and b), due mainly to axial deformation of the columns, with a relatively small shear component from racking; at another extreme, a medium-rise rigid frame deflects in a mainly shear mode due to double-bending of the members (Fig. 15.2a and b); in a third extreme case, of a pair of shear walls with a system of very flexible coupling beams, the walls deflect independently in a flexural mode, with virtually no composite action (Fig. 15.3a and b). In practice, the typical coupled wall usually adopts an intermediate combined shear-flexure mode with a flexural configuration in the lower region and a shear configuration in the upper (Fig. 15.4a and b). Theoretically, however, depending on the relative stiffnesses of the walls and beams, the coupled wall is capable of behaving in one of the three extreme ways described above, and in any number of intermediate positions with different proportions of flexural and shear mode behavior. For this reason the coupled wall may be considered as the generic high-rise bent, and coupled-wall theory as a generalized theory for shear-flexural cantilevers.

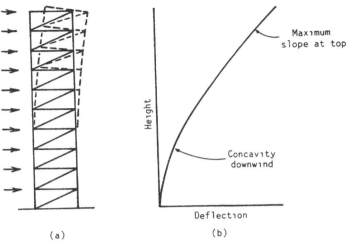

Fig. 15.1 (a) Laterally loaded tall braced frame; (b) flexural mode of deflection.

It is shown in this chapter how coupled-wall theory may be used to estimate the deflection behavior of individual shear-flexure bents of different types, and how the theory may be extended to represent approximately the behavior of multibent structures that combine different types of bent.

15.1 COUPLED WALL THEORY

Referring to Fig. 15.5a, it can be shown that the deflection profile of a coupled shear wall subjected to a uniformly distributed horizontal load is governed by the

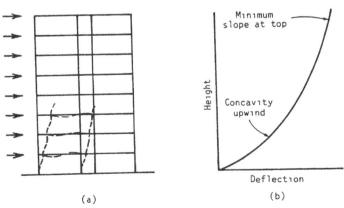

Fig. 15.2 (a) Laterally loaded rigid frame; (b) shear mode of deflection.

374 GENERALIZED THEORY

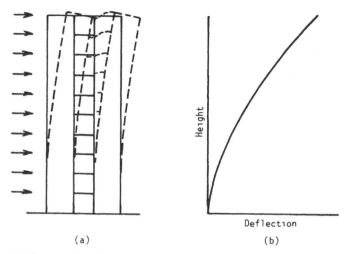

Fig. 15.3 (a) Shear walls with very low stiffness coupling beams; (b) flexural mode of deflection.

differential equation

$$\frac{d^4y}{dx^4} - (k\alpha)^2 \frac{d^2y}{dx^2} = \frac{w}{EI}\left[1 - (k\alpha)^2 \frac{(k^2-1)}{k^2}\frac{x^2}{2}\right] \quad (15.1)$$

where

$$\alpha^2 = \frac{12I_c l^2}{b^3 hI} \quad (15.2)$$

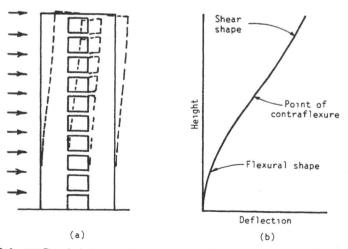

Fig. 15.4 (a) Coupled shear wall structure; (b) flexural-shear mode of deflection.

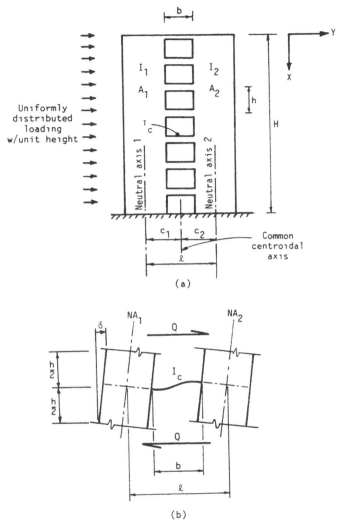

Fig. 15.5 (a) Coupled-wall bent; (b) story-height segment of coupled wall.

and

$$k^2 = 1 + \frac{AI}{A_1 A_2 l^2} \tag{15.3}$$

in which

$$I = I_1 + I_2 \tag{15.4}$$

and

$$A = A_1 + A_2 \tag{15.5}$$

Considering in Fig. 15.5b a story-height segment of the coupled wall, and representing the racking rigidity of the coupled wall by the symbol (GA), it can be shown that

$$(GA) = \frac{Qh}{\delta} = \frac{12EI_c l^2}{b^3 h} \qquad (15.6)$$

Substituting from Eq. (15.6) into Eq. (15.2) gives

$$\alpha^2 = \frac{(GA)}{EI} \qquad (15.7)$$

The parameter α indicates the ratio of the shear rigidity of the bent to the sum of the flexural rigidities of the walls.

Assuming boundary conditions of zero displacement and inclination at the base, and zero external moment and shear at the top, the solution to Eq. (15.1) leads to an expression for the horizontal displacement at a distance x from the top of the structure:

$$y(x) = \frac{wH^4}{EI} \left\{ \left[\frac{1}{8} - \frac{1}{6}\left(\frac{x}{H}\right) + \frac{1}{24}\left(\frac{x}{H}\right)^4 \right] \frac{(k^2-1)}{k^2} + \frac{1}{k^2} \left[\frac{1-(x/H)^2}{2(k\alpha H)^2} \right. \right.$$
$$\left. \left. + \frac{\cosh k\alpha(H-x) - 1 - k\alpha H(\sinh k\alpha H - \sinh k\alpha x)}{(k\alpha H)^4 \cosh k\alpha H} \right] \right\} \qquad (15.8)$$

A physical interpretation of this equation is made easier if, instead of writing it in terms of I, the sum of the individual wall inertias, it is written in terms of the gross moment of inertia I_g of the bent, as though the structure behaved fully compositely, where

$$I_g = I + \sum A_i c_i^2 \qquad (15.9)$$

in which A_i is the sectional area of wall i, and c_i is the distance of the centroid of wall i from the common centroid of the wall sectional areas. Now

$$c_1 = \frac{A_2 l}{A} \quad \text{and} \quad c_2 = \frac{A_1 l}{A} \qquad (15.10)$$

Substituting these into Eq. (15.9) gives

$$I_g = I + \frac{A_1 A_2^2 l^2}{A^2} + \frac{A_1^2 A_2 l^2}{A^2} \qquad (15.11)$$

$$= I + \frac{A_1 A_2 l^2}{A} \qquad (15.12)$$

which, after substituting from Eq. (15.3), gives

$$I = I_g \frac{(k^2 - 1)}{k^2} \qquad (15.13)$$

Substituting this into Eq. (15.8) gives the deflection equation in terms of the gross moment of inertia I_g, thus

$$y(x) = \frac{wH^4}{EI_g} \left\{ \left[\frac{1}{8} - \frac{1}{6}\left(\frac{x}{H}\right) + \frac{1}{24}\left(\frac{x}{H}\right)^4 \right] + \frac{1}{(k^2 - 1)} \left[\frac{1 - (x/H)^2}{2(k\alpha H)^2} \right. \right.$$

$$\left. \left. + \frac{\cosh k\alpha(H - x) - 1 - k\alpha H (\sinh k\alpha H - \sinh k\alpha x)}{(k\alpha H)^4 \cosh k\alpha H} \right] \right\}$$

$$(15.14)$$

By rearranging Eq. (15.13) and substituting from Eq. (15.9) it can be shown that

$$k^2 = 1 + \frac{EI}{\sum EA_i c_i^2} \qquad (15.15)$$

demonstrating that k depends on the ratio of the sum of the walls' individual flexural rigidities to the overall flexural rigidity of the walls associated with their axial deformation.

Equation (15.14), which describes the deflection of the coupled wall, is written in terms of three independent, nondimensional parameters, αH, k, and x/H. The first of these parameters expresses the relative racking shear and individual-flexural rigidities of the structure, the second expresses the relative overall- and individual-flexural rigidities, and the third represents the level of the deflection. In Eq. (15.14), the function in the braces, which is written in terms of the parameters, controls the shear-flexure deflected shape of the structure, while the term preceding the braces, wH^4/EI_g, which refers to the size of structure and value of the loading, controls its magnitude.

15.2 PHYSICAL INTERPRETATION OF THE DEFLECTION EQUATION

An inspection of Eq. (15.14) shows it to consist of parts that, when considered separately, reduce its apparent complexity. The first part

$$y(x) = \frac{wH^4}{EI_g}\left[\frac{1}{8} - \frac{1}{6}\left(\frac{x}{H}\right) + \frac{1}{24}\left(\frac{x}{H}\right)^4 \right] \qquad (15.16)$$

is the deflection equation for a flexural cantilever of rigidity EI_g subjected to a uniformly distributed load of intensity w. The second part of Eq. (15.14)

$$y(x) = \frac{wH^4}{EI_g} \frac{1}{k^2 - 1} \frac{1 - (x/H)^2}{2(k\alpha H)^2} \qquad (15.17)$$

is less easily identifiable. By substituting for I_g and α^2 from Eqs. (15.13) and (15.7), however, it can be shown that

$$y(x) = \frac{1}{k^4} \frac{wH^2}{2(GA)} \left[1 - \left(\frac{x}{H}\right)^2 \right] \qquad (15.18)$$

For structures with axially very stiff columns, for which I_g, the gross inertia, is large relative to I, the sum of the individual inertias, the value of k may be seen from Eq. (15.15) to be approximately unity. Then Eq. (15.18) becomes

$$y(x) = \frac{wH^2}{2(GA)} \left[1 - \left(\frac{x}{H}\right)^2 \right] \qquad (15.19)$$

which is the deflection equation of a shear cantilever of rigidity (GA) subjected to a uniformly distributed load w.

The remaining part of Eq. (15.14)

$$y(x) = \frac{wH^4}{(k^2 - 1)EI_g}$$
$$\cdot \left[\frac{\cosh k\alpha(H - x) - 1 - k\alpha H(\sinh k\alpha H - \sinh k\alpha x)}{(k\alpha H)^4 \cosh k\alpha H} \right]$$

$$(15.20)$$

cannot be explained so simply, but it expresses the effect of the walls' axial deformations on the shear interaction between the walls.

15.3 APPLICATION TO OTHER TYPES OF BENT

The above theory, which was developed with reference to the deflections of a coupled wall bent, represents in essence the behavior of a cantilever that deforms by variable amounts of overall flexure, single-curvature flexure of the vertical elements, and racking shear. A consideration of other types of bent, for example, braced frames, rigid frames, and wall–frames, shows that these too can be perceived in similar terms. Therefore, it is evident that the theory developed for coupled walls should apply also to other types of bent. Studies have shown that the coupled wall theory can indeed be applied in a generalized way to other types of shear-flexure cantilever bents [15.1, 15.2]. This is achieved by determining for the particular bent the values of the rigidities EI, ΣEAc^2, and (GA), using procedures appropriate to the type of bent, in order to obtain the values αH and k for substitution in the deflection equation [Eq. (15.14)].

15.3.1 Determination of Rigidity Parameters

The determination of the rigidities is made for different types of bent as follows:

Sum of the Individual Flexural Rigidities EI. In most of the types of bent this is obtained by summing the individual flexural rigidities of the columns, shear walls, and cores. That is,

$$EI = EI \text{(columns)} + EI \text{(walls)} + EI \text{(cores)} \quad (15.21)$$

Braced frames are an exception to the way that EI is evaluated in that, if the story-height column segments are not continuous at floor levels, their moments of inertia are effectively zero, causing k and αH to have unit and infinite values, respectively. Because these values cannot be used satisfactorily in Eq. (15.20), it is recommended for the columns that a nominal nonzero value of I be used equal to, say, $0.001 \Sigma Ac^2$. This will yield from Eq. (15.15) a value for k^2 of 1.001 and from Eq. (15.7) a fictitious but conforming large value of αH. Using these two values in Eq. (15.14) will give practically the same results as those for the limiting case in which I is zero.

Sum of the Axial Flexural Rigidities, $\Sigma (Ac^2)$. This is obtained for a coupled wall, rigid frame, and braced frame by finding first the centroid of the sectional areas of the columns and walls of the bent, and then taking the sum of their second moments about the common centroid, that is

$$\sum_{i}^{n} (Ac^2)_i = A_1 c_1^2 + A_2 c_2^2 + \cdots + A_i c_i^2 + \cdots + A_n c_n^2 \quad (15.22)$$

in which A_i is the sectional area of column or wall i, and c_i is the distance of the centroid of column or wall i from the common centroid of all the columns and walls.

Racking Shear Rigidity (GA). The racking shear rigidity is the property whose method of determination is the most particular to the type of bent. It is dependent on the deformations of the web members as the structure racks under the action of shear. The methods of determining (GA) are summarized below for various types of bent.

Coupled Walls (Fig. 15.5a). The racking rigidity is dependent on the double-curvature bending of the connecting beam, and is given by Eq. (15.6)

$$(GA) = \frac{12EI_c l^2}{b^3 h}$$

Rigid Frames (Fig. 15.2a). The racking rigidity of rigid frames is related to the double-curvature bending of the columns and girders and is given by

$$(GA) = \frac{12E}{h\left(\dfrac{1}{C} + \dfrac{1}{G}\right)} \qquad (15.23)$$

in which $C = \Sigma^n I_c/h$ for the n columns in a story of the bent, and $G = \Sigma^{n-1} I_g/L$ for the $n - 1$ spans in a floor of the bent.

Braced Frames. The racking rigidity of a braced frame is usually dependent on the axial deformations of the braces and girders. In certain types of frame, however, the racking rigidity also involves bending of the girder. The shear rigidities of some of the more common types of bracing can be derived from the deflection formulas for braced frames, given in Table 6.1, in conjunction with Eq. (15.6). Referring to the figures in Table 6.1:

1. Single-diagonal bracing

$$(GA) = Eh \bigg/ \left(\frac{d^3}{L^2 A_d} + \frac{L}{A_g}\right) \qquad (15.24)$$

2. Double-diagonal bracing

$$(GA) = 2Eh \bigg/ \left(\frac{d^3}{L^2 A_d}\right) \qquad (15.25)$$

3. K-bracing

$$(GA) = Eh \bigg/ \left(\frac{2d^3}{L^2 A_d} + \frac{L}{4A_g}\right) \qquad (15.26)$$

4. Story-height knee bracing

$$(GA) = Eh \bigg/ \left[\frac{d^3}{2m^2 A_d} + \frac{m}{2A_g} + \frac{h^2(L - 2m)^2}{12 I_g L}\right] \qquad (15.27)$$

5. Offset diagonal bracing

$$(GA) = Eh \bigg/ \left[\frac{d^3}{(L-2m)^2 A_d} + \frac{(L-2m)}{A_g} + \frac{h^2 m^2}{3 I_g L} \right] \quad (15.28)$$

When the EI, ΣAc^2, and (GA) properties have been determined for a particular bent, they are used to determine the characteristic nondimensional parameters of the bent, thus

$$\alpha H = H \sqrt{\frac{(GA)}{EI}} \quad (15.29)$$

and

$$k = \sqrt{1 + \frac{I}{\Sigma Ac^2}} \quad (15.30)$$

15.3.2 Calculation of Deflection

Having determined αH and k for a bent, the deflection at a level x from the top is obtained by substituting the values of αH, k, and αx in Eq. (15.14).

It is possible also to determine the story drift δ_i at a level i in the structure by using

$$\delta_i = \left(h \frac{dy}{dx} \right)_i \quad (15.31)$$

in which h_i is the height of story i and $(dy/dx)_i$ is determined from the first derivative of Eq. (15.14).

$$\frac{dy}{dx}(x) = \frac{wH^3}{EI_g} \left\{ -\frac{1}{6} + \frac{1}{6} \left(\frac{x}{H} \right)^3 + \frac{1}{k^2 - 1} \left[\frac{-x/H}{(k\alpha H)^2} \right. \right.$$

$$\left. \left. + \frac{-\sinh k\alpha(H-x) + k\alpha H \cosh k\alpha x}{(k\alpha H)^3 \cosh k\alpha H} \right] \right\} \quad (15.32)$$

Certain actions, such as wall moments and shears and hence frame moments and shears, can be determined also, by using the second and third derivatives of Eq. (15.14) [15.3].

15.4 APPLICATION TO MIXED-BENT STRUCTURES

When a number of parallel bents are combined in a building structure that is symmetric on plan about the axis of loading, the bents are constrained by the inplane rigidity of the slabs to deflect identically. If the bents are of different types, it may be assumed that each individual bent, of whatever type, acts as a shear-flexure cantilever with the characteristics of a coupled wall. The total structure may be considered, therefore, as a set of mutually constrained coupled walls. A rigorous continuum analysis of such a combination of different coupled walls is too complex to allow a practical solution and, therefore, an approximate approach to a solution is necessary. The proposed method has been used for the analysis of a wide variety of mixed-bent structures, and comparisons with finite element analyses have shown it to be reasonably accurate.

The continuum approximate method involves using the stiffness parameters of the individual bents to determine the overall parameters αH and k for the total structure, and then applying the deflection solution [Eq. (15.14)] for a single coupled-wall bent to obtain the deflections of the total structure. It is implicit in the process of calculating the overall αH and k values that a structure, such as that shown in Fig. 15.6a and b, is modeled by first reducing the bents to a set of equivalent single-bay coupled walls having an arbitrarily chosen standard distance between the walls' centroidal axes (Fig. 15.6c). The set of coupled walls is then considered to be lumped into a single coupled-wall bent (Fig. 15.6d) to which the single coupled-wall theory, as presented earlier in the chapter, may be applied.

The procedure for evaluating the overall parameters involves finding first the parameters EI, (GA), and ΣEAc^2 for each individual bent. The overall structure parameters (suffix T) are then determined from the individual bent parameters (suffix B) in the following ways:

$$\alpha_T H = H \sqrt{\frac{(GA)_T}{(EI)_T}} \tag{15.33}$$

in which

$$(EI)_T = \Sigma (EI)_B \tag{15.34}$$

and

$$(GA)_T = \Sigma (GA)_B \tag{15.35}$$

Shear walls that are not connected in their plane by beams to other vertical components may be regarded as coupled walls with zero coupling stiffness. Consequently, their values of EI are included in $(EI)_T$ [Eq. (15.34)], but their contribution to $(GA)_T$ [Eq. (15.35)] is zero.

The second principal parameter k_T is given by

$$k_T = \sqrt{1 + \frac{\Sigma (EI)_B}{\Sigma \left| \Sigma (EAc^2) \right|_B}} \tag{15.36}$$

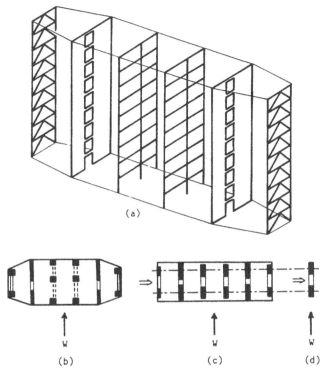

Fig. 15.6 (a) Mixed-bent symmetrical structure; (b) plan of mixed-bent structure; (c) bents of structure reduced to equivalent coupled walls; (d) set of equivalent coupled walls consolidated into a single equivalent coupled-wall bent.

In Eq. (15.36), noncoupled shear walls contribute their values of EI to the numerator, but nothing to the denominator. The gross flexural rigidity is given by

$$(EI_g)_T = (EI)_T + \Sigma \left[\Sigma (EAc^2) \right]_B \quad (15.37)$$

Having evaluated $\alpha_T H$, k_T, and $(EI_g)_T$, they may be substituted in the deflection equation [Eq. (15.14)] to obtain the profile of the structure, and in Eqs. (15.31) and (15.32) to obtain the story drift.

15.5 ACCURACY OF THE METHOD

The theory is based on the assumption that the considered structure is uniform throughout its height, and that its discrete story-by-story arrangement may be represented by a continuum. It has been found [15.1] for the continuum representation that, for structures of 10 stories or more, the error in deflections due to the assumption of a continuum is less than 5%, which is acceptable for most practical purposes. For structures whose properties reduce significantly up the height, a uniform continuum solution would not be appropriate. If the reduction of properties over the height is not severe, a uniform continuum solution may be used, subject to the following modifications. For shear wall or braced frame structures with reductions in their gross EI value of $1:\frac{1}{3}$ or $1:\frac{1}{4}$ from the base to the top, calculate the top deflection for the uniform structure with the base properties of the real structure, and increase the resulting deflection by 10 or 15%, respectively. For a typical wall–frame structure with values of αH in the range of 1 to 4 and with a reduction in wall and column inertias of $1:\frac{1}{3}$ or $1:\frac{1}{4}$, increase the top deflection calculated for the corresponding base-property uniform structure by 20 or 25%, respectively. For a rigid frame structure with a reduction in values of (GA) of $1:\frac{1}{3}$ or $1:\frac{1}{4}$, increase the top deflection calculated for the corresponding base-property uniform structure by 40 or 50%, respectively.

In addition to inaccuracies caused by the use of continuum theory to represent a discrete member structure, other sources of inaccuracy are introduced by using a coupled wall and its theory to represent different types of bent, and by lumping together various types of bent in a mixed-bent structure. Generally, the more closely a bent resembles a coupled wall, the more accurate the method. Coupled walls, single-bay rigid frames, and braced frames are well represented by the theory. Multibay rigid frames are less so, especially if the bays are of unequal width. Generally, within a mixed-bent structure, the greater the difference between the free deflection characteristics of the various bents, the less well is the structure represented by the lumped model. In a large number of test analyses of uniform structures with a variety of practically proportioned bents having a wide range of values of αH and k, the deflection results showed maximum discrepancies of 10% from those obtained by finite element analyses and, in the majority of cases, significantly less than 10%.

For tall wall–frame structures in particular, this generalized method, which takes the axial deformation of the frame columns into account, is significantly more accurate than the basic wall–frame method given in Chapter 11, which neglects column axial deformations. This point is illustrated, together with the general use of the method, in the following numerical example.

15.6 NUMERICAL EXAMPLE

A floor plan of a 30-story, 375-ft (114.3-m)-high structure is shown in Fig. 15.7. A uniformly distributed load of 30 lb/ft^2 (1.44 kN/m^2) and an elastic modulus of $E = 4.0 \times 10^6$ lb/in.^{2} (2.76×10^{10} N/m^2) are assumed. The five-bent symmetrical plan includes two types of rigid frame (Bents I and III), and a pair of assemblies (Bents II) in each of which a central shear wall is connected by beams to exterior columns. The methods of calculating the parameters for the bents are as referred to in Sections 15.3.1 and 11.2.5. The values of the bent parameters

15.6 NUMERICAL EXAMPLE

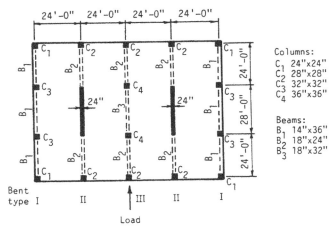

Fig. 15.7 Plan of example mixed-bent structure.

and the overall structure parameters are given in Table 15.1. Substituting these parameters in Eq. (15.14) gave the deflection profile.

The structure can be categorized as a wall–frame. The approximate continuum method developed specifically for a wall–frame in Chapter 11 does not account for axial deformations of the columns whereas the present generalized method does. Consequently, two cases of the example structure were analyzed: Case 1 in which axial deformations of the columns were allowed, and Case 2 in which axial deformations were prevented by assigning arbitrarily high sectional areas to the columns.

TABLE 15.1 Values of Structural Parameters

	EI	(GA)	EAc^2
Bent I	$2.301 \times 10^5 \cdot E$ lb in.^{2}	$31.951 \cdot E$ lb	$2.973 \times 10^8 \cdot E$ lb in.^{2}
Bent II	$7.597 \times 10^7 \cdot E$ lb in.^{2}	$27.458 \cdot E$ lb	$3.260 \times 10^8 \cdot E$ lb in.^{2}
Bent III	$3.824 \times 10^5 \cdot E$ lb in.^{2}	$20.849 \cdot E$ lb	$3.992 \times 10^8 \cdot E$ lb in.^{2}
Total (2-Bents I and II plus 1-Bent III)	$1.528 \times 10^8 \cdot E$ lb in.^{2}	$139.667 \cdot E$ lb	$16.458 \times 10^8 \cdot E$ lb in.^{2}

$$\alpha H = H \sqrt{\frac{(GA)_T}{(EI)_T}} = 375 \times 12 \sqrt{\frac{139.667 \cdot E}{1.528 \times 10^8 \cdot E}} = 4.302$$

$$k = \sqrt{1 + \frac{(EI)_T}{\sum (EAc^2)_T}} = \sqrt{1 + \frac{1.528 \times 10^8 \cdot E}{16.458 \times 10^8 \cdot E}} = 1.045$$

386 GENERALIZED THEORY

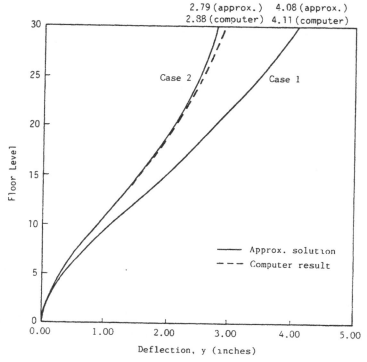

Fig. 15.8 Deflections of example structure.

Finite element analyses for the structure, considered as an assembly of discrete members, were also run for the two cases. The results of the four analyses are shown in Fig. 15.8. The results of the approximate method are shown to compare reasonably closely with those of the finite element analyses, and to account well for the axial deformation of the columns.

SUMMARY

A generalized method is presented for estimating the deflections of structures that are uniform with height and that do not twist.

The method is based on the assumption that a tall lateral load resisting bent, whether a rigid frame, a braced frame, or a wall-frame, can be represented in its behavior by an equivalent coupled wall. It is also assumed that the system of replacement coupled walls that comprises the equivalent total structure can be represented by a single equivalent coupled wall. Coupled wall theory is then used to estimate the deflected shape of the structure.

The characteristic differential equation for the deflection of a coupled wall subjected to uniformly distributed loading is given, and its solution for the deflection shape written in terms of the gross inertia of the structure and three nondimensional parameters αH, k, and x/H. The deflection solution is interpreted to show that it consists of three parts: one that represents the effect of overall flexure, a second that represents the effect of racking shear, and a third part that is related to the interaction between the overall flexure and the shear effects.

The methods for evaluating the stiffness parameters for different types of bent, and their method of combination to obtain the overall structure parameters, are given. The use of these in determining the structure deflections and story drifts is explained.

A numerical example for a realistic tall wall-frame structure is then used to illustrate the application of the method and to indicate its accuracy.

REFERENCES

15.1. Stafford Smith, B., Kuster, M., and Hoenderkamp, J. C. D. "A Generalized Approach to the Deflection Analysis of Braced Frame, Rigid Frame and Coupled Wall Structures." *Can. J. Civ. Engineer.* **8**(2), June 1981, 230-240.

15.2. Stafford Smith, B., Kuster, M., and Hoenderkamp, J. C. D., "Generalized Method for Estimating Drift in High Rise Structures." *J. Struct. Engineer. ASCE*, Structural Division, **110**(7), July 1984, 1549-1562.

15.3 Hoenderkamp, J. C. D. and Stafford Smith, B. "Simplified Analysis of Symmetric Tall Building Structures Subject to Lateral Loads." *Proc. 3rd Int. Conf. on Tall Buildings*, Hong Kong and Guangzhou, December 1984, pp. 28-36.

CHAPTER 16

Stability of High-Rise Buildings

The increasing height and greater structural efficiency of tall buildings have led to their having smaller reserves of stiffness and, consequently, stability. A check on the effects of this reduction in stability has become an important part of the building design process.

In considering stability, that of the structure as a whole, as well as that of individual members that make up the building, must be examined. However, the design for stability of individual columns is the same for high-rise buildings as for low-rise structures, and this aspect is usually covered by national Design Code requirements. This discussion on stability is, therefore, concerned with the whole structure, or with whole stories of the structure, rather than with individual members.

In its overall behavior a high-rise structure resembles a cantilever column of moderate slenderness ratio. It differs, however, from a typical structural column, which is essentially a flexural element, by including the possibility of a significant, or even a dominant, shear flexibility. Consequently, the potential modes of overall buckling include not only a flexural mode (Fig. 16.1a) but, alternatively, a shear mode (Fig. 16.1b), or, quite possibly, a combination of both these modes (Fig. 16.1c). Furthermore, these mode shapes might occur not only in transverse buckling of the structure, but in torsional or transverse-torsional forms of buckling.

The total gravity load on a high-rise structure is usually a small proportion of the load that would be required to cause overall buckling. Consequently, the possibility of collapse in this way is remote. The more serious stability consideration concerns the second-order effects of gravity loading acting on transverse displacements caused by horizontal loading, or acting on initial vertical misalignments in the structure. The vertical eccentricity of the gravity loading causes increases in the transverse displacements and in the member moments. In an extreme case this so-called P-Delta effect may be sufficient to initiate collapse. Usually, however, the P-Delta effects are either small and may be neglected, or of only moderate magnitude, in which case they can be accommodated by small increases in the sizes of the members. Nevertheless, in the design of any high-rise building, it is prudent to assess whether P-Delta effects may be significant and, if so, to account for them in the analysis and design.

In this chapter, methods of analysis for overall buckling and for P-Delta effects are presented.

16.1 OVERALL BUCKLING ANALYSIS OF FRAMES: APPROXIMATE METHODS

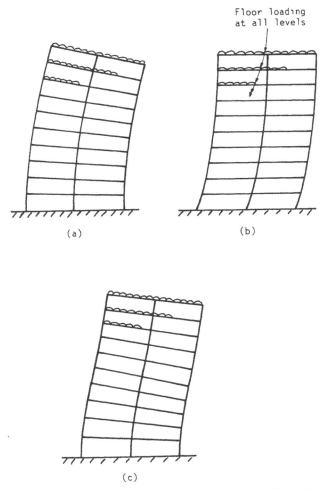

Fig. 16.1 (a) Flexural buckling; (b) shear buckling; (c) combined flexural-shear buckling.

16.1 OVERALL BUCKLING ANALYSIS OF FRAMES: APPROXIMATE METHODS

Methods for the determination of the overall buckling load are included because first, it indicates an upper bound for the critical gravity load, second, it allows an assessment of the relative vulnerability of the building to transverse buckling or torsional buckling, and third, it may be used, in a structure for which an approximate P-Delta analysis is appropriate, to evaluate an amplification factor for the displacements and moments.

16.1.1 Shear Mode

This mode of buckling occurs in moment resistant frames as a result of story sway associated with double bending of the columns and girders. Any effects of axial deformations of the columns are neglected in this approximate method. It is shown in Section 16.3.5 that the story drift in a frame, including the second-order effects of gravity loading [16.1], can be estimated by

$$\delta_i^* = \frac{1}{1 - (P_i \delta_i / h_i Q_i)} \delta_i \qquad (16.1)$$

in which, with suffix i referring to story i, δ_i is the first-order story drift caused by the external shear Q_i, P_i is the gravity loading carried by the columns in the story, and h_i is the story height.

The loss of stability is indicated approximately by a zero denominator in Eq. (16.1), in which case the displacement δ_i^* becomes infinite. Then

$$\frac{P_{icr} \delta_i}{h_i Q_i} = 1 \qquad (16.2)$$

which gives the critical load in the shear mode as

$$P_{icr} = \frac{Q_i h_i}{\delta_i} \qquad (16.3)$$

It is shown in Chapter 7, Eq. (7.19), that the lateral stiffness of story i may be written as

$$\frac{Q_i}{\delta_i} = \frac{12E}{h_i^2 \left(\dfrac{1}{C_i} + \dfrac{1}{G_i} \right)} \qquad (16.4)$$

in which $C_i = \Sigma(I_c/h)_i$, for which the summation is carried out over all columns (of inertias I_c and height h) in story i, and $G_i = \Sigma(I_g/L)_i$, for which the summation includes all the girders (of inertias I_g and lengths L) in the floor at the top of story i.

Substituting Eq. (16.4) into (16.3) gives the following expression for the critical load in a typical story i entirely in terms of the story members' dimensions and properties

$$P_{icr} = \frac{12E}{h_i \left(\dfrac{1}{C_i} + \dfrac{1}{G_i} \right)} \qquad (16.5)$$

16.1 OVERALL BUCKLING ANALYSIS OF FRAMES: APPROXIMATE METHODS

Special consideration should be given to the first story of a frame, which, if it has rigid base connections, can be shown from Eq. (7.20) to have a buckling load equal to

$$P_{0cr} = \frac{12E[1 + (C_1/6G_1)]}{h_1[(1/C_1) + (2/3G_1)]} \qquad (16.6)$$

or, if the frame has effectively pinned base connections, can be shown from Eq. (7.21) to have a buckling load equal to

$$P_{0cr} = \frac{12E}{h_1[(4/C_1) + (3/2G_1)]} \qquad (16.7)$$

16.1.2 Flexural Mode

This mode presumes that the entire structure buckles as a flexural cantilever by axial deformations of the columns. The greater the slenderness of a structure, the more vulnerable it becomes to instability in the flexural mode as opposed to the shear mode.

The buckling load is a function of the moment of inertia of the cantilever, which is taken as the second moment of the column sectional areas about their common centroid. Assuming this moment of inertia to vary in the frame from I_0 at the base to $I_0(1 - \beta)$ at the top, in order to allow for the reduction in the sizes of the columns up the height, an energy analysis with a slight modification to calibrate for the uniform member case yields [16.1]

$$P_{0cr} = \frac{7.83 E I_0}{H^2}(1 - 0.2974\beta) \qquad (16.8)$$

where P_{0cr} is the critical total gravity load on the structure and H is the total height of the structure.

16.1.3 Combined Shear and Flexural Modes

For cases in which a combination of shear and flexural modes may contribute to buckling, an analogy is drawn with the case of the buckling of a vertical cantilever with a gravity load at its top, for which the following solution exists

$$\frac{1}{P_{cr}} = \frac{1}{P_f} + \frac{1}{P_s} \qquad (16.9)$$

in which P_{cr}, P_f, and P_s are the critical loads for the combined, flexural, and shear modes of buckling, respectively.

Applying the analogy to the case of distributed gravity loading gives

$$\frac{1}{P_{0cr}} = \frac{1}{P_{0f}} + \frac{1}{P_{0s}} \qquad (16.10)$$

in which P_{0cr}, P_{0f}, and P_{0s} are the critical loads in the base story for the combined, flexural, and shear modes of buckling, respectively.

This very approximate approach is suggested as being useful for the preliminary stages of design and for assessing the importance of the flexural mode relative to the usually dominant shear mode of buckling [16.1].

16.2 OVERALL BUCKLING ANALYSIS OF WALL-FRAMES

Equations (16.5) to (16.10) provide very approximate estimates of the overall buckling load of a structure in the shear, flexure, and combined shear-flexure modes. A more rigorous analysis for plan-symmetric, uniform wall-frame structures provides solutions for the buckling loads of frame structures at one extreme, shear wall structures at the other, and any combination of shear walls and frames in between [16.2, 16.3].

16.2.1 Analytical Method

The method assumes the properties of the structure to be uniform, and the applied gravity loading to be distributed uniformly throughout the height (Fig. 16.2a). Representing the walls collectively by a flexural cantilever, the frames by a shear cantilever, and their connections by a stiff linking medium distributed uniformly over the height (Fig. 16.2b), differential equations for equilibrium were formulated

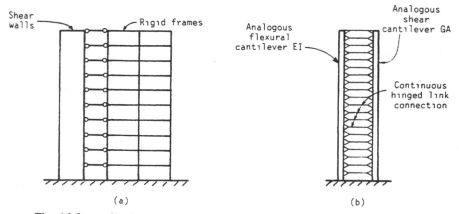

Fig. 16.2 (a) Uniform wall-frame structure; (b) analogous continuum model.

16.2 OVERALL BUCKLING ANALYSIS OF WALL-FRAMES

and solved to determine the critical buckling load. Solutions to these differential equations were obtained for a wide practical range of frame-to-wall relative stiffnesses [16.3].

Although the analyses are complex and too lengthy to justify presentation here, their results permit both the transverse and the torsional critical loads of a wall-frame structure to be calculated relatively simply by the procedure outlined below.

Consider the doubly symmetric structure in Fig. 16.3, in which the walls and rigid frames are aligned with the principal X and Y axes.

1. Determine the total flexural rigidities, $(EI)_t$, of the walls in the X and Y directions using, respectively

$$(EI)_{tx} = \Sigma (EI)_x \quad (16.11a)$$

$$(EI)_{ty} = \Sigma (EI)_y \quad (16.11b)$$

2. Determine the total shear rigidities, $(GA)_t$, of the frames in the X and Y directions using

$$(GA)_{tx} = \Sigma (GA)_x \quad (16.12a)$$

$$(GA)_{ty} = \Sigma (GA)_y \quad (16.12b)$$

where the shear rigidity of an individual frame is obtained for a typical story i from

$$(GA) = \frac{12E}{h_i[(1/C) + (1/G)]_i} \quad (16.13)$$

in which C_i and G_i are as defined for Eq. (16.4).

Walls and frames that are skew to the X and Y axes can be accommodated in the analysis by including their respective components of stiffness in the above totals.

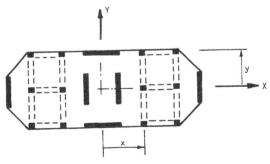

Fig. 16.3 Plan of symmetric wall-frame structure.

3. Determine the torsional rigidities $(EI_\omega)_t$ for the walls and $(GJ)_t$ for the frames.

For the walls:

$$(EI_\omega)_t = \sum EI_x y^2 + \sum EI_y x^2 \qquad (16.14a)$$

For the frames:

$$(GJ)_t = \sum GA_x y^2 + \sum GA_y x^2 \qquad (16.14b)$$

in which x is the distance from a wall or frame aligned in the Y direction to the center of twist, and y is the corresponding distance of a wall or frame aligned in the X direction.

Since torsional buckling is influenced not only by the plan distribution of the structural components but also by that of the gravity loading, a weight distribution parameter is required and is defined by

$$R = \frac{\sum pr^2}{\sum p} \qquad (16.15)$$

in which the floor loading is represented as a set of point loads p at distances r from the center of rotation.

4. The transverse and torsional stiffnesses obtained from Eqs. (16.11), (16.12), and (16.14) are then used to obtain the following transverse and torsional characteristic parameters:

$$(\alpha H)_x = H \sqrt{\frac{(GA)_{tx}}{(EI)_{tx}}} \qquad (16.16a)$$

$$(\alpha H)_y = H \sqrt{\frac{(GA)_{ty}}{(EI)_{ty}}} \qquad (16.16b)$$

$$(\alpha H)_\theta = H \sqrt{\frac{(GJ)_t}{(EI_\omega)_t}} \qquad (16.16c)$$

5. The three parameters (αH) above are used to find the corresponding coefficients s_x, s_y, and s_θ, from Table 16.1, that enable the calculation of the critical loads.

The critical gravity loads for transverse buckling are given by

$$P_{0crx} = \frac{s_x (EI)_{tx}}{H^2} \qquad (16.17a)$$

$$P_{0cry} = \frac{s_y (EI)_{ty}}{H^2} \qquad (16.17b)$$

16 2 OVERALL BUCKLING ANALYSIS OF WALL-FRAMES

TABLE 16.1 Coefficients s_1, s_1, s_θ for Wall-Frame Instability [Ref. 16.3]

αH	s_1, s_1, s_θ	αH	s_1, s_1, s_θ	αH	s_1, s_1, s_θ
0.00	7.84	3.40	36.4	6.80	97.0
0.10	7.90	3.50	37.8	6.90	99.2
0.20	7.97	3.60	39.3	7.00	101.4
0.30	8.14	3.70	40.8	7.10	103.6
0.40	8.33	3.80	42.3	7.20	105.8
0.50	8.61	3.90	43.8	7.30	108.1
0.60	8.94	4.00	45.3	7.40	110.4
0.70	9.28	4.10	46.9	7.50	112.7
0.80	9.74	4.20	48.5	7.60	115.0
0.90	10.3	4.30	50.1	7.70	117.4
1.00	10.8	4.40	51.7	7.80	119.7
1.10	11.4	4.50	53.3	7.90	122.1
1.20	12.1	4.60	55.0	8.00	124.6
1.30	12.8	4.70	56.7	8.10	127.0
1.40	13.5	4.80	58.4	8.20	129.5
1.50	14.3	4.90	60.1	8.30	132.0
1.60	15.2	5.00	61.8	8.40	134.5
1.70	16.1	5.10	63.6	8.50	137.1
1.80	17.0	5.20	65.4	8.60	139.6
1.90	18.0	5.30	67.2	8.70	142.2
2.00	19.0	5.40	69.0	8.80	144.8
2.10	20.0	5.50	70.9	8.90	147.5
2.20	21.1	5.60	72.8	9.00	150.2
2.30	22.2	5.70	74.7	9.10	152.8
2.40	23.4	5.80	76.6	9.20	155.6
2.50	24.6	5.90	78.5	9.30	158.3
2.60	25.8	6.00	80.5	9.40	161.1
2.70	27.0	6.10	82.5	9.50	163.8
2.80	28.3	6.20	84.5	9.60	166.7
2.90	29.6	6.30	86.5	9.70	169.5
3.00	30.9	6.40	88.6	9.80	172.3
3.10	32.2	6.50	90.6	9.90	175.2
3.20	33.6	6.60	92.7	10.00	178.1
3.30	35.0	6.70	94.9		

and the critical gravity load for torsional buckling is given by

$$P_{0cr\theta} = \frac{s_\theta (EI_\omega)_t}{RH^2} \qquad (16.17c)$$

These critical loads will be shown to be useful also for evaluating an amplification factor to give an estimate of the P-Delta effects.

396 STABILITY OF HIGH-RISE BUILDINGS

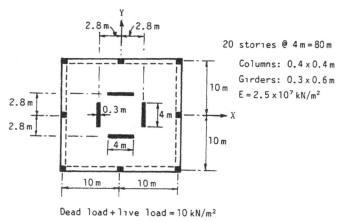

Dead load + live load = 10 kN/m²

Fig. 16.4 Plan of example structure.

16.2.2 Example: Stability of Wall–Frame Structure

Figure 16.4 shows the doubly symmetric plan of a 20-story, 80-m-high, reinforced concrete building consisting of shear walls and rigid frames. It is required to determine the magnitudes of the gravity loading that would cause lateral buckling and torsional buckling of the structure.

Member Properties

$$\text{Inertia of a single wall about its strong axis} = \frac{0.3 \times 4^3}{12} = 1.6 \text{ m}^4$$

$$\text{Inertia of a single column} = \frac{0.4 \times 0.4^3}{12} = 0.002 \text{ m}^4$$

$$\text{Inertia of a girder} = \frac{0.3 \times 0.6^3}{12} = 0.005 \text{ m}^4$$

$$\text{Modulus of elasticity } E = 2.5 \times 10^7 \text{ kN/m}^2$$

Translational Parameters. Because the structure is symmetric and identical in plan about its X and Y axes, only one direction of transverse buckling will be assessed. Considering X-direction buckling, assume the two walls and two frames aligned in the X-direction resist buckling, with a negligible contribution from the Y-direction components.
For the walls:

$$(EI)_{t_1}(2 \text{ walls}) = 2 \times 2.5 \times 10^7 \times 1.6 = 8.0 \times 10^7 \text{ kNm}^2$$

16.2 OVERALL BUCKLING ANALYSIS OF WALL-FRAMES

For the frames:

$$(GA)_{t,t}(2 \text{ frames}) = 2 \times \frac{12E}{h[(1/C) + (1/G)]}$$

where

$$C = \sum \frac{I_c}{h} = \frac{3 \times 0.002}{4} = 0.0015$$

$$G = \sum \frac{I_g}{L} = \frac{2 \times 0.005}{10} = 0.001$$

$$\therefore (GA)_{t,t}(2 \text{ frames}) = \frac{2 \times 12 \times 2.5 \times 10^7}{4[(1/0.0015) + (1/0.001)]} = 90\,000 \text{ kN}$$

Then

$$(\alpha H)_t = H \sqrt{\frac{(GA)_{t,t}}{(EI)_{t,t}}} = 80 \sqrt{\frac{9.0 \times 10^4}{8.0 \times 10^7}} = 2.68$$

Torsional Parameters. Torsional buckling will be resisted by the four walls and four frames acting in their planes and rotating about the center of the structure. For the walls:

$$(EI_\omega)_t = 2(EI)_t y^2 + 2(EI)_t x^2$$
$$= 2.5 \times 10^7 (2 \times 1.6 \times 2.8^2 + 2 \times 1.6 \times 2.8^2)$$
$$= 1.25 \times 10^9 \text{ kNm}^4$$

For the frames:

$$(GJ)_t = 2(GA)_t y^2 + 2(GA)_t x^2$$
$$= 90{,}000 \times 10^2 + 90{,}000 \times 10^2 = 1.8 \times 10^7 \text{ kNm}^2$$

Then

$$(\alpha H)_\theta = H \sqrt{\frac{(GJ)_t}{(EI_\omega)_t}} = 80 \sqrt{\frac{1.8 \times 10^7}{1.25 \times 10^9}} = 9.6$$

Weight Distribution Parameter. Dividing the floor plan at a typical level into 25, 4m × 4m regions, each carrying 160 kN gravity load, and taking the distance from the center of each region to the center of the structure as r.

$$\sum pr^2 = 256{,}000 \text{ kNm}^2 \quad \text{and} \quad \sum p = 25 \times 160 = 4000 \text{ kN}$$

Hence

$$R = \frac{\sum pr^2}{\sum p} = \frac{256{,}000}{4000} = 64 \text{ m}^2$$

For the gravity load to cause lateral buckling, P_{0cr_1}: The value $(\alpha H)_1 = 2.68$ in Table 16.1 gives by interpolation $s_1 = 26.8$. Then

$$P_{0cr_1} = \frac{s_1(EI)_{t_1}}{H^2} = \frac{26.8 \times 8.0 \times 10^7}{80^2} = 33.5 \times 10^4 \text{ kN}$$

Because of symmetry, the critical load for lateral buckling in the Y direction P_{0cr_1} will be identical.

For the gravity load to cause torsional buckling, $P_{0cr\theta}$: The value $(\alpha H)_\theta = 9.6$ in Table 16.1 gives $s_\theta = 166.7$. Then

$$P_{0cr\theta} = \frac{s_\theta(EI_\omega)_t}{RH^2} = \frac{166.7 \times 1.25 \times 10^9}{64 \times 80^2} = 50.9 \times 10^4 \text{ kN}$$

The actual maximum value of the total loading over 20 stories is

$$P_0 = 20 \times 4000 = 8.0 \times 10^4 \text{ kN}$$

which leaves adequate margins of safety against overall buckling in both the translational and torsional modes.

16.3 SECOND-ORDER EFFECTS OF GRAVITY LOADING

16.3.1 The P-Delta Effect

A first-order computer analysis of a building structure for simultaneously applied gravity and horizontal loading results in deflections and forces that are a direct superposition of the results for the two types of loading considered separately. Any interaction between the effects of gravity loading and horizontal loading is not accounted for by the analysis.

In reality, when horizontal loading acts on a building and causes it to drift, the resulting eccentricity of the gravity loading from the axes of the walls and columns produces additional external moments to which the structure responds by drifting further. The additional drift induces additional internal moments sufficient to equilibrate the gravity load moments. This effect of the gravity loading P acting on the horizontal displacements Δ is known as the P-Delta effect.

The second-order P-Delta additional deflections and moments are small for typical high-rise structures, with a magnitude usually of less than 5% of the first-order values. If the structure is exceptionally flexible, however, the additional forces

16.3 SECOND-ORDER EFFECTS OF GRAVITY LOADING   399

might be sufficient to require consideration in the members' design, or the additional displacement might cause unacceptable total deflections that require the structure to be stiffened. In an extreme case of lateral flexibility combined with exceptionally heavy gravity loading, the additional forces from the P-Delta effect might cause the strength of some members to be exceeded with the possible consequence of collapse. Or, the additional P-Delta external moments may exceed the internal moments that the structure is capable of mobilizing by drift, in which case the structure would collapse through instability. Such failures would occur at gravity loads less than the critical overall buckling load predicted in the previous section.

Although a translational P-Delta effect is the most obvious case to consider, a torsional P-Delta mode is possible and should also be assessed. The torsional mode occurs when a building twists, and its walls and frames displace at each floor about some center of rotation. As a result the gravity loading, which is distributed over the building plan, is vertically misaligned with the axes of the resisting elements causing, in effect, an additional torque. The building responds by twisting more until the additional internal resisting torque and the external P-Delta torque are in equilibrium. Since the P-Delta torque and the torsional resistance of the structure depend on the plan locations of the gravity loading and of the walls and frames, these locations must be included in the parameters of a stability analysis. The more widely dispersed the vertical bents are from the center of rotation, the more effective they are in resisting torsion and the P-Delta torsional effects.

The methods of P-Delta analysis to be described include the following: a very approximate method in which a constant amplification factor is applied to all the results of a first-order analysis, a more accurate method involving an iteration of the first-order analysis with the primary lateral force augmented by increments whose effects are equivalent to those of the gravity loads, a second iterative method in which the gravity loads are applied to the laterally deflecting structure, then a so-called "direct" method for rigid frame structures in which iterations are avoided by making a direct second-order adjustment of the displacements and moments, and, finally, two methods in which the structure is modeled so that a stiffness matrix analysis incorporates both the first-order and second-order effects.

16.3.2 Amplification Factor P-Delta Analysis

It has been shown [16.4] for a vertical cantilever displaced laterally by a uniformly distributed horizontal load that the addition of a concentrated vertical load P at the free end of the cantilever (Fig. 16.5a) increases the horizontal displacements by an amplification factor F, where

$$F = \frac{1}{1 - (P/P_{cr})} \qquad (16.18)$$

in which P_{cr} is the concentrated vertical load at the top to cause buckling of the cantilever.

STABILITY OF HIGH-RISE BUILDINGS

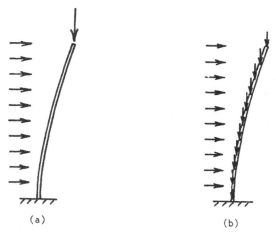

Fig. 16.5 (a) Cantilever with lateral and concentrated axial loading; (b) cantilever with lateral and distributed axial loading.

The final displacements Δ^* are given, therefore, in terms of the initial displacements Δ by

$$\Delta^* = F\Delta = \frac{1}{1 - (P/P_{cr})} \Delta \qquad (16.19)$$

Since the amplification factor is a constant over the height of the structure subjected to load P, the increase in deflection is proportional to the initial displacements at all levels.

Extending the amplification factor method to the tall building structure in which the gravity loading is distributed throughout the height (Fig. 16.5b), P is replaced by P_0, the total gravity load, and P_{cr} becomes P_{0cr}, the overall buckling load, so that the equation for the total drift is taken to be

$$\Delta^* = \frac{1}{1 - (P_0/P_{0cr})} \Delta \qquad (16.20)$$

The P-Delta effect causes an increase not only in drift but also in internal moments. Therefore an initial set of moments M in a structure, calculated by a first-order analysis, would be increased by second-order effects to a set of final moments M^*, where

$$M^* = \frac{1}{1 - (P_0/P_{0cr})} M \qquad (16.21)$$

An analysis of the total drift and moments in the structure at the limit state under consideration would require:

1. a first-order analysis of the structure for the factored horizontal loading only;
2. the evaluation of the amplification factor, using the factored gravity loading and the critical buckling load obtained either from an approximate method, using Eq. (16.5), (16.8), or (16.10), or from a more rigorous analytical method, using Eqs. (16.17);
3. increasing the drift and moments in the structure, resulting from the horizontal loading first-order analysis, by the amplification factor derived in (2); and
4. adding the results of (3), for horizontal loading, to the forces in the structure determined for gravity loading.

The above procedure should be used to assess the P-Delta effects about the two major bending axes of the structure as judged necessary for each of the directions. To assess the torsional P-Delta effects on the structure, the same procedure can be used with a torsional amplification factor applied to the forces and displacements caused by torque. The value of P_{0cr} to be used for torsion should be determined from Eq. (16.17c).

16.3.3 Iterative P-Delta Analysis

In cases of heavy gravity loading or of a flexible structure, the accuracy of the amplification factor method deteriorates and it becomes necessary to use a more accurate, second-order, method of analysis. In the iterative second-order method, an initial first-order analysis of the structure is made with the external horizontal loading. The resulting horizontal deflections are then used in conjunction with the gravity loading to compute at each floor level an equivalent increment of horizontal load. This increment is added to the initial horizontal load and the analysis is repeated. The resulting increased deflections are then used in conjunction with the gravity loads to compute another set of equivalent horizontal increments, which again are added to the initial horizontal load for a reanalysis. The iterations are continued until increases in the deflections become negligible.

To determine the increment of horizontal load with an effect equivalent to the eccentric gravity load, consider first a single-story column (Fig. 16.6a) collectively representing the vertical structural components in story i of a multistory structure. The horizontal loading on the structure, with a shear Q_i in story i, has caused a story drift δ_i. Assuming the column to be straight over the story height, therefore neglecting any effects due to its bending, the total gravity load P_i, carried by the columns and walls in story i with an eccentricity δ_i, causes an additional moment at the bottom of the column equal to $P_i \delta_i$ (Fig. 16.6b).

The equilibrium of the column could be produced alternatively by replacing the vertical load P_i by an additional increment of shear having the same moment at the column base (Fig. 16.6c), that is,

$$\delta Q_i h_i = P_i \delta_i \qquad (16.22)$$

402 STABILITY OF HIGH-RISE BUILDINGS

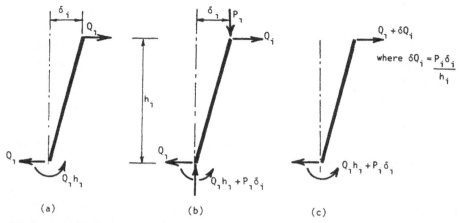

Fig. 16.6 (a) Column deflected by shear; (b) axial load added to column deflected by shear; (c) equivalent augmented shear.

from which the equivalent increment of shear is given by

$$\delta Q_i = \frac{P_i \delta_i}{h_i} \qquad (16.23)$$

Consider now the resultant effect of the shear increments in successive stories (Fig. 16.7). The increment of shear at the top of story i, and the opposite shear at

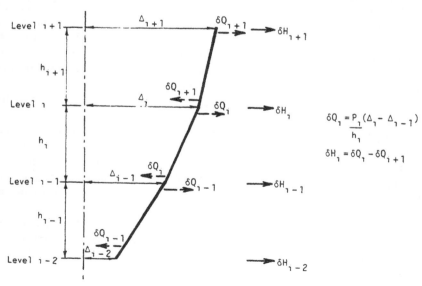

Fig. 16.7 Equivalent additional lateral load at successive floor levels.

the bottom of the story is given by

$$\delta Q_i = \frac{P_i}{h_i}(\Delta_i - \Delta_{i-1}) \qquad (16.24)$$

The resultant additional increment of horizontal load to be applied at floor level i is then

$$\delta H_i = \delta Q_i - \delta Q_{i+1} \qquad (16.25)$$

This increment is added to the original horizontal load at level i for the next reanalysis, as described earlier.

It can be shown that the set of horizontal increments causes, at any level of the structure, the same additional moment as would the set of gravity loads acting at their respective eccentricities from the original vertical line of the structure.

The iterative method is applicable to all types of frame, wall, and other forms of building structure.

16.3.4 Iterative Gravity Load P-Delta Analysis

In the Iterative Method of Section 16.3.3, the requirement of having to repeatedly evaluate the increments of horizontal load at many floors can be tedious. This is avoided in the following simpler and more realistic iterative gravity load method of P-Delta analysis [16.5].

After a first-order horizontal load analysis of the structure using a frame analysis program has been made, the gravity loads are applied to the unloaded structure deflected by the first-order values of drift, Δ_i, to obtain an increment of drift δ_{i1} (Fig. 16.8a). The gravity loads are then applied to the structure deflected by the increments δ_{i1} to obtain another increment in drift, δ_{i2} (Fig. 16.8b). The procedure is repeated until the additional drift increment δ_{in} is negligible (Fig. 16.8c). The final drift at story i, Δ_i^* including the P-Delta effect, is the sum of the first-order drift and all the increments of drift.

$$\Delta_i^* = \Delta_i + \delta_{i1} + \delta_{i2} + \delta_{i3} + \cdots \qquad (16.26)$$

The iterative process is required because when the vertical loads are applied, they are not being applied to the final deflected shape. The final moment at story i, M_i^*, including the P-Delta effect, is obtained by adding the first-order moment, M_i, and the increments δM_i, obtained from the iterative analysis; that is

$$M_i^* = M_i + \delta M_{i1} + \delta M_{i2} + \delta M_{i3} + \cdots \qquad (16.27)$$

In practice, the method can be simplified by adding a full-height, axially rigid, fictitious column with a flexural stiffness equivalent to zero, and connecting it to the structure either by internodal horizontal constraints or by axially rigid links

404 STABILITY OF HIGH-RISE BUILDINGS

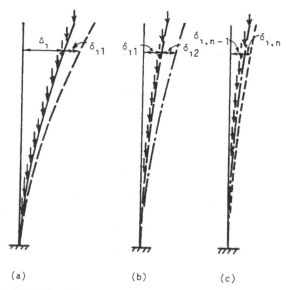

Fig. 16.8 (a-c) Gravity loading applied to successive increments of lateral deflection.

(Fig. 16.9). Then the total gravity loading at each floor is applied only to corresponding levels of the deflected fictitious column in the same iterative manner as described above. In this way, only the coordinates of the fictitious column, and not the entire structure, need to be altered. The results obtained by this gravity load method are identical to those given by the iterative incremental lateral load method, while the analysis can be performed in less than one-third of the time.

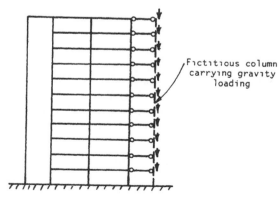

Fig. 16.9 Alternative approach with gravity loading applied to deflected fictitious column.

16.3.5 Direct P-Delta Analysis

The iterative analysis described in Section 16.3.3 can be reduced for rigid frame structures to a first-order analysis plus a direct second-order adjustment.

From the first-order analysis, using horizontal loading only, the shear stiffness of story i of a rigid frame structure can be expressed as

$$K_{si} = \frac{Q_i}{\delta_i} \qquad (16.28)$$

The P-Delta effect at the final deflected state can now be represented by the initial external shear Q_i and increment δQ_i, as defined in Eq. (16.23), to give an effective total shear

$$Q_i^* = Q_i + \delta Q_i = Q_i + \frac{P_i \delta_i^*}{h_i} \qquad (16.29)$$

in which δ_i^* is the final drift in story i, including the P-Delta effect.

Consequently, the final drift in story i can be expressed as [16.6]

$$\delta_i^* = [Q_i + (P_i \delta_i^*/h_i)]/K_{si} \qquad (16.30)$$

that is,

$$\delta_i^* = [Q_i + (P_i \delta_i^*/h_i)]/[Q_i/\delta_i] \qquad (16.31)$$

then

$$\delta_i^* = \frac{1}{1 - (P_i \delta_i / Q_i h_i)} \delta_i \qquad (16.32)$$

This resulting drift in story i includes the P-Delta second-order effect. The total drift at floor level n, including the P-Delta effect, can then be obtained from

$$\Delta_i^* = \sum_{i=1}^{n} \delta_i^* \qquad (16.33)$$

The increase in the first-order story drift δ_i due to the P-Delta effect is shown by Eq. (16.32) to be a nonlinear function of the gravity loading P_i acting in a story. Therefore, it is essential in a P-Delta analysis to use the magnitude of loading corresponding to the limit state in question. In contrast to the simpler amplification factor method, the increases in drift and moments calculated for gravity loading by this method are not a constant proportion of the corresponding first-order values at different levels of the structure.

The increased moments M_i^* in the members of story i, including the P-Delta effect, can be estimated similarly from the moments M_i of the first-order analysis

$$M_i^* = \frac{1}{1 - (P_i \delta_i / Q_i h_i)} M_i \qquad (16.34)$$

Second-order P-Delta results for drift and moments are obtained, therefore, by applying Eqs. (16.32) and (16.34), respectively, to the results of a horizontal loading first-order analysis. To these augmented results should be added the results of an analysis for gravity loading. Because the derivation of Eqs. (16.32) and (16.34) depends on Eq. (16.28), which is valid only for structures that deform predominantly in shear, for example, rigid frames, the method is not applicable to structures that deform mainly in flexure, that is, those including shear walls.

16.4 SIMULTANEOUS FIRST-ORDER AND P-DELTA ANALYSIS

In the following method for the P-Delta analysis of building structures, the second-order effects are accounted for by modifying the structural model so that, when analyzed for the actual horizontal loading, the resulting values of drift and member forces include the P-Delta effects. A single computer run is then sufficient to analyze the structure and include the P-Delta effects. The modification to the structural model consists of attaching to it a fictitious column. This causes the horizontal displacement terms of the first-order stiffness matrix of the structure to be automatically augmented so as to incorporate the effects of gravity loading acting on the lateral displacements. The first-order stiffness matrix is therefore converted into a second-order matrix. An explanation of the method follows.

16.4.1 Development of the Second-Order Matrix

Referring to Fig. 16.7 and Eqs. (16.24) and (16.25), which relate to the iterative method of analysis, the increment of horizontal load at level i, equivalent in effect to the gravity loading, can be expressed as

$$\delta H_i = \frac{-P_i \Delta_{i-1}}{h_i} + \left(\frac{P_i}{h_i} + \frac{P_{i+1}}{h_{i+1}} \right) \Delta_i - \frac{P_{i+1} \Delta_{i+1}}{h_{i+1}} \qquad (16.35)$$

The set of equations for the equivalent increments at all the story levels can be written in matrix form, thus

16.4 SIMULTANEOUS FIRST-ORDER AND P-DELTA ANALYSIS

$$\begin{Bmatrix} \delta H_1 \\ \delta H_2 \\ \delta H_3 \\ \rule{0pt}{10pt}\overline{} \\ \delta H_i \\ \overline{} \\ \delta H_N \end{Bmatrix} = \begin{bmatrix} \dfrac{P_1}{h_1}+\dfrac{P_2}{h_2} & \dfrac{-P_2}{h_2} & & & & & \\ \dfrac{-P_2}{h_2} & \dfrac{P_2}{h_2}+\dfrac{P_3}{h_3} & \dfrac{-P_3}{h_3} & & & & \\ & \dfrac{-P_3}{h_3} & \dfrac{P_3}{h_3}+\dfrac{P_4}{h_4} & \dfrac{-P_4}{h_4} & & & \\ \hline & & & & & & \\ & & \dfrac{-P_i}{h_i} & \dfrac{P_i}{h_i}+\dfrac{P_{i+1}}{h_{i+1}} & \dfrac{-P_{i+1}}{h_{i+1}} & & \\ \hline & & & & & \dfrac{-P_N}{h_N} & \dfrac{P_N}{h_N} \end{bmatrix} \begin{Bmatrix} \Delta_1^* \\ \Delta_2^* \\ \Delta_3^* \\ \overline{} \\ \Delta_i^* \\ \overline{} \\ \Delta_N^* \end{Bmatrix}$$

(16.36)

or

$$\{\delta H\} = [K_G]\{\Delta^*\} \tag{16.37}$$

where the matrix K_G, which is square, tridiagonal, and symmetric about the leading diagonal, is the geometric stiffness matrix and is a function of the gravity loading. The matrix $\{\Delta^*\}$ is the vector of the total lateral displacements, which includes the P-Delta effect.

Referring next to the terms of a building structure's stiffness matrix corresponding to horizontal displacements of the floor levels:

$$\begin{Bmatrix} H_1 \\ H_2 \\ H_3 \\ \overline{} \\ H_i \\ \overline{} \\ H_N \end{Bmatrix} = \begin{bmatrix} K_{11} & K_{12} & & & & & \\ K_{21} & K_{22} & K_{23} & & & & \\ 0 & K_{32} & K_{33} & K_{34} & & & \\ \hline & & & & & & \\ & & & K_{i,i-1} & K_{ii} & K_{i,i+1} & \\ \hline & & & & & K_{N,N-1} & K_{NN} \end{bmatrix} \begin{Bmatrix} \Delta_1 \\ \Delta_2 \\ \Delta_3 \\ \overline{} \\ \Delta_i \\ \overline{} \\ \Delta_N \end{Bmatrix} \tag{16.38}$$

that is

$$\{H\} = [K]\{\Delta\} \tag{16.39}$$

in which K, the first-order lateral stiffness matrix, is also tridiagonal.

The P-Delta effects of the gravity loading can be represented by adding an equivalent increment of horizontal loading to the actual loading. This can be stated in matrix form as

$$\{H\} + \{\delta H\} = [K]\{\Delta^*\} \tag{16.40}$$

Therefore, substituting from Eq. (16.37) into (16.40)

$$\{H\} + [K_G]\{\Delta^*\} = [K]\{\Delta^*\} \tag{16.41}$$

and rearranging, gives

$$\{H\} = [K - K_G]\{\Delta^*\} \tag{16.42}$$

in which $\{H\}$ is the matrix of the actual horizontal loading, $[K - K_G]$ is the second-order stiffness matrix incorporating the modifications due to gravity loading, and $\{\Delta^*\}$ represents the final horizontal displacements, including the P-Delta effects. If the original lateral stiffness matrix K is reduced by K_G, a single first-order analysis will produce results for deflections and member forces that will incorporate the P-Delta second-order effects.

Writing the first three rows of matrix Eq. (16.42) indicates how this modification can be made.

$$\begin{Bmatrix} H_1 \\ H_2 \\ H_3 \end{Bmatrix} = \begin{bmatrix} K_{11} - \left(\dfrac{P_1}{h_1} + \dfrac{P_2}{h_2}\right) & K_{12} + \dfrac{P_2}{h_2} & \\ K_{21} + \dfrac{P_2}{h_2} & K_{22} - \left(\dfrac{P_2}{h_2} + \dfrac{P_3}{h_3}\right) & K_{23} + \dfrac{P_3}{h_3} \\ & K_{32} + \dfrac{P_3}{h_3} & K_{33} - \left(\dfrac{P_3}{h_3} + \dfrac{P_4}{h_4}\right) & K_{34} + \dfrac{P_4}{h_4} \end{bmatrix} \begin{Bmatrix} \Delta_1^* \\ \Delta_2^* \\ \Delta_3^* \end{Bmatrix}$$

$$(16.43)$$

A simple and convenient physical way of modifying the first-order stiffness matrix is by adding to the structure a fictitious, full structure-height column with either a negative shear area or a negative inertia, as explained below.

16.4.2 Negative Shear Area Column

Consider, for example, a wall–frame structure (Fig. 16.10a) with an added fictitious shear column, that is, a column with infinite flexural rigidity but deformable in shear. The terms of the second column in the stiffness matrix of the fictitious

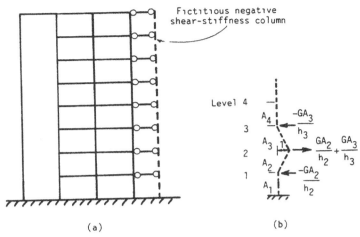

Fig. 16.10 (a) Structure with fictitious shear column; (b) lateral stiffness coefficients of shear column.

shear column, corresponding to unit horizontal displacement at level 2 (Fig. 16.10b) are

$$k_{12} = \frac{-GA_2}{h_2}; \quad k_{22} = \frac{GA_2}{h_2} + \frac{GA_3}{h_3}; \quad k_{32} = \frac{-GA_3}{h_3} \quad (16.44)$$

Referring to the second columns of the stiffness matrices in Eqs. (16.38) and (16.43), the increments of stiffness required to convert the terms of the first-order stiffness matrix of the structure to the corresponding terms of the second-order stiffness are

$$k_{12} = \frac{P_2}{h_2}; \quad k_{22} = -\left(\frac{P_2}{h_2} + \frac{P_3}{h_3}\right); \quad k_{32} = \frac{P_3}{h_3} \quad (16.45)$$

It is evident from Eqs. (16.44) and (16.45) that the fictitious column will satisfy these requirements if

$$\frac{-GA_2}{h_2} = \frac{P_2}{h_2}; \quad \text{that is, if } A_2 = \frac{-P_2}{G} \quad (16.46)$$

and

$$\frac{-GA_3}{h_3} = \frac{P_3}{h_3}; \quad \text{that is, if } A_3 = \frac{-P_3}{G} \quad (16.47)$$

Therefore, the whole stiffness matrix will be modified appropriately if the fictitious column is assigned in each level i a shear area

$$A_i = \frac{-P_i}{G} \qquad (16.48)$$

and an infinite, or extremely large, flexural rigidity.

In Eq. (16.48), P_i is the total axial force carried by the walls and columns in story i, and G is the shear modulus to be assigned to the fictitious column.

Thus, by adding the fictitious column, which is connected at each floor level to the model of the structure by either internodal horizontal constraints or hinged-end axially rigid links, and assigning the column in each story to have a negative shear area in accordance with Eq. (16.48), a single first-order analysis can be made that fully incorporates the second-order P-Delta effects.

The resulting values of drift and member forces in the structure will include the second-order P-Delta effects. The total shear in the structure at any level, however, will slightly exceed the external horizontal shear at that level, with the excess being caused by the gravity loading acting on the inclined columns and walls.

This method is suitable for use with structural analysis programs that allow the shear area of members to be specified, and that accept negative member properties.

16.4.3 Negative Flexural Stiffness Column

If the available analysis program allows negative member properties but does not have a shear area option, an alternative approach is to use a fictitious negative inertia column with its floor level nodes restrained against rotation, as shown diagrammatically in Fig. 16.11a.

The stiffness terms of an ordinary flexural column, corresponding to a unit horizontal displacement at level 2 (Fig. 16.11b) are

$$k_{12} = \frac{-12EI_2}{h_2^3}; \qquad k_{22} = \frac{12EI_2}{h_2^3} + \frac{12EI_3}{h_3^3}; \qquad k_{32} = \frac{-12EI_3}{h_3^3} \qquad (16.49)$$

The fictitious column will generate the required modifying terms of Eq. (16.45) for automatic incorporation into the second-order stiffness matrix if

$$\frac{-12EI_2}{h_2^3} = \frac{P_2}{h_2}; \quad \text{that is, if } I_2 = \frac{-P_2 h_2^2}{12E} \qquad (16.50)$$

and if

$$\frac{-12EI_3}{h_3^3} = \frac{P_3}{h_3}; \quad \text{that is, if } I_3 = \frac{-P_3 h_3^2}{12E} \qquad (16.51)$$

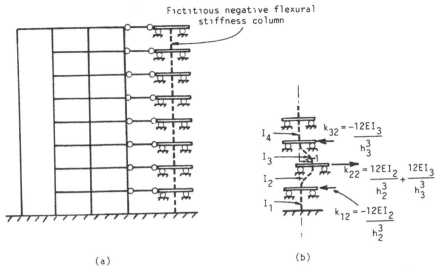

Fig. 16.11 (a) Structure with fictitious flexural column; (b) stiffness coefficients of flexural column.

Therefore, in a typical story i, the fictitious column must be assigned the negative value of inertia

$$I_i = \frac{-P_i h_i^2}{12E} \qquad (16.52)$$

Similarly to the negative shear area column, the negative inertia column is attached to the original model of the structure at each floor level by either internodal horizontal constraints or axially rigid hinged-end links. The effect of the rigid horizontal arms in Fig. 16.11a, representing the restraint on the fictitious column against rotation at the floor levels, should be specified in the analysis program simply by a rotational restraint on the column at each node.

16.5 TRANSLATIONAL-TORSIONAL INSTABILITY

The second-order methods of P-Delta analysis have been concerned so far with translational effects. In the case of an unsymmetrical building, combined translational and torsional P-Delta effects are possible. These can be analyzed directly and accurately, as a combined first- and second-order analysis of the problem, by adding to the structural model a fictitious column with negative transverse and torsional properties [16.9]. Each story is augmented by a fictitious column with negative shear areas in the two major directions of translation, and a negative torsional constant about its axis. The shear areas are, as before, functions of the gravity loading, while the torsion constant accounts also for the plan distribution of the loading. The translational shear areas are assigned to the column in story i as

$$A_{ix} = A_{iy} = \frac{-P_i}{G} \qquad (16.53)$$

If the computer program does not allow the assignment of shear areas, a negative inertia column with its ends restrained against vertical plane rotations may be used as an alternative. In that case

$$I_{ix} = I_{iy} = \frac{-P_i h_i^2}{12E} \qquad (16.54)$$

The fictitious column is located at the centroid of the total gravity loading above story i (Fig. 16.12) and its negative torsion constant J_i is found from

$$\frac{1}{h_i}\sum [P_{ij}(d_{xij}^2 + d_{yij}^2)] = \frac{1}{h_i} P_i r_i^2 = \frac{-GJ_i}{h_i} \qquad (16.55)$$

in which P_{ij} is the gravity load in column or wall j in story i, d_{xij} and d_{yij} are its Y and X distances from the center of gravity loading in story i, and r_i is the radius of inertia of the total load P_i about the center.

Hence, in addition to its two negative values of A (or I) the column is assigned a torsion constant

$$J_i = \frac{-P_i r_i^2}{G} \qquad (16.56)$$

If the computer program does not allow the specification of the torsion constant for beam elements, this may be accommodated by splitting the negative stiffness column into two half-columns. If the column is specified as a negative shear area column, using Eq. 16.53, the shear areas of the half-columns would be

$$A_{ix} = A_{iy} = \frac{-P_i}{2G} \qquad (16.57)$$

or, if the column is specified as a negative flexural rigidity column using Eq. (16.54), the properties of the half-columns would be

$$EI_i = \frac{-P_i h_i^2}{24} \qquad (16.58)$$

16.5 TRANSLATIONAL-TORSIONAL INSTABILITY

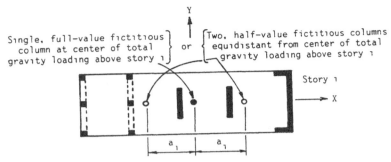

Fig. 16.12 Plan of structure for transverse and torsional P-Delta analysis.

The two half-columns would be located at a distance a_t on each side of the center of total gravity loading (Fig. 16.12) where a_t is given by

$$a_t = \sqrt{\frac{\sum P_{ij}(d_{yij}^2 + d_{xij}^2)}{P_t}} \qquad (16.59)$$

A story-to-story regular floor plan would allow the fictitious column or columns to be continuous throughout the structure. If, however, the position of the center of the gravity loading changes from one floor to another, continuity of the fictitious column, or columns, should be provided by rigid horizontal beams at the floor levels (Fig. 16.13).

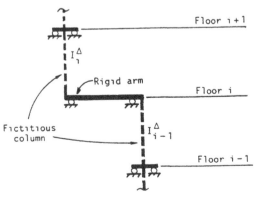

Fig. 16.13 Connection between fictitious columns in successive stories.

16.6 OUT-OF-PLUMB EFFECTS

When walls or columns are constructed out-of-plumb, the gravity loads acting on the vertical misalignment cause drift and moment P-Delta effects. The normally allowed erection tolerances restrict the out-of-plumb effects to negligible proportions, and the improbability that all columns in a single story, or that all successive stories lean by the maximum amount in the same direction, reduces even further the importance of the problem.

It is prudent, however, when the usual P-Delta effects are small, to check whether the out-of-plumb P-Delta effects are larger. If they are larger, and of significance, the out-of-plumb P-Delta effects should then be used in designing the structure.

The out-of-plumb effects can be accounted for by analyzing the structure for equivalent lateral loads δH_i, as in the Iterative Method. The first values of δH_i should be obtained by using the out-of-plumb displacements, based on the allowable tolerances, and Eqs. (16.24) and (16.25). If the analysis is to be made by computer, an alternative and more rapid method of analysis would be the iterative gravity load method (Section 16.3.4), with the gravity loads applied initially to a fictitious column deflected by the allowable tolerances.

The erection tolerances used to estimate δ_i vary between Codes of Practice, but $h/1000$ is a typical value. The West German Code for reinforced concrete, DIN 1045 (1972), for example, is more detailed [16.10] in specifying

$$\delta_i = \pm \frac{h_i}{100\sqrt{H \text{ (meters)}}} \qquad (16.60)$$

or

$$\delta_i = \pm \frac{h_i}{55\sqrt{H \text{ (feet)}}} \qquad (16.61)$$

where H is the total height of the structure. These values are more severe than $h/1000$ for buildings of less than 100 m height, and less severe for taller buildings.

16.7 STIFFNESS OF MEMBERS IN STABILITY CALCULATIONS

When checking stability effects for the loading at the factored design load level, the effective flexural rigidities (EI) of members in reinforced concrete structures may be substantially less than the rigidities calculated from the instantaneous loading values of E and the gross section values of I. The effect of sustained loading on the elastic modulus of concrete, and of cracking from shrinkage or bending on the moment of inertia, both contribute to a significant reduction in the flexural rigidity. The accuracy of the stability calculations is as vulnerable to an overestimation of the stiffness of the members as to approximations in the method of analysis.

Because the nonlinearity of P-Delta calculations requires them to be carried out for factored design loads, it is recommended that the estimated value of EI for reinforced concrete members should be that for the stage just before the onset of yielding. Studies of the stiffness of reinforced concrete members [16.11, 16.12] have led to some detailed recommendations [16.10] for reductions in the stiffness of individual members. In tall buildings, however, with many thousands of members at different states of loading, it would be impractical to adjust individually the properties of the members. Rather, it is recommended that

1. to allow for sustained loading, the initial value of E for concrete should be reduced by the factor 0.8;
2. to allow for cracking:
   a. the gross value of I for beams should be reduced by the factor 0.5;
   b. if appropriate, the gross value of I for columns should be reduced by the factor 0.8.

16.8 EFFECTS OF FOUNDATION ROTATION

A flexible foundation will affect the overall stability of a building by reducing the effective lateral stiffness of the vertical cantilever structure. It will also increase the deflections from horizontal loading and hence increase the P-Delta effect.

It has been proposed [16.13] that the amplification factor [Eq. (16.18)] may be modified to account for foundation flexibility by reducing the effective stiffness of the cantilever structure, and hence reducing P_{0cr}, in the following way.

In Fig. 16.14, the top deflection of a uniform flexural cantilever subjected to a

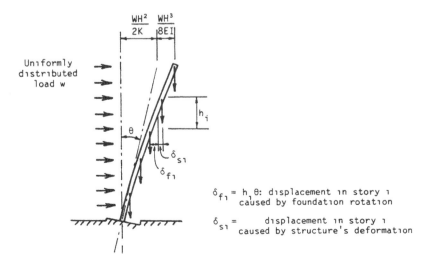

Fig. 16.14 Structure with flexible foundation.

uniformly distributed total load W is given by

$$\Delta_T = \frac{WH^3}{8EI} \qquad (16.62)$$

If the foundation rotational flexibility, defined as the rotation per unit moment, is K, then the top deflection is increased to

$$\Delta_T = \frac{WH^3}{8EI} + \frac{WH^2}{2K} \qquad (16.63)$$

which is the same as the top deflection of a rigidly based cantilever with an effective flexural rigidity of

$$(EI)_c = \frac{\mu}{\mu + 4} EI$$

where

$$\mu = KH/EI \qquad (16.64)$$

Assuming that the reduction in the effective EI causes a proportional reduction in P_{0cr}, the term P_0/P_{0cr} in the denominator of Eq. (16.20) becomes

$$\frac{\mu + 4}{\mu} \frac{P_0}{P_{0cr}} \qquad (16.65)$$

where P_{0cr} is determined as for a fixed base structure. The modified term from Eq. (16.65) should be used to calculate the amplification factor for use in Eqs. (16.20) and (16.21).

If a second-order P-Delta analysis is used, of the type in which factors for individual stories are applied to a set of first-order results, the first-order results and hence the second-order factors will automatically include the effects of foundation flexibility, provided these effects were specified for the first-order analysis.

Similarly, if a negative stiffness column is used in a computer analysis to include P-Delta effects, the specification of a flexible foundation by vertical or rotational springs under the real column and wall bases will automatically take care of the problem.

SUMMARY

The chapter considers the effect of gravity loading on the overall stability of a structure, and the second-order P-Delta effects of gravity loading, which increase the lateral deflections and moments due to transverse loading.

The overall buckling load defines an upper bound to the gravity loading that a structure can carry. It may also be used to assess the need for a P-Delta analysis and for making an approximate estimate of the P-Delta effects. The approximate buckling loads of structures that fail in either a shear mode or a flexural mode, or a combination of the two modes, can be determined from simple formulas. The buckling loads of plan-symmetric wall-frame structures with uniform properties can be determined from an analytical approach using stiffness parameters. This method covers the full range, from the shear mode of buckling for a frame at one extreme, to the flexural mode of buckling of a wall at the other, with any intermediate combination of walls and frames. The method also allows an estimate of the critical load for torsional buckling.

The second-order effect of gravity loading on the first-order displacements and moments due to transverse loading, that is, the P-Delta effect, is the most common aspect of building instability that has to be considered in building design. An approximate assessment of the increase in displacements and moments due to P-Delta effects can be made by an amplification factor applied throughout the structure. In cases of flexible structures or of heavy gravity loading, giving rise to significant P-Delta effects, a more accurate, second-order analysis should be made. This could be by a story-by-story adjustment of the results of a first-order horizontal load analysis, using in each story a factor based on the gravity loading in that story. Alternatively, it could be a direct stiffness matrix analysis of the structural model, modified by fictitious columns with negative stiffness properties to simulate the effective reduction in lateral stiffness caused by gravity loading. Because a second-order analysis is nonlinear, the loading should correspond to the limit state under consideration. At this higher loading, the gross properties of reinforced concrete members should be reduced to allow for cracking while, for sustained loading, the modulus of elasticity of the concrete should also be reduced to allow for creep effects.

A flexible foundation increases the P-Delta effects of gravity loading. The erection of a structure out of plumb also gives rise to P-Delta effects from gravity loading. The probable maximum values of these can be calculated from the erection tolerances specified in the local Design Code. If these exceed the usual P-Delta effects, and if they are significant, they should be included in the design calculations.

REFERENCES

16.1 Goldberg, J. E. "Approximate Methods for Stability and Frequency Analysis of Tall Buildings." *Proc. Regional Conf. on Tall Buildings*, Madrid, Spain, September 1973, pp. 123-146.

16.2 Gluck, J. and Gellert, M. "On the Stability of Elastically Supported Cantilever with Continuous Lateral Restraint." *Int. J. Mech. Sci.* **13**, 1971, 887-891.

16.3 Rosman, R. "Stability and Dynamics of Shear-Wall Frame Structures." *Build. Sci.* **9**, 1974, 55-63.

16.4 Timoshenko, S. P. and Gere, J. M. *Theory of Elastic Stability*. Engineering Society Monograph, McGraw-Hill, New York, 1961.

16.5 Gaiotti, R. and Stafford Smith, B. "P-Delta Analysis of Building Structures." *J. Struct. Engineer.* ASCE, **115**(4), April 1989, 755-770.

16.6 Goldberg, J. "Approximate Methods in Stress and Stability Analysis of Tall Building Frames." *Proc. Regional Conf. on Tall Buildings*, Bangkok, Thailand, January 1974, pp. 177-194.

16.7 Nixon, D., Beaulieu, D., and Adams, P. F. "Simplified Second-Order Frame Analysis." *Can. J. Civil Engineer.* **12**(4), 1975, 602-605.

16.8 Rutenberg, A. "A Direct P-Delta Analysis Using Standard Plane Frame Computer Programs." *Computers Structures* **14**(1-2), 1981, 97-102.

16.9 Rutenberg, A "Simplified P-Delta Analysis for Asymmetric Structures." *J. Struct. Div. ASCE*, **108**, ST9, September 1982, 1995-2013.

16.10 "Structural Design of Tall Concrete and Masonry Buildings." *Monograph on Planning and Design of Tall Buildings*, ASCE, **CB** 1978, 371-376.

16.11 Kordina, K. "Cracking and Crack Control." *Planning and Design of Tall Buildings*, Proc. ASCE—IABSE Int. Conf., 1972, Vol. III, No. 24-D2.

16.12 Hage, S. E. "The Second-Order Analysis of Reinforced Concrete Frames." *Struct. Engineer. Report No. 49*, University of Alberta, Edmonton, 1974.

16.13 Dicke, D. "Structural Design of Tall Concrete and Masonry Buildings." *Monograph on Planning and Design of Tall Buildings*, ASCE, **CB**, 1978, 369.

CHAPTER 17

Dynamic Analysis

Advances in structural design concepts and analytical techniques, combined with the availability of newer and more efficient materials and construction methods for buildings, have led to significant reductions in their structural weight and stiffness. In addition, heavy masonry cladding and partition walls, which were very effective in increasing the stiffness of structural frames, are less frequently used. Consequently, the more typical light and flexible modern tall building is much more responsive to dynamic exciting forces than its earlier counterpart. The resulting dynamic stresses may be much greater than static values, while induced motions may disturb the comfort and equanimity of the building's occupants.

Tall building motions may be classified as static or dynamic. The first refers to the motions produced by slowly applied forces such as gravitational or thermal effects, or the long period component of wind. Dynamic motions refer to those caused by time-dependent dynamic forces, notably seismic accelerations, short period wind loads, blasts, and machinery vibrations, the first two usually being of the greatest concern. Although the deformations arising from static forces may be of possible detriment to the integrity of the structure, unless they lead to the collapse of the building they are unlikely to provoke any reaction from the occupants.

Dynamic wind pressures produce sinusoidal or narrow-band random vibration motions of the building, which will generally oscillate in both along-wind and cross-wind directions, and possibly rotate about a vertical axis. The magnitudes of the three displacement components will depend on the velocity distribution and direction of the wind, and on the shape, mass, and stiffness properties of the structure. In certain cases, the effects of cross-wind motions of the structure may be greater than those due to the along-wind motions.

The ground shaking which occurs in an earthquake may be described as a series of virtually multi-directional random acceleration pulses. The ground movements will generally produce simultaneous translations along and rocking about the two orthogonal horizontal axes, as well as displacement along and torsion about the vertical axis of the structure. For normal buildings, however, it is generally sufficient for design purposes to consider only translatory accelerations in the two horizontal directions, together with the associated vertical axis torsional actions. An accelerogram recorded during the occurrence of an earthquake shows an irregularly timed sequence of both positive and negative peaks of acceleration having varying amplitudes. The response of a building to such acceleration-time histories may be determined from an elastic analysis, unless the inertial forces are sufficiently large to cause inelastic deformations or localized failures.

The intensity of the ground shaking to which the building may be subjected during its lifetime can be estimated from the recorded earthquake history in the area in which it is situated. Continuing records are used to produce maps showing regions of relative possible seismic hazards, and these can be extended and refined as knowledge of such events is accumulated. When designing a tall building to resist seismic forces, the design loads may be determined from a dynamic analysis of the building's response to time-history base accelerations, based on an actual recorded local event, or an artificially generated time-history. Such a time-consuming rigorous approach may be simplified by the use of earthquake response spectra, which, although requiring less computational effort, yield acceptably similar results for peak responses.

The seismic response of the building will depend on the dynamic properties of the structure, the ground motion at the foundation, and the mode of soil-structure interaction. The motion of a very stiff building will be almost identical to the ground motion, but that of a flexible structure will be quite different. The response will depend on the proximity of the natural frequencies of the structure to that of the predominant ground-motion frequency, the damping inherent in the structure, the foundation behavior, the ductility of the structure, and the duration of the earthquake.

The nature and magnitude of both wind and earthquake loading on buildings have been discussed in Chapter 3. Although both are dynamic and transient in character, it has been shown that for design purposes they may be frequently replaced by equivalent static loads, which are chosen to represent their probable worst magnitudes. The equivalent static loads for wind effects will be based on a statistical knowledge of the likely occurrence and magnitude of wind velocities and pressures, and for earthquakes on the time-history of accelerations. For the majority of tall buildings, the quasistatic loadings are adequate for design purposes, and have proved satisfactory in most situations. A dynamic analysis is required only when the building is relatively flexible or, because of its shape, structural arrangement, mass distribution, foundation condition, or use, is particularly sensitive to wind or seismic accelerations. Then consideration has to be given to both the stress levels that occur and the accelerations that may affect the comfort of the occupants.

This chapter considers the particular circumstances under which the designer may need to undertake a study of the dynamic response, and examines briefly the techniques available for the analysis. The field of structural dynamics is very extensive, and is well covered by existing textbooks. Consequently, this chapter will highlight briefly only the major techniques that are important for tall structures. Finally, consideration is given to the human response to dynamic motions and its effect on structural design.

17.1 DYNAMIC RESPONSE TO WIND LOADING

A complete description of the wind loading process relies on a proper definition of the wind climate from meteorological records, together with an understanding of atmospheric boundary layers, turbulence properties and the variation of wind speed with height, the aerodynamic forces produced by the interaction of the building with the turbulent boundary layer, and the dynamic response of the structure to the wind forces. A detailed consideration of all these effects is beyond the scope of this chapter, and the interested reader is referred to specialist texts on the topics concerned [e.g., 17.1].

The object of this section is to consider briefly the problem of deciding when a dynamic analysis of a tall building is necessary and how to determine the peak dynamic response of the structure to fluctuating wind forces.

17.1.1 Sensitivity of Structures to Wind Forces

The principal structural characteristics that affect the decision to make a dynamic design analysis are the natural frequencies of the first few normal modes of vibration and the effective size of the building. When a structure is small, the whole building will be loaded by gusts so that the full range of frequencies from both boundary layer turbulence and building-generated turbulence will be encountered. On the other hand, when the building is relatively large or tall, the smaller gusts will not act simultaneously on all parts, and will tend to offset each other's effects, so that only the lower frequencies are significant.

If the structure is stiff, the first few natural frequencies will be relatively high, and there will be little energy in the spectrum of atmospheric turbulence available to excite resonance. The structure will thus tend to follow any fluctuating wind forces without appreciable amplification or attenuation. The dynamic deflections will not be significant, and the main design parameter to be considered is the maximum loading to which the structure will be subjected during its lifetime. Such a structure is termed "static," and it may be analyzed under the action of static equivalent wind forces, as described in Chapter 3.

If a structure is flexible, the first few natural frequencies will be relatively low, and the response will depend on the frequency of the fluctuating wind forces. At frequencies below the first natural frequency, the structure will tend to follow closely the fluctuating force actions. The dynamic response will be attenuated at frequencies above the natural frequency, but will be amplified at frequencies at or near the natural frequency; consequently the dynamic deflections may be appreciably greater than the static values. The lateral deflection of the structure then becomes an important design parameter, and the structure is classified as "dynamic." In such structures, the dynamic stresses must also be determined in the design process. Furthermore, the accelerations induced in dynamic structures may be important with regard to the comfort of the occupants of the building and must be considered.

When a structure is very flexible, its oscillations may interact with the aerodynamic forces to produce various kinds of instability, such as vortex-capture resonance, galloping oscillations, divergence, and flutter. In this exceptional case, the potential for disaster is so great that the design must be changed or the aerodynamic effects modified to ensure that this form of unstable behavior cannot occur.

It is thus important for the engineer to be able to determine in the early design stages if the structure is static or dynamic, particularly in view of the comfort criteria for the occupants. To rectify an unacceptable dynamic response after the structure has been built will, if at all possible, generally be difficult and very expensive.

Unfortunately, it is not yet possible, particularly in the early design stages, to assess accurately whether a dynamic analysis will be required, although several empirical guidelines are available in Design Codes. For example, the Australian Code [17.2] defines a dynamic building as one in which

1. the height exceeds five times the least plan dimension and
2. the natural frequency in the first mode of vibration is less than 1.0 Hz.

In the Canadian Code [17.3] dynamic buildings are those whose height is greater than four times their minimum effective width, or greater than 120 m in height. Such empirical guidelines should be considered applicable only to traditional forms of building structure, and may be inappropriate to apply to more radical innovations.

A more sophisticated, but necessarily more complicated, approach for structural classification has recently been devised [17.4]. It allows a judgment on whether a structure is potentially dynamic, that is, not stiff enough to be analyzed by static methods alone, but sufficiently stiff to prevent aerodynamic instability. The procedure requires a knowledge of the damping ratio and the lowest natural frequency of vibration of the structure. In the early stages of design, it is not possible to calculate accurately the fundamental natural frequency, but empirical formulas are available to allow an assessment to be made (cf. Section 17.2.3). The question of damping ratios is considered in Section 17.2.4.

17.1.2 Dynamic Structural Response due to Wind Forces

The prediction of the structural response involves two stages: (1) the prediction of the occurrence of various mean wind speeds and their associated directions, and (2) given the occurrence of that wind, the prediction of the maximum dynamic response of the structure. The former requires an assessment of the wind climate of the region, adjusted to take account of the local topography of the site, and of the local wind characteristics (mean velocity profile and turbulence structure). The steady pressures and forces due to the mean wind, and the fluctuating pressures on the exterior, may then be determined. The properties of the mean wind can be conveniently expressed only in statistical terms.

Although the design of cladding may be strongly influenced by local pressures, the response of the building as a whole depends on the integrated values over the different faces of the building.

The exciting forces on a structure due to wind actions tend to be random in amplitude and spread over a wide range of frequencies. The structure's response is dominated by the action of its resonant response to wind energy available in the narrow bands close to the natural frequencies of the structure. The major part of the exciting energy will generally be at frequencies much lower than the fundamental natural structural frequency, and the amount of energy decreases with increasing frequency. Consequently, for design purposes, it is usually necessary to consider the structure's response only in the fundamental modes; the contribution from higher modes is rarely significant.

The fluctuations in response of a structure can be considered as those associated with the mean wind speed, and those associated with the turbulence of the wind, which are predominantly dynamic in character. Consequently, it is convenient to describe windspeeds, forces, deflections, etc. in terms of an hourly mean value together with the average maximum fluctuation likely to occur in an hour. When these are added, the sum can be used as an average hourly maximum, or peak response, to define equivalent static design data.

The peak value can be expressed statistically in terms of the number of standard deviations by which the peak exceeds the mean value. For design purposes, the conventional practice is to define the peak value of the variable, $x(\max)$ say, by the relationship

$$x(\max) = \bar{x} + g_p \sigma \qquad (17.1)$$

where $x(\max)$, $\bar{x}$, and σ are the peak, mean, and standard deviations, respectively, of the variable x concerned, referred to a record period of 1 hour, and g_p is the "peak" factor.

When considering the response of a tall building to wind actions, both along-wind and cross-wind motions must be considered. These arise from different forcing mechanisms, the former being due primarily to buffeting effects caused by turbulence, while the latter is due primarily to alternate-side vortex shedding. The cross-wind response may be of particular importance with regard to the comfort of the occupants.

If knowledge of the mass, stiffness, and damping properties of the structure is available, a time-history of its response to any applied forcing function can be achieved by an integration of the dynamic equation of motion. However, such a detailed history involving a lengthy analysis is not essential since only the peak responses (moments, deflections, accelerations, etc.) are required for design purposes. Fortunately, the experience gained in large numbers of wind tunnel tests on models for prototype designs, coupled with the very considerable research effort and advances in both experimental and analytical wind engineering techniques, have allowed the inclusion in Design Codes of simplified methods for the dynamic analysis of tall buildings subjected to wind forces. Two such approaches are described in the succeeding sections.

17.1.3 Along-Wind Response

The pioneering work of Davenport [17.5] and Vickery [17.6] has shown that the along-wind response of most structures is due almost entirely to the action of the incident turbulence of the longitudinal component of the wind velocity, superimposed on a mean displacement due to the mean drag. The resulting analytical meth-

424 DYNAMIC ANALYSIS

ods, using spectral and spatial correlation considerations to predict the structural response, have been developed to such a level that they are now employed in modern design wind Codes. The work has led to the development of the gust factor method for the prediction of the building response.

The gust factor method is based on the assumption that the fundamental mode of vibration of a structure has an approximately linear mode shape. In essence, the aim of the method is to determine a gust factor G that relates the peak to mean response in terms of an equivalent static design load, or load effect $\overline{Q}$, such that,

$$\text{Design value, } Q(\max) = G\overline{Q} \qquad (17.2)$$

where $\overline{Q}$ defines the mean value of the quantity concerned.

For example, if the mean pressures acting on the face of a tall building are summed to give the mean base overturning moment $\overline{M}$, the design dynamic base overturning moment $M(\max)$ will be obtained by multiplying $\overline{M}$ by the gust factor G.

$$M(\max) = G\overline{M} \qquad (17.3)$$

As described in Section 17.2, similar approaches are used to determine the peak response due to random earthquake loading.

A schematic representation of Davenport's design procedure is shown in Fig. 17.1. The procedure is a combination of two parts: the first involves the modeling of the wind forces, and the second involves the use of structural dynamic analysis to determine the resulting response. In the diagram, the force spectrum is found from the wind velocity spectrum, represented by an algebraic expression based on

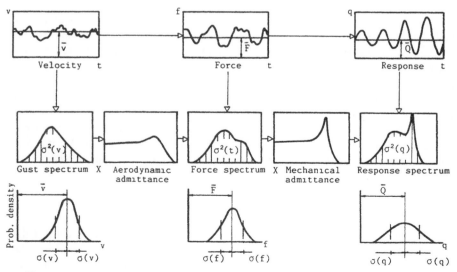

Fig. 17.1 Graphic representation of the design process (after Davenport [17.5]).

observations of the real wind, through the aerodynamic admittance, which relates the size of the gust disturbance to the size of the structure. The aerodynamic admittance may be determined theoretically [17.3], from an empirical formula [17.8], or measured experimentally in a wind tunnel. To find the response of the structure in this mode to the force spectrum, it is necessary to know the mechanical admittance, which is a function of the natural frequency, the damping, and the stiffness of the structure. The mechanical admittance has a sharp resonance peak at the natural frequency, similar in form to the dynamic magnification curve found in the response of dynamic systems. As a result of the peak in the mechanical admittance function, the response has a peak at the natural frequency, the amplitude of which is determined by the damping present. For the orders of damping found in most buildings, this peak usually contains most of the area in the response spectrum. For this reason, most of the fluctuations take place at or near the natural frequency. The area under the loading effect spectrum is taken as the sum of two components, the area under the broad hump of the diagram, which must be integrated numerically for each structure, and the area under the resonance peak, for which a single analytic expression is available. These background and resonant excitation components are represented in Eq. (17.4) by B and R, respectively, combined vectorially to give the peak response.

In Davenport's analysis, the response of a tall slender building to a randomly fluctuating wind force is determined by treating it as a rigid spring-mounted cantilever whose dynamic properties are specified by the fundamental natural frequency n_0 and an appropriate damping ratio β. Consequently, only a single linear mode of vibration need be considered.

Expressions for the gust factor are given in a number of publications. In this section, the approach followed is that of the NBCC [17.3], which is representative of the different formulations. All formulations are of the same form, and the associated design curves have similar forms.

The gust factor can be regarded as a relationship between the wind gusts and the magnification due to the structural dynamic properties. As such, it will depend on the properties of the structure (height H, and breadth/height ratio W/H), fundamental natural frequency n_0, and critical damping ratio β, the mean design wind speed V, and the particular location of the building (i.e., whether it is sited in the center of a large city, in suburbs or wooded areas, or in flat open country).

It may be shown [17.5] that the gust factor G may be expressed as

$$G = 1 + g_p r (B + R)^{1/2} \qquad (17.4)$$

In Eq. (17.4), g_p is a peak factor that accounts for the time history of the excitation, and is determined from the duration time T over which the mean velocity is averaged and the fundamental frequency of vibration n_0; in practice, T is taken as 3600 sec (1 hour). r is a roughness factor, which depends on the location and height of the building (Fig. 17.2); B is the excitation due to the background turbulence or gust energy, which depends on the height and aspect ratio of the building (Fig. 17.3); and R is the excitation by the turbulence resonant with the structure, which depends on the size effect S, the gust energy ratio at the natural

426 DYNAMIC ANALYSIS

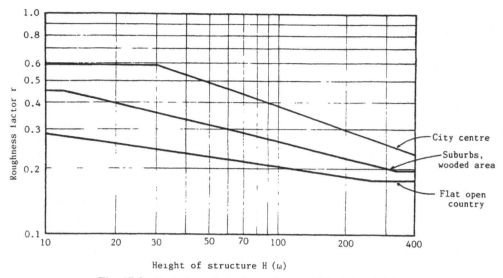

Fig. 17.2 Variation of roughness factor with building height.

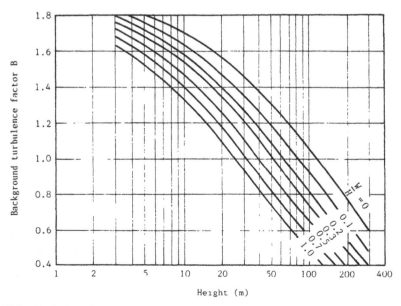

Fig. 17.3 Variation of background turbulence factor with height and aspect ratio of building.

frequency of the structure, F, and the critical damping ratio β, that is,

$$R = \frac{SF}{\beta} \qquad (17.5)$$

The size reduction factor S depends on the aspect ratio W/H, the natural frequency n_0, and the mean wind velocity at the top of the structure, V_H, as shown in Fig. 17.4. The gust energy ratio F is a function of the inverse wavelength, n_0/V_H, as shown in Fig. 17.5.

If resonant effects are small, then R will be small compared to the background turbulence B, and vice versa.

The peak factor g_p in Eq. (17.4) gives the number of standard deviations by which the peak load effect is expected to exceed the mean load effect, and is shown in Fig. 17.6 as a function of the average fluctuation rate v given by

$$v = \frac{n_0}{\sqrt{(1 + B/R)}} \qquad (17.6)$$

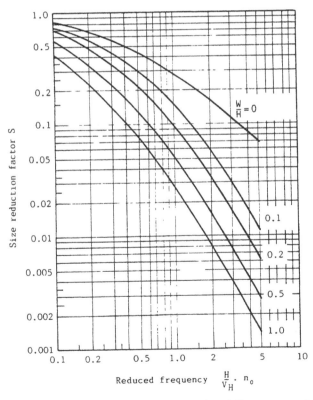

Fig. 17.4 Variation of size reduction factor with reduced frequency and aspect ratio of building.

428 DYNAMIC ANALYSIS

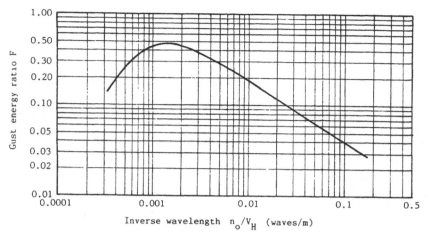

Fig. 17.5 Variation of gust energy ratio with inverse wavelength.

In the above formulas, the variables V_H, n_0, and β must relate to the along-wind direction.

Substitution of the known values of g_p, r, B, and R into Eq. (17.4) then produces the desired value of the gust factor.

Once the gust factor G has been determined, the peak dynamic forces and displacements may be determined by multiplying the values due to the mean wind loading by G.

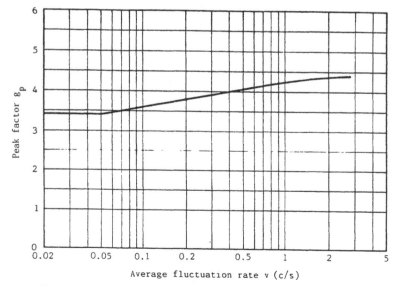

Fig. 17.6 Variation of peak factor with average fluctuation rate.

Peak Along-Wind Accelerations. As discussed in Section 17.3, the most important criterion for the comfort of the building's occupants is the peak acceleration they experience. It is thus important to be able to estimate at an early stage in design the likely maximum accelerations in both the along-wind and across-wind directions.

The maximum acceleration a_D in the along-wind direction may be estimated from the expression [17.3]

$$a_D = 4\pi^2 n_0^2 g_p r \sqrt{R}\left(\frac{\Delta}{G}\right) \quad (17.7)$$

where Δ = the maximum wind-induced deflection at the top of the building in the along-wind direction (m).

The natural frequency n_0 and damping ratio β must be again in the along-wind direction. The other symbols have been defined previously in connection with Eq. (17.4).

17.1.4 Cross-Wind Response

The cross-wind excitation of tall buildings is due predominantly to vortex shedding. However, generalized empirical methods of predicting the response have been difficult to derive, even assuming that the motions are due entirely to wake excitation, because of the effects of building geometry and density, structural damping, turbulence, operating reduced frequency range, and interference from upstream buildings. The last effect can alter significantly the cross-wind motions. Consequently, as yet, the most accurate method of determining the cross-wind structural response has been from tests on an aeroelastic model in a wind tunnel.

The work of Saunders, Melbourne, and Kwok [17.9, 17.10], using the results of empirical wind tunnel data, has led to an approximate analysis that can take into account the most important variables concerned. The technique employed to calculate the response due to wake excitation is to solve the equation of motion for a lightly damped structure in modal form with the forcing function mode generalized in spectral format [17.10].

Although it is generally found that the maximum lateral wind loading and deflection are in the along-wind direction, the maximum acceleration of the building, which is particularly important for human comfort, may often occur in the cross-wind direction. Across-wind accelerations are likely to exceed along-wind accelerations if the building is slender about both axes, such that the geometric ratio $\sqrt{WD}/H$ is less than one-third, where D is the along-wind plan dimension [17.3].

Based on a wide range of turbulent boundary layer wind tunnel studies, a tentative formula is given in the NBCC for the peak acceleration a_w at the top of the building, namely,

$$a_w = n_0^2 g_p [WD]^{1/2} \left(\frac{a_r}{\rho g \sqrt{\beta}}\right) \quad (17.8)$$

where

ρ = average density of the building (kg/m^3)
$a_r = 78.5 \times 10^{-3} [V_H/(n_0\sqrt{WD})]^{3.3}$ (Pa)
g = acceleration due to gravity (m/sec^2)

Because of the relative sensitivities of the expressions in Eqs. (17.7) and (17.8) to the natural frequencies, it is recommended that the latter be determined using fairly rigorous analytical methods, and that approximate formulas (cf. Section 17.2.3) be used with caution.

17.1.5 Worked Example

To illustrate the calculations involved in the estimation of the peak wind load effects, the example is considered of a tall square office building in the center of a large city. The following data are given:

Height $H = 180$ m
Breadth B = Depth $D = 30$ m
Estimated fundamental natural frequency $n_0 = 0.2$ Hz
Estimated critical damping ratio $\beta = 0.015$
Mean wind speed at top of building $V_H = 35$ m/sec
Estimated maximum deflection at top of building [17.3] $\Delta = 0.36$ m
Estimated average building density $\rho = 175$ kg/m^3

The required calculations are as follows:

1. Gust Factor
   From Fig. 17.2, roughness factor $r = 0.305$
   Aspect ratio $W/H = 30/180 = 0.167$
   $\therefore$ From Fig. 17.3, background turbulence factor $B = 0.64$
   Reduced frequency $n_0 \dfrac{H}{V_H} = \dfrac{0.2 \times 180}{35} = 1.029$
   $\therefore$ From Fig. 17.4, size reduction factor $S = 0.11$
   Inverse wavelength $n_0/V_H = 0.2/35 = 0.00571$
   $\therefore$ From Fig. 17.5, gust energy ratio $F = 0.27$
   $\therefore$ Resonant turbulence factor $R = \dfrac{SF}{\beta} = \dfrac{0.11 \times 0.27}{0.015} = 1.98$

That is, the resonant turbulence exitation is greater than the background turbulence excitation.

From Eq. (17.6), average fluctuation rate $\nu = \dfrac{0.2}{\sqrt{1 + (0.64/1.98)}}$
$= 0.174$

∴ From Fig. 17.6, peak factor $g_p = 3.75$
∴ From Eq. (17.4), gust factor $G = 1 + 3.75 \times 0.307 \sqrt{0.64 + 1.98}$
$$= 2.85$$

That is, the peak dynamic forces and displacements are obtained by multiplying all static values due to the mean wind loading by 2.85.

2. Along-wind acceleration

From Eq. (17.7), $a_D = 4\pi^2 \times (0.2)^2 \times 3.75 \times 0.305 \sqrt{1.98} \times \dfrac{0.36}{2.85}$
$$= 0.32 \text{ m/sec}^2 \ (3.3\% \text{ g})$$

3. Across-wind acceleration

$$a_r = 78.5 \times 10^{-3} \left(\dfrac{35}{0.2 \times 30}\right)^{3.3} = 26.448$$

∴ From Eq. (17.8), $a_w = (0.2)^2 \times 3.75 \times 30 \times \dfrac{26.448}{175 \times 9.81 \times \sqrt{0.015}}$
$$= 0.57 \text{ m/sec}^2 \ (5.8\% \text{ g})$$

The peak across-wind accelerations are therefore considerably greater than the peak along-wind accelerations in this case.

Great progress has been made over the last three decades in understanding the wind climate and the effect of turbulent winds on tall buildings. It has been possible only to touch on the subject in this section and the interested reader is referred to [17.11] and to the Proceedings of the various International Conferences on Wind Effects on Buildings and Structures for a more comprehensive treatment of the subject.

17.2 DYNAMIC RESPONSE TO EARTHQUAKE MOTIONS

This section describes how tall building structures are excited by ground motions during earthquakes, and how a response spectrum analysis may be used to obtain realistic estimates of the peak structural response.

17.2.1 Response of Tall Buildings to Ground Accelerations

For a structure subjected to ground accelerations $\ddot{u}_g$ in some particular direction, the governing equation of motion will always be of the form [17.14]

$$\mathbf{M}\ddot{\mathbf{u}} + \mathbf{C}\dot{\mathbf{u}} + \mathbf{K}\mathbf{u} = -\mathbf{M}\mathbf{I}\ddot{u}_g(t) \qquad (17.9)$$

in which $\mathbf{M}$, $\mathbf{C}$, and $\mathbf{K}$ are, respectively, the mass matrix, the viscous damping matrix, and the stiffness matrix, $\mathbf{u}$ is the displacement vector, and $\mathbf{I}$ is a unit vector.

432 DYNAMIC ANALYSIS

The fundamental problem is to determine the displacement response $u(t)$; internal forces and other response quantities of interest can subsequently be derived from the displacement response.

The masses to be used in the analysis of a building should be based on the known dead loads, plus probable values of the live loads, applying appropriate reductions to take account of the probability that not all floors will be fully loaded at any one time, as discussed in Chapter 3. The stiffness of the structure may be evaluated from the load-displacement relationships determined from the techniques described in earlier chapters. In steel structures, the flexibility of the joints, and, in concrete structures, the loss of stiffness due to cracking should be included when evaluating the stiffness of the structure.

Although the mass and stiffness matrices for the structure can be calculated from the dimensions and elastic properties of the structural and nonstructural elements, it is not possible to determine the damping matrix in a similar manner. Energy dissipation in a tall building is due to the combined effects of a number of sources, such as friction at structural joints and between structural and nonstructural components, inherent damping in the material, and microcracking in concrete. It is not yet possible to define quantitatively the local energy dissipating mechanisms to allow a direct evaluation of the damping matrix. Consequently, damping in the structure is normally estimated on an overall basis in terms of modal damping ratios, with values obtained from measurements on similar structures serving as a guide. The evaluation of damping ratios is discussed in Section 17.2.4.

Except in very special situations, only the three translational components (two horizontal and one vertical) are considered in the design calculations.

Time-histories of earthquake ground motions as measured by strong-motion accelerographs are now available for a number of earthquakes (e.g., Fig. 17.7). The accelerograph records the three orthogonal components of ground acceleration, each of which may be integrated to yield the corresponding velocity- and displacement-time histories. The earthquake accelerogram, or acceleration-time trace, can

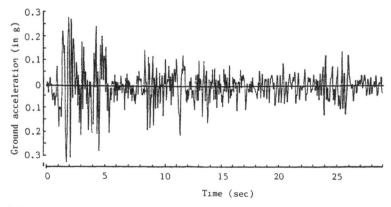

Fig. 17.7 Typical earthquake accelerogram. North-South component of ground acceleration recorded at El Centro (approximately 4 miles from fault).

be interpreted directly to obtain estimates of peak ground acceleration, duration of strong ground shaking, and frequency content.

The accelerogram can be digitised, so that the time-history is defined by the numerical ordinates of the accelerogram at time intervals spaced sufficiently closely to define accurately the ground accelerations. Various procedures have been developed [e.g., 17.12, 17.13] to allow the equation of motion [Eq. (17.9)] to be integrated numerically for the input time-history of accelerations [$\ddot{u}_g(t)$]. The solution gives the time-history of velocities and displacements of the structure. Once the displacement response history $\mathbf{u}(t)$ has been determined, the equivalent lateral force $\mathbf{F}(t)$ at any specific time can be obtained from the lateral stiffness matrix $\mathbf{K}$ as

$$\mathbf{F}(t) = \mathbf{K}\mathbf{u}(t) \qquad (17.10)$$

allowing the total base shear and overturning moment to be evaluated.

The complete history of any response quantity such as horizontal deflection, velocity, acceleration, base shear and moment, and equivalent lateral forces can thus be determined for any input base accelerogram.

A rigorous three-dimensional dynamic analysis of a tall building is a formidable task, particularly if the motions are sufficiently strong to cause significant yielding in the structure, and it would be an impractical requirement in the design of most buildings. Fortunately, for design purposes, it is not necessary to have a complete history of the structural response to any recorded earthquake or simulated design earthquake, since only the maximum response to the earthquake need be established.

Two techniques, the equivalent lateral force procedure and the modal analysis method, have been devised to allow the peak dynamic response to be determined directly. Both techniques lead to the consideration of lateral forces on the building in the direction of the ground motion, the main difference between the two lying in the magnitude and distribution of the lateral forces over the height of the building.

In the simpler equivalent static lateral force procedure, the magnitude of the forces is based on an estimate of the fundamental period of the structure, as well as on other factors such as the seismicity of the area, operational importance factor, the energy absorption capacity of the structural system, and the dead weight of the building, as discussed in Section 3.3.1. The total base shear derived from the design formula [Eq. (3.3)] is applied partly as a concentrated load at the top, to represent the effects of higher modes of vibration, and the remainder distributed over the height of the building in proportion to the product of the mass at each level and its height above the base, giving an approximately triangular distribution with the apex at the base. A static structural analysis will then give a response that should be a close estimate to the peak dynamic response.

In the more accurate modal analysis procedure, described in the next section, the lateral forces are based on the properties of the building's natural modes of vibration that are likely to be excited, which are functions of the distributions of mass and stiffness over the height. The earthquake properties are input through a response spectrum.

The two procedures have similar capabilities and are subject to similar limitations, and the direct results allow an estimation of the effects of lateral forces in the direction under consideration, that is, the story shears and moments, lateral deflections, and story drifts. The equivalent static lateral force procedure is the more suitable for preliminary design calculations, and is normally required even if a more refined technique is used for the final analysis. It requires a knowledge of the fundamental natural frequency of vibration of the building, for which various empirical formulas have been devised to allow an estimate to be made for preliminary design purposes, as discussed in Section 17.2.3. The more refined modal analysis procedure requires a knowledge of the mass and stiffness properties of the building; therefore, a preliminary design for the structure must be made before the method can be implemented. The modal analysis procedure has the advantage that it provides a more accurate lateral force distribution on the structure.

The simpler equivalent static lateral force procedure will generally be sufficiently accurate for the analysis of regular uniform buildings that are not exceptionally tall. For very tall or important buildings, buildings with significant irregularities in plan or elevation, buildings with setbacks, major discontinuities in mass or resistance, with large or irregular eccentricities between the centers of mass and stiffness, the modal analysis procedure should be used.

If there is doubt about whether the lateral force procedure is adequate, a quick calculation based on the equivalent lateral force procedure may be employed to determine whether a modal analysis is also advisable.

The sequence of calculations in such an assessment is as follows [17.14]:

1. Calculate the lateral forces and story shears from the equivalent lateral force procedure.
2. Approximately dimension the structural members.
3. Calculate the lateral deflections δ_i of the designed structure due to the lateral forces from Step 1.
4. Calculate new sets of lateral forces F_i and corresponding story shears from the formula

$$F_i = V_0 \frac{w_i \delta_i}{\sum_{j=1}^{N} w_j \delta_j}$$

in which V_0 is the base shear, w_i is the building mass lumped at the ith floor level, δ_i is the lateral deflection at the ith level, and N is the number of stories in the building.

5. If at any story the recalculated story shear from Step 4 differs from the original value Step in 1 by more than 30%, a modal analysis is necessary. If the difference is less than 30%, the modal analysis is unnecessary, and the structure should be designed using the story shears from Step 4.

Natural Frequencies and Modes of Vibration. The dynamic response of a structure to any exciting force is dependent on its vibration characteristics, defined by the natural frequencies ω_n and modes of vibration φ_n ($n = 1$ to N, for a system with N degrees of freedom). In the case of an undamped structure, and in the absence of any exciting force, the equation of motion becomes

$$\mathbf{M\ddot{u} + Ku} = 0 \qquad (17.11)$$

Assuming that the free vibration motion is simple harmonic, the displacement may be expressed in the form

$$\mathbf{u} = \varphi \sin(\omega t + \theta) \qquad (17.12)$$

where θ is the phase angle.

Substitution of Eq. (17.12) into Eq. (17.11) yields the governing eigenvalue equation,

$$\mathbf{K}\varphi = \omega^2 \mathbf{M}\varphi \qquad (17.13)$$

Many procedures have been developed for the numerical solution of the eigenproblem [17.12, 17.13] to give the N natural frequencies ω_n and the modes of vibration φ_n, and computer programs for this purpose are now widely available.

17.2.2 Response Spectrum Analysis

This section discusses the derivation of earthquake design response spectra, describes briefly the modal method of analysis of the dynamic behavior of structures, and shows how the two may be combined to estimate the peak response of the structure to earthquake ground motions.

Earthquake Design Response Spectra. Earthquake accelerograms show the irregularity of the ground accelerations as a function of time. Although these provide basic information about the nature of the ground motions, the structural engineer requires a more meaningful characterization for design purposes. This is provided by a response spectrum, which can be defined as a graphic representation of the maximum response of a damped single-degree-of-freedom (SDOF) mass-spring system with continuously varying natural periods to a given ground excitation.

The SDOF mass-spring system employed is represented in Fig. 17.8. The mass M is connected to the support by a spring of stiffness K acting in parallel with a dashpot to simulate the viscous damping in the system, equal to $C\dot{u}$. The support is assumed to displace by an amount $y(t)$ and the mass by $x(t)$, the relative displacement $u(t)$ being equal to $(x - y)$.

The equation of motion is given by,

$$M\ddot{u}(t) + C\dot{u}(t) + Ku = -M\ddot{y}(t) = -p(t) \qquad (17.14)$$

where p is the effective support excitation loading.

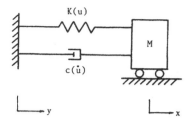

Fig. 17.8 Single degree of freedom damped spring mass oscillator system.

Equation (17.14) may be expressed alternatively in the form

$$\ddot{u}(t) + 2\beta\omega\dot{u}(t) + \omega^2 u(t) = -\ddot{y}(t) \tag{17.15}$$

in which $\omega = (K/M)^{1/2}$ is the natural frequency of vibration and $\beta = C/2M\omega$ is the fraction of critical damping, or damping ratio, where C is the damping coefficient and the critical damping ($2M\omega$) is the minimum value of C that results in a nonvibrating response.

The solution of Eq. (17.15) at time t is [17.12]

$$u(t) = -\frac{1}{\omega\sqrt{1-\beta^2}} \int_0^t [\ddot{y}(\tau) \exp[-\omega\beta(t-\tau)] \\ \cdot \sin\omega\sqrt{1-\beta^2}\,(t-\tau)] \, d\tau \tag{17.16}$$

in which τ is a dummy variable of integration.

Thus, before it is possible to determine the relative displacement-time history, it is necessary to know the acceleration-time history of the support, the natural frequency ω of the system, and the fraction of critical damping β, which is a measure of the structure's energy dissipative qualities. For any input acceleration $\ddot{y}$, the solution will yield the maximum absolute value of relative displacement u, termed the spectral displacement S_d, which will be a function of the natural frequency (or period) and damping factor.

The maximum pseudo relative velocity S_v and maximum absolute pseudo acceleration S_a are then given by

$$\begin{aligned} S_v &= \omega S_d \\ S_a &= \omega^2 S_d \end{aligned} \tag{17.17}$$

The pseudo acceleration is identical to the maximum acceleration when there is no damping, which, for normal levels of structural damping, is practically the same as the maximum acceleration.

As a result of the relationships described in Eqs. (17.16) and (17.17), the complete response spectrum may be represented on a tripartite logarithmic plot of the form shown in Fig. 17.9. The velocity is plotted on the vertical axis, while the displacement and acceleration are plotted on axes at $-45°$ and $+45°$ to the vertical, with the natural frequency plotted along the horizontal axis, all to a logarithmic scale. Alternatively, the horizontal axis may be used to show the period

17.2 DYNAMIC RESPONSE TO EARTHQUAKE MOTIONS   437

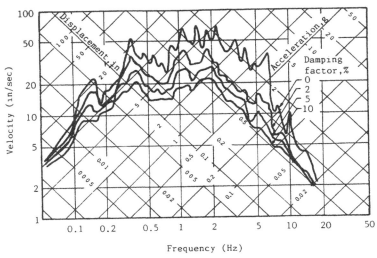

Fig. 17.9 Response spectra. El Centro earthquake, N-S direction (after Newmark and Hall [17.15]).

of vibration, the inverse of the frequency, which has the effect of turning the diagram from "front to back." The response spectra shown refer to the well-documented El Centro earthquake of May 1940, N-S direction, and are for different damping factors [17.15]. The jagged response is a plot of the maximum response of different oscillators to a given accelerogram, and is therefore a description of a particular ground motion. The value of the response spectrum is that it provides a more significant and meaningful measure of the effect of an earthquake motion than just a single value, such as the peak acceleration, does.

Although the actual response spectra for earthquake motions are quite irregular, they have the general shape of a trapezoid when plotted in tripartite logarithmic form as in Fig. 17.9. For design purposes, the actual response spectrum is normally smoothed to produce a curve that consists only of straight line portions, as shown in Fig. 17.10. The smoothing is performed on a statistical basis, in recognition of the fact that the detailed response spectrum of any future earthquake is unknown. The linear form is also more appropriate because of the difficulty of calculating exactly what the period of a tall building will be during strong shaking, especially as a nonlinear response is highly probable.

To establish the ground-motion characteristics of the design earthquake, and the maximum probable earthquake for any specific location, it is necessary to study the seismic history of the region concerned. This will allow an estimation of both the design and the probable maximum earthquake in terms of their magnitudes and distances from the particular building site. It may be possible to average the response spectra of more than one earthquake to give a more meaningful design input, and it is also possible to use probabilistic theory [17.12] to construct simulated accelerograms and design response spectra. Design spectra may be prescribed in the local or national design code concerned, such as in the NBCC [17.3].

438 DYNAMIC ANALYSIS

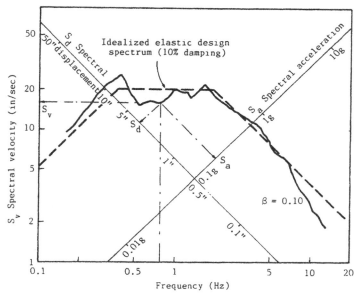

Fig. 17.10 Idealized design response spectrum.

If the natural frequency of a structure is calculated, and the degree of damping present is estimated, other corresponding important design parameters such as the maximum displacement and maximum acceleration can be obtained directly from the response spectrum diagram.

For a building with a known frequency of oscillation ω, and an estimated damping ratio β, a response spectrum diagram such as that shown in Fig. 17.10 can be used with the modal method of analysis, described in the next section, to determine the peak response of the structure to the design earthquake.

Modal Analysis Procedure. In general, the set of governing dynamic equations of motion [Eq. (17.9)] must be solved simultaneously by available computational procedures [17.12, 17.13] to determine all displacements u that define the motions of the structure. This approach can be avoided by using the computationally more efficient modal method of analysis. The method, which is based on linear elastic structural behavior, employs the superposition of a limited number of modal peak responses, as determined from a prescribed response spectrum, and with appropriate modal combination rules it will yield results that compare closely with those from a time-history analysis.

This method of analysis is based on the fact that for certain forms of damping that are reasonable approximations for many buildings, the equations of motion can be uncoupled so that the response in each natural mode of vibration can be calculated independently of the others [17.12]. Each mode will respond with its own particular displacement profile, the mode of vibration φ_n, its own frequency, the natural frequency of vibration ω_n, and with its own modal damping, the damping ratio β_n.

17.2 DYNAMIC RESPONSE TO EARTHQUAKE MOTIONS

In the structural idealization, the mass is usually lumped at the floor levels. Only one degree of freedom per floor, the horizontal deflection for which the structure is being analyzed, is used, and so the matrices involved are of the same order as the number of stories N in the building.

The uncoupling of the N equations yields the typical equation of motion for the nth natural mode as [17.16]

$$\ddot{Y}_n + 2\beta_n \omega_n \dot{Y}_n + \omega_n^2 Y_n = -\frac{L_n}{M_n} \ddot{u}_g(t) \qquad (17.18)$$

where

$$L_n = \sum_{j=1}^{N} m_j \varphi_{jn}$$

and the modal mass

$$M_n = \sum_{j=1}^{N} m_j \varphi_{jn}^2$$

and m_j is the mass at the jth floor level.

Equation (17.18) is of the same form as that for the single-degree of freedom dynamic system [Eq. (17.15)] with natural frequency ω_n and damping ratio β_n, excited by a ground excitation $(L_n/M_n)\ddot{u}_g(t)$.

In Eq. (17.18), the displacement function Y_n is the normal or generalized coordinate, or modal amplitude, for the nth natural mode, used to simplify the equations of motion [Eq. (17.9)] [17.12]. The geometric coordinate u_n in Eq. (17.9) is equal to the product of the generalized coordinate Y_n and the mode-shape vector φ_n.

Thus,

$$Y_n(t) = -\frac{L_n}{M_n} \frac{1}{\omega_n \sqrt{1-\beta_n^2}} \int_0^t [\ddot{u}_g(\tau) \exp[-\beta_n \omega_n (t-\tau)] \\ \cdot \sin \omega_n \sqrt{1-\beta_n^2}\,(t-\tau)]\,d\tau \qquad (17.19)$$

In Eq. (17.19), it is convenient to define the quotient L_n/M_n as the participation factor, γ_n, for the nth mode. Physically, the participation factor then represents the effective contribution of the mass for the particular mode considered.

The contribution of the nth mode to the modal displacement $u_j(t)$ at the jth floor is then equal to the product of the amplitude generalized coordinate and the mode shape,

$$u_{jn}(t) = Y_n(t)\varphi_{jn} \qquad (17.20)$$

To determine the dynamic story shears and moments, it is convenient to introduce the concept of equivalent lateral forces, defined as the static external forces p that would produce the same structural displacements u. Hence, at any time t, the equivalent forces corresponding to the modal displacements $u_n(t)$ will be, from

Eqs. (17.10) and (17.13),

$$P_n(t) = Ku_n(t) = K\varphi_n Y_n(t) = \omega_n^2 M \varphi_n Y_n(t) \tag{17.21}$$

The equivalent lateral force at the jth floor level is then,

$$p_{jn}(t) = \omega_n^2 m_j \varphi_{jn} Y_n(t) \tag{17.22}$$

in which Y_n is given by Eq. (17.19)

The contributions from each mode may then be summed to give the total equivalent force P at each floor level. For example, at level j,

$$P_j = \sum_{n=1}^{N} p_{jn}(t) \tag{17.23}$$

The internal dynamic shears and moments at any level can then be obtained by summing all the story forces and the moments of these forces above the level concerned.

In a similar manner, the displacement at any level may be obtained by combining the responses from each mode at that position. The drift U_N at the top of the building is then

$$U_N = \sum_{n=1}^{N} u_{Nn} \tag{17.24}$$

and the interstory drift is given by the difference between the total displacements of the floors above and below the level concerned.

The great attraction of this method is that an independent analysis can be made of a single-degree-of-freedom system for each natural mode of vibration. The response generally needs to be determined for only the first few modes since the total response to earthquakes is primarily due to the lowest modes of vibration. Sufficiently accurate design values of forces and deformations in tall buildings should be achieved by combining no more than about six modes in each component direction. Three would probably suffice for medium-rise buildings. Checks may be carried out since the relative influence of each successive mode on the important design parameters may be examined during the calculations.

The earthquake response is obtained by combining the contributions of all the modes of vibration involved, and this can be used to give a complete time-history of the structural actions. However, only the evaluation of the peak response is of importance in design, and this may be derived directly from the design response spectrum.

Design Response Spectrum Analysis. Since in the modal analysis the response of the structure in each mode of vibration is derived from a single-degree-of-freedom system, the maximum response in that mode can be obtained directly from the earthquake design response spectrum.

17.2 DYNAMIC RESPONSE TO EARTHQUAKE MOTIONS

The maximum response in the nth mode can be expressed in terms of the ordinates of the displacement S_{dn}, pseudo velocity S_{vn}, and pseudo acceleration S_{an}, which correspond to the frequency ω_n and damping ratio β_n. The three quantities are related by, from Eqs. (17.17)

$$S_{an} = \omega_n S_{vn} = \omega_n^2 S_{dn} \qquad (17.25)$$

Expressed in terms of the modal participation factor γ_n, the maximum values of the modal response quantities then become, from Eqs. (17.18) to (17.24):
Maximum modal displacement

$$\overline{Y}_n = \gamma_n S_{dn} \qquad (17.26)$$

Maximum displacement at jth floor

$$\overline{u}_{jn} = \gamma_n S_{dn} \varphi_{jn} \qquad (17.27)$$

Maximum interstory drift in jth story

$$\overline{\Delta}_{jn} = \gamma_n S_{dn} (\varphi_{jn} - \varphi_{j-1,n}) \qquad (17.28)$$

Maximum value of equivalent lateral force at jth floor $\overline{P}_{jn}$

$$\overline{P}_{jn} = \gamma_n S_{an} m_j \varphi_{jn} \qquad (17.29)$$

In Eqs. (17.26) to (17.29), a bar above a particular variable is used to denote the maximum value of the quantity concerned.

The maximum values of the internal forces in the building, particularly the story shear and moments, are then obtained by a static analysis of the structure, taking due account of the senses of these equivalent forces. Although they act in the same direction for the lowest natural modes, they may act in opposite directions in the higher modes.

The maximum base shear $\overline{V}_{0n}$ and base moment $\overline{M}_{0n}$ will then be, using Eq. (17.29).

$$\overline{V}_{0n} = \sum_{j=1}^{N} \overline{P}_{jn} = \gamma_n S_{an} \sum_{j=1}^{N} m_j \varphi_{jn} = \gamma_n L_n S_{an} \qquad (17.30)$$

$$\overline{M}_{0n} = \sum_{j=1}^{N} h_j \overline{P}_{jn} = \gamma_n S_{an} \sum_{j=1}^{N} h_j m_j \varphi_{jn} \qquad (17.31)$$

where h_j is the distance from the jth floor to the base.

The maximum modal response can thus be expressed in terms of the displacements or accelerations, evaluated for the particular frequency and damping ratio for the mode, from the design response spectrum.

Equations (17.25), (17.27), and (17.29) show that if the displacements and forces are both expressed in terms of the spectral displacement S_{dn}, the forces are multiplied by the square of the natural frequency. Consequently, the higher modes will be of greater significance in defining the forces in the structure than they are in the deflections, and it will be necessary to include more modal components to evaluate the forces to the same degree of accuracy as the deflections.

The total response R of the building to earthquake motions is the sum of the individual responses r_n of the natural modes. However, the maximum total response $\bar{R}$ is not generally equal to the absolute sum of the maximum modal responses, $\bar{r}_n$, since they will not normally occur simultaneously. Such a sum would, however, give an upper bound to the maximum likely total response.

A more realistic design estimate of the maximum response is to combine the modal maxima according to the square root of the sum of the squares (SRSS) method,

$$\bar{R} = \sqrt{(\Sigma \bar{r}_n^2)} \qquad (17.32)$$

The maximum values of displacements, interstory drifts, story shears, and moments may all be evaluated using Eq. (17.32).

This formula will generally give realistic estimates of peak response for structures in which the natural frequencies of vibration are well separated, a property that is usually valid for idealized building structures in which lateral displacements in one plane are considered. If this is not the case, and some natural frequencies are so close that the motions may be coupled together, a more realistic combination, such as the complete quadratic combination method, should be undertaken [17.17].

The maximum estimated response due to earthquake motions may thus be derived from the following procedure:

1. Establish the response spectrum from the ground motions.
2. Calculate the mass and stiffness matrices M and K, and estimate the modal damping ratios β_n.
3. Determine the first few natural frequencies ω_n and the modes of vibration φ_n.
4. Calculate the maximum response of the structure in each individual mode as follows:
    a. For a natural frequency ω_n and damping ratio β_n, determine the ordinates S_{dn} and S_{an} of the displacement and acceleration response spectra.
    b. Calculate the floor displacements from Eq. (17.27) and the story drifts from Eq. (17.28).
    c. Calculate the equivalent story-level lateral forces from Eq. (17.29).
    d. Calculate the internal shears and moments from statics for a cantilever structure subjected to these lateral forces.
5. Calculate the peak value of the major design actions (displacement, drift, story shear, and moment) by combining the maximum modal values according to Eq. (17.32).

It is necessary to consider only the modes that contribute most to the response of the structure. Since most of the energy of vibration is contained in the lower modes of vibration, it is generally sufficient to consider no more than six modes in each horizontal direction [17.14]. A convenient rule is to include a sufficient number of modes r so that an effective modal mass e of at least 90% of the total mass of the building is represented by the modes chosen, that is,

$$e = \frac{\sum_{n=1}^{r} \gamma_n L_n / 100}{\sum_{j=1}^{N} m_j} \not< 90$$

Worked Example. To illustrate the method of analysis, the simple idealized three-story shear building of Fig. 17.11 is considered. It is assumed that the columns are uniform throughout the height, and that the mass of the building may be lumped equally at the three floor levels. For simplicity, it is assumed that the floors are rigid, and that the axial deformations of the columns can be neglected. The movement of the structure can then be expressed in terms of the three horizontal deflections u_1, u_2, and u_3 at the floor levels. For undamped free vibrations, the mass and stiffness matrices in Eq. 17.11 become

$$\mathbf{M} = \begin{bmatrix} W & 0 & 0 \\ 0 & W & 0 \\ 0 & 0 & W \end{bmatrix} \quad \text{and} \quad \mathbf{K} = \begin{bmatrix} k & -k & 0 \\ -k & 2k & -k \\ 0 & -k & 2k \end{bmatrix}$$

in which $k = 12EI/h^3$, where EI is the sum of the flexural rigidities of the columns, and h is the story height. A value of k/W of 250 is assumed.

On substituting the mass and stiffness matrices into Eq. 17.13, and solving the resulting eigenvalue equation, the three natural frequencies, periods, and corresponding modes of vibration are found to be

$\omega_1 = 7.036$ rad/sec $\omega_2 = 19.717$ rad/sec $\omega_3 = 28.491$ rad/sec

$n_1 = 1.12$ Hz $n_2 = 3.14$ Hz $n_3 = 4.53$ Hz

$T_1 = 0.89$ sec $T_2 = 0.32$ sec $T_3 = 0.22$ sec

$$\varphi_1 = \begin{bmatrix} 1 \\ 0.802 \\ 0.445 \end{bmatrix} \quad \varphi_2 = \begin{bmatrix} -1 \\ 0.555 \\ 1.247 \end{bmatrix} \quad \varphi_3 = \begin{bmatrix} 1 \\ -2.247 \\ 1.802 \end{bmatrix}$$

For convenience, the top floor displacement is taken as unity for each of the three mode shapes.

The NBCC design earthquake response spectrum used, normalized to a peak ground acceleration of 1 g, is shown in Fig. 17.12. It is further assumed that the building is located in a seismic zone for which the peak ground acceleration is taken to be 0.1 g, and that the structure has a damping ratio of 3%. Consequently, the design spectral displacements, velocities, and accelerations corresponding to

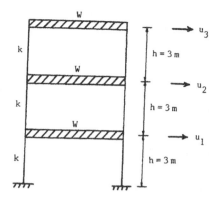

Fig. 17.11 Example structure.

the value of the first three natural periods may be obtained by multiplying the ordinates of S_d, S_v, and S_a from Fig. 17.12 by a factor of 0.1. The values obtained are as follows:

Mode	1	2	3
Natural frequency (Hz)	1.12	3.14	4.53
S_d (m)	0.035	0.010	0.0053
S_v (m/sec)	0.25	0.20	0.15
S_a (m/sec^2)	1.76	3.94	4.24
	(0.18 g)	(0.4 g)	(0.43 g)

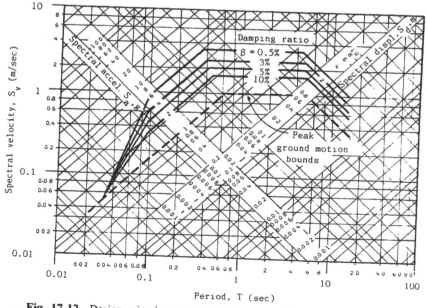

Fig. 17.12 Design seismic response spectrum used for worked example.

17.2 DYNAMIC RESPONSE TO EARTHQUAKE MOTIONS

The sequence of calculations is as follows:

Modal Participation Factors. From Eq. 17.18, for mode 1,

$$L_1 = \sum_{j=1}^{3} m_j \varphi_{j1} = W1 + W0.802 + W0.445 = 2.247W$$

$$M_1 = \sum_{j=1}^{3} m_j \varphi_{j1}^2 = W(1)^2 + W(0.802)^2 + W(0.445)^2 = 1.841W$$

∴ Participation factor

$$\gamma_1 = \frac{L_1}{M_1} = \frac{2.247W}{1.841W} = 1.221$$

Similar calculations for modes 2 and 3 yield

$$L_2 = 0.802W \quad M_2 = 2.863W \quad \gamma_2 = 0.280$$
$$L_3 = 0.555W \quad M_3 = 9.296W \quad \gamma_3 = 0.0597$$

Horizontal Deflections. Due to mode 1, the maximum horizontal deflections at the three story levels become, from Eq. 17.27

$$u_{31} = \gamma_1 S_{d_1} \varphi_{31} = 1.221 \times 0.035 \times 1 \quad\;\; = 0.0427 \text{ m}$$
$$u_{21} = \gamma_1 S_{d_1} \varphi_{21} = 1.221 \times 0.035 \times 0.802 = 0.0343 \text{ m}$$
$$u_{11} = \gamma_1 S_{d_1} \varphi_{11} = 1.221 \times 0.035 \times 0.445 = 0.0190 \text{ m}$$

The corresponding values for modes 2 and 3 become

$$u_{32} = -0.0028 \text{ m} \quad u_{22} = 0.0016 \text{ m} \quad u_{12} = 0.0035 \text{ m}$$
$$u_{33} = 0.0003 \text{ m} \quad u_{23} = -0.0007 \text{ m} \quad u_{13} = 0.0006 \text{ m}$$

Due to all three modes, the maximum horizontal deflection at the top of the building can then be estimated to be, from Eq. 17.32,

$$\bar{u}_3 = \sqrt{0.0427^2 + 0.0028^2 + 0.0003^2} = 0.043 \text{ m}$$

In this case, the second and third modes have little influence on the maximum top deflection.

The horizontal deflections may be used to estimate the corresponding magnitudes of the interstory drifts for all three stories of the structure. For mode 1, the interstory drifts d_{j1} for the stories j become

446 DYNAMIC ANALYSIS

Story 1. $d_{11} = 0.0190 - 0 = 0.0190$ m

Story 2. $d_{21} = 0.0343 - 0.0190 = 0.0153$ m

Story 3. $d_{31} = 0.0427 - 0.0343 = 0.0084$ m

The corresponding values for modes 2 and 3 are

$d_{12} = 0.0035$ m $\quad d_{22} = -0.0019$ m $\quad d_{32} = -0.0044$ m

$d_{13} = 0.0006$ m $\quad d_{23} = -0.0013$ m $\quad d_{33} = 0.0010$ m

The modal values may again be combined to give an estimate of the maximum probable interstory drifts d_i. For the first story

$$d_1 = \sqrt{0.0190^2 + 0.0035^2 + 0.0006^2} = 0.0193 \text{ m}$$

Similarly, for the second and third stories,

$$d_2 = 0.0155 \text{ m} \quad d_3 = 0.0095 \text{ m}$$

Equivalent Lateral Forces. To calculate the magnitudes of the forces involved it is assumed that each mass $W = 30,000$ kg.

From Eq. 17.29, the maximum value of the equivalent lateral force at roof level due to mode 1 is

$$\bar{P}_{31} = \gamma_1 S_{a1} m_3 \varphi_{31} = 1.221 \times 1.76 \times 30,000 \times 1 \text{ N} = 64.47 \times 10^3 \text{ N}$$

$$= 64.47 \text{ kN}$$

Similarly, at levels 1 and 2,

$$\bar{P}_{21} = 1.221 \times 1.76 \times 30,000 \times 0.802 \text{ N} = 51.7 \text{ kN}$$

$$\bar{P}_{11} = 1.221 \times 1.76 \times 30,000 \times 0.445 \text{ N} = 28.69 \text{ kN}$$

Similarly, for modes 2 and 3, the corresponding values become

$\bar{P}_{32} = -33.1$ kN $\quad \bar{P}_{22} = 18.37$ kN $\quad \bar{P}_{12} = 41.27$ kN

$\bar{P}_{33} = 7.59$ kN $\quad \bar{P}_{23} = -17.06$ kN $\quad \bar{P}_{13} = 13.68$ kN

Base Shears and Moments. Due to mode 1, the maximum base shear and moment become, from Eq. 17.30

$$\bar{V}_{01} = 64.47 + 51.70 + 28.69 = 144.86 \text{ kN}$$

$$\bar{M}_{01} = 64.47 \times 9 + 51.70 \times 6 + 28.69 \times 3 = 976.5 \text{ kNm}$$

The corresponding values due to modes 2 and 3 become, taking due account of the senses of the forces,

$$\overline{V}_{02} = 26.54 \text{ kN} \qquad \overline{V}_{03} = 4.21 \text{ kN}$$

$$\overline{M}_{02} = -63.87 \text{ kNm} \qquad \overline{M}_{03} = 6.99 \text{ kNm}$$

The maximum probable base shears and moments due to all modes then become

$$\overline{V}_0 = \sqrt{144.86^2 + 26.54^2 + 4.21^2} = 147.3 \text{ kN}$$

$$\overline{M}_0 = \sqrt{976.5^2 + 63.87^2 + 6.99^2} = 978.6 \text{ kNm}$$

Lateral-Torsional Coupling. As noted previously, the usual procedure in the spectrum analysis of tall buildings is to consider the response of the structure to two independent orthogonal motions. Strictly speaking, this approach is valid only if the centers of mass and resistance are coincident, although the errors may be acceptable if the eccentricity is small.

However, coupling between translational motions and rotational motions about a vertical axis will likely occur if either the eccentricity between the centers of mass and resistance are large, or the natural frequencies of lateral and torsional vibration are close together. For such structures, independent analyses of two orthogonal translatory motions may not be sufficiently accurate, and all three degrees of freedom—two translations and one rotation—should be included at each floor level. The modal method of analysis may then be applied to the system with three degrees of freedom at each floor level. In so doing, it must be borne in mind that modes which are predominantly torsional can be excited by translational components of ground motion, and that a particular mode can be excited by both horizontal components. The resulting maxima should then be combined by the more general formula of Newmark and Rosenblueth [17.13].

Vertical Components of Ground Motion. Although horizontal ground motions are normally dominant, it may not be possible to neglect the vertical accelerations, which may be large compared to the gravitational acceleration in regions of high seismicity. The dynamic vertical response may be estimated by the use of a vertical response spectrum to obtain the equivalent vertical forces for which the building must be designed.

The peak response due to horizontal and vertical ground motions may then be estimated from the square root of the sum of the squares of the individual maxima.

Nonlinear Inelastic Behavior. Although a nonlinear analysis can always be achieved from a step-by-step piecewise-linear elastic analysis, with the calculation divided up into a series of time steps, and with the current stiffness of the structure appropriate to the stress level concerned being used during each step, such a procedure would be too cumbersome for design purposes.

Although the modal method of analysis is strictly applicable only to elastic systems, a simple procedure that produces results of acceptable accuracy [17.15]

448 DYNAMIC ANALYSIS

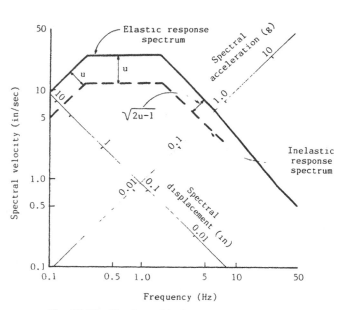

Fig. 17.13 Elastic and inelastic response spectra.

uses a linear analysis in conjunction with an inelastic response spectrum. The approach is reasonably accurate for building structures in which the deformation is limited to ductility factors of about 5 or 6.

The inelastic design spectrum can be derived from the elastic spectrum and the allowable ductility factor μ, defined as the ratio of the ultimate or maximum displacement μ_m to the effective elastic limit deflection u_y, that is,

$$\mu = u_m/u_y \qquad (17.33)$$

The elastic design spectrum of Fig. 17.13 may be divided into three regions—the displacement region (ω between 0.1 and 0.4 Hz), velocity region (ω between 0.4 and 4 Hz), and acceleration region (ω between 4 and 7.7 Hz). If the elastic-plastic load-displacement relationship is of bilinear form, as shown in Fig. 17.14,

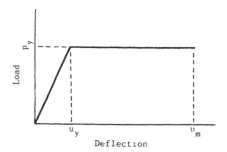

Fig. 17.14 Idealized load-displacement-curve.

it can be shown [17.14] that the inelastic spectrum may be obtained by multiplying the values in the displacement and velocity regions by a factor $1/\mu$, and the values in the acceleration region by a factor $1/(2\mu - 1)^{1/2}$. The resulting inelastic design spectrum is shown by the broken line in Fig. 17.13 for a ductility factor of 2.

With this approach, is possible to define design spectra to take account of inelastic structural behavior.

Soil–Structure Interaction. If a structure is founded on rock, the motion experienced by the base is essentially that of the original soil, and the calculations may be carried out as if the base were rigid. However, a compliant soil adds flexibility to the system, but it also absorbs energy due to the propagation of stress waves in the soil, and the amount of damping can be considerably increased. In this case, soil–structure interaction may have a significant effect on structural response, and should be considered in the analysis.

The simplest method of including the foundation flexibility is to represent the soil under the base by a group of springs to simulate the stiffness and a series of dashpots to simulate the damping. The earthquake motions may then be applied to the base of the springs and dashpots. If it is assumed that the springs are of constant stiffness and damping is uniform in each mode, a response spectrum analysis may be used to determine the peak response [17.17]. More complex methods of analysis, including perhaps a time-history response, will be required if a more accurate nonlinear representation of the soil is incorporated in the model.

Comparison between Design Methods for Dynamic Responses due to Wind and Earthquake Forces. The similarity between the design methods for the dynamic analysis of structures subjected to turbulent wind forces or earthquake ground motions is illustrated in Table 17.1 [17.8].

17.2.3 Empirical Relationships for Fundamental Natural Frequency

It has been shown earlier that a knowledge of the fundamental natural frequency of a building is necessary to determine the design loading and response to either turbulent wind or earthquake actions. The estimation of this parameter can pose difficulties in the early design stages when the final structural form has not been decided in detail.

Many studies reported in the literature have attempted to derive empirical formulas to allow an estimation of the fundamental natural frequency, or period, of a building, based only on simple geometric parameters. Two widely used formulas for the fundamental frequency n_0 are

1.
$$n_0 = \frac{\sqrt{D}}{0.091H} \text{ (Hz)} \quad (17.34)$$

where D is the base direction (in meters) in the direction of motion considered, and H is the height of the building (in meters).

TABLE 17.1 Comparison between Design Procedures against Wind and Earthquake Actions

Design for Wind Response	Design for Earthquake Response
Climate (wind speeds, directions, storm types)	Zone (seismicity, energy release, frequency of occurrence)
↓	↓
Exposure (terrain roughness, topography, local obstructions)	Local geological and geotechnical features (faults, soils, foundations)
↓	↓
Time-History of Wind Field	**Time-History of Earthquake Accelerations**
↓	↓
Structure: aerodynamics properties (geometry: height, width, breadth)	Structure: geometric properties (configuration, framing, weight)
↓	↓
Time-History of Aerodynamic Forces	**Time-History of Inertial Forces**
↓	↓
Structure: mechanical properties (stiffness, mass, damping)	Structure: mechanical properties (stiffness, mass, damping)
↓	↓
Time-History of Structural Response	**Time-History of Structural Response**
↓	↓
stress, deflection, acceleration, pressure	stress, deflection, acceleration, damage
|	|
Design Criteria	**Design Criteria**

It has been suggested [17.15] that this formula is particularly applicable to reinforced concrete shear wall buildings and braced steel frames.

2.
$$n_0 = \frac{10}{N} \text{ (Hz)} \qquad (17.35)$$

where N is the number of stories in the building.

It is suggested in the NBCC [17.3] that this formula should be used when the lateral force-resisting system consists of a moment-resisting space frame that resists the entire lateral forces, and the frame is not enclosed or adjoined by more rigid elements that would tend to prevent the frame from resisting lateral forces.

Another formula that has also been commonly used is,

$$n_0 = \frac{1}{C_T H^{3/4}} \qquad (17.36)$$

where C_T is equal to 0.035 or 0.025 for steel or concrete structures, respectively, and H is the building height (in feet). This formula is also most appropriate when moment-resisting frames are the sole lateral-load-resisting element in the building [17.15].

The first two formulas were tested by Ellis [17.18] against the measured natural frequencies of 17 buildings, ranging in height from 7 to 44 stories. He found that errors greater than $\pm 50\%$ were not abnormal, but that the simpler formula [Eq. (17.35)] generally gave more accurate results.

He tried different simple predictor formulas and compared the results with the actual measured frequencies of 163 buildings of rectangular plan-form. As a result of his study, the following formula was recommended:

$$n_0 = \frac{46}{H} \qquad (17.37)$$

where H is the building height (in meters).

It may be noted that he also suggested that the frequency of the first orthogonal translational mode can be estimated as $58/H$, and the frequency of the first torsional mode as $72/H$. Care must also be taken when using these formulas, particularly the torsional mode predictor.

Once a preliminary design has been achieved and the stiffness of the building is known, a more accurate estimate of the fundamental natural frequency may be determined from established procedures. A reasonably accurate approximate formula based on Rayleigh's method [17.15] is

$$n_0 = \frac{1}{2\pi} \left(\frac{g \Sigma F_i u_i}{\Sigma W_i u_i^2} \right)^{1/2} \qquad (17.38)$$

in which W_i is the weight of the ith floor, u_i is the calculated static horizontal deflection at the ith level due to a set of equivalent lateral loads F_i at the floor levels, and g is the gravitational acceleration. Any reasonable distribution of loads F_i may be selected, but it is convenient to use the statically equivalent forces due to wind or earthquake actions discussed in Chapter 3.

17.2.4 Structural Damping Ratios

As discussed earlier, structural damping is normally estimated in terms of a percentage of the critical damping ratio, with values obtained from measurements on full-scale similar structures serving as a guide. The amount of damping in a structure will depend on the magnitude of the displacements and stress levels involved, and different values will generally be quoted for service and ultimate conditions. Typical values for reinforced concrete (RC) and steel framed buildings are given in Table 17.2 [17.4].

More complex formulas for the above structural forms, expressing the structural damping ratio in terms of the fundamental natural frequency and the drift index, are also given in Ref. [17.4].

452 DYNAMIC ANALYSIS

TABLE 17.2 Typical Structural Damping Ratios

Form of Construction	Damping Ratio β	
	Service	Ultimate
RC core, cantilever floor, light weight cladding	0.016	0.022
RC columns, slab floors, few internal walls	0.030	0.070
RC frame, few internal walls	0.030	0.070
RC frame, shear walls	0.030	0.080
RC shear core and columns, some internal walls	0.040	0.120
RC frame, some internal walls	0.040	0.120
RC all forms, many internal walls	0.050	0.160
Steel frame, no internal walls	0.005	0.007
Steel frame, few internal walls	0.025	0.060
Steel frame, many internal walls	0.040	0.150

17.3 COMFORT CRITERIA: HUMAN RESPONSE TO BUILDING MOTIONS

From the public's point of view, a building structure should remain stationary, and so movements that in other circumstances might be quite acceptable can in a tall building induce a wide range of responses, ranging from anxiety to acute nausea in its occupants. Motions that have psychological or physiological effects on the occupants may thus result in an otherwise acceptable structure becoming an undesirable building, with a resulting reputation that may produce difficulties in renting the floor space. It is thus not sufficient to provide a structure capable of resisting the stresses induced by the design loading, with sufficient stiffness to prevent excessive movement and damage to nonstructural elements; the designer must ensure also that there are no undesirable motions that could adversely affect the occupants.

It would be prohibitively expensive to construct a building that would not move perceptibly in the worst storm or during a severe earthquake. Consequently, since some motion is inevitable, the goal is to determine levels of motion and rates of occurrence that are both economic and acceptable to the building occupants.

This section examines briefly the criteria involved and discusses the design implications of the provision of human comfort.

17.3.1 Human Perception of Building Motion

The most common causes of vibration in structures are wind, earthquakes, machinery, nearby industrial plants, and the various types of transportation. The motions resulting from these causes can vary greatly in duration and intensity, and there are a variety of mechanisms by which the apparent motion may be exaggerated.

17.3 COMFORT CRITERIA: HUMAN RESPONSE TO BUILDING MOTIONS

The perception of building movement depends largely on the degree of stimulation of the body's central nervous system, the sensitive balance sensors within the inner ears playing a crucial role in allowing both linear and angular accelerations to be sensed. Human response to building vibration is influenced by many factors, such as the movement of suspended objects, and the noise due to turbulent wind or fretting between building components. If the building twists, objects at a distance viewed by the occupants may appear to move slightly, and relative movements of adjacent buildings vibrating out of phase will be exaggerated. False cues may result from wind forces causing flexing of glass in windows. Human reaction will be affected by any fear for the structural soundness of the building, and by any previous experience of this or similar situations.

17.3.2 Perception Thresholds

There are as yet no generally accepted international standards for comfort criteria, although they are under active consideration, and the engineer must base his design criteria on an assessment of published data. In recent years, a considerable amount of research has been carried out into the important physiological and psychological parameters that affect human perception to motion and vibration in the low frequency range of 0-1 Hz encountered in tall buildings. These parameters include the subject's expectancy and experience, the activity he is pursuing, his body posture and orientation, visual and acoustic cues, and the amplitude, frequency, and accelerations for both the translational and rotational motions to which he is subjected. A considerable amount of data is now available, but only the most important general consequences for design are considered here.

It is now generally agreed that acceleration is the predominant parameter in determining the nature of human response to vibration [17.19], and this, with a knowledge of the frequency of oscillation, can define all other relevant parameters of sinusoidal vibration.

Based on a series of studies in which subjects were called on to undertake a range of manual tasks when subjected to a variety of building motions, a set of thresholds for tall building motions has been proposed. Figure 17.15 shows a typical interaction curve indicating upper bounds of accelerations for different degrees of hindrance to task performance as affected by the period of vibration [17.20]. The boundaries indicated by the letters refer to the following limits: (A) perceptible threshold; (B) desk work and psychological limit; (C) ambulatory limit; (D) building motion limit. The numbers refer to the different perception levels defined in Table 17.3, which illustrates how human behavior and motion perception are affected by different ranges of acceleration [17.20]. Corresponding tentative proposals [17.21] for perception thresholds, showing the comfort limits expressed in terms of the accelerations, displacement amplitudes, and periods of vibration, are shown in Fig. 17.16.

Based on an examination of the available data, curves have been produced [17.19, 17.22] to give recommended upper limits of acceleration for various frequencies depending on the building use. Since individual perception thresholds

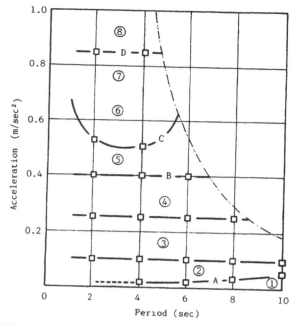

Fig. 17.15 Building motion criteria for human response (after Yamada and Goto [17.20]).

TABLE 17.3 Human Perception Levels

Range	Acceleration (m/sec^2)	Effect
1	<0.05	Humans cannot perceive motion
2	0.05–0.10	Sensitive people can perceive motion; hanging objects may move slightly.
3	0.1–0.25	Majority of people will perceive motion; level of motion may affect desk work; long-term exposure may produce motion sickness
4	0.25–0.4	Desk work becomes difficult or almost impossible; ambulation still possible
5	0.4–0.5	People strongly perceive motion; difficult to walk naturally; standing people may lose balance.
6	0.5–0.6	Most people cannot tolerate motion and are unable to walk naturally
7	0.6–0.7	People cannot walk or tolerate motion.
8	>0.85	Objects begin to fall and people may be injured

17.3 COMFORT CRITERIA HUMAN RESPONSE TO BUILDING MOTIONS

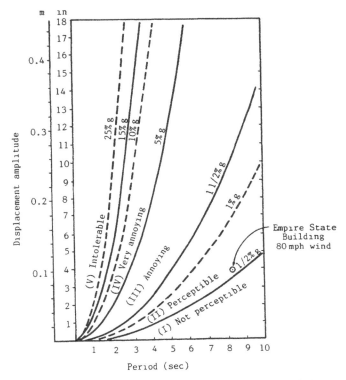

Fig. 17.16 Human comfort curves (after Chang [17.21]).

tend to follow a lognormal probability distribution, there will, theoretically at least, always be a certain number of people who will notice any building motion however small the displacements. The curves are therefore based on the accepted criterion that not more than 2% of those occupying the parts of the building where motions are greatest complain about them. The figures are based on the motion caused by the peak 10 minutes of the worst wind storm with a return period of 5 years or more. Since human response to motion depends on previous experience of similar events, different criteria should be used for storm-induced building motions than are used for regularly recurring motions; the criteria for the latter should be related to perception thresholds.

The curves shown in Fig. 17.17 may be used as a guide to the limits of motion perception and comfort criteria for horizontal building movements. The criteria are based on the root-mean-square (rms) value of the translational acceleration. The appropriate curves E1 to E5 for design use for different adverse comment levels for the maximum 10 minutes of the worst storm in a 1- or 5-year return period are indicated in Table 17.4 [17.19]. The curves refer to a critical working area, such as a hospital operating theatre. For less critical areas, the curves may be multiplied by the factors shown in Table 17.5.

456 DYNAMIC ANALYSIS

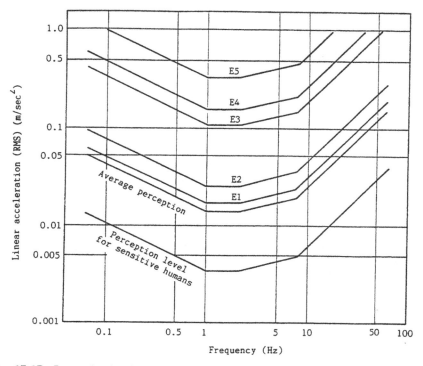

Fig. 17.17 Perception levels and comfort criteria for human response to building motions (after Irwin [17.19]). (To be used in association with Table 17.4.)

The curves are based on the assumption that the building is properly clad and insulated acoustically so that noises caused by wind and building motions are not transmitted to the occupants. If this is not the case, the acceptable magnitudes must be reduced by an appropriate factor [17.19] since the effect of noise is to exaggerate the feelings caused by the motion.

The NBCC [17.3] gives an alternative useful guideline, recommending accel-

TABLE 17.4 Appropriate Curves and Adverse Comment Levels for the Maximum 10 Minutes of the Worst Storm in a 1- or 5-Year Return Period

Application	Buildings			Structures Manned by Trained Personnel		
Curve (Fig. 17.17)	E1	E2	E3	E4		E5
Return period in years	1	5	1	1	5	1	—
Percentage adverse comment	2	2	12	2	2	12	Hand holds or restraint harness required

After Irwin [17.19].

TABLE 17.5 Environmental Factors for Use with Comfort Criteria Curves.

Place	Time	Steady Conditions	Impulsive Events
Critical area	Any	1	1
Residential	Day	2 to 4	60 to 90[a]
	Night	1.4	20
Office	Any	4	128[a]
Workshop	Any	8	128[a]

[a] With prior warning.
After Irwin [17.19].

eration limitations of 1–3% of gravity (0.09 to 0.27 m/sec^2) once in every 10 years, the two figures being more appropriate for apartment and office blocks respectively.

If the building is subjected simultaneously to two orthogonal horizontal translational acceleration components, (expressed in m/sec^2 rms values), they must be summed vectorially, and any rotational acceleration component included to give the final acceleration to be considered.

17.3.3 Use of Comfort Criteria in Design

In the design stages, the probable motion of a planned structure can be predicted from a dynamic analysis. In the case of a prestigious or unusual structure, it may be necessary to perform a wind-tunnel investigation on an aeroelastic model of the proposed design. The predicted values should be verified subsequently by measured motions of the built structure so that possible problems in service may be foreseen.

The predicted accelerations and periods of vibration may then be compared with the threshold curves to ascertain whether any problems are likely to be encountered.

If unacceptable levels of motion are predicted, a number of solutions of varying complexity may be employed. Although static deflections may be reduced by the provision of extra stiffness, it has only a slight influence on the reduction of accelerations.

This can be shown by considering the general equation of motion for a structure

$$M\ddot{u} + C\dot{u} + Ku = Pf(t) \qquad (17.39)$$

from which it may be deduced that the acceleration is proportional to $u_{max}\omega^2$, where u_{max} is the peak displacement and ω is the circular frequency of motion. Increasing the stiffness K of the structure by a factor λ reduces u_{max} by the same factor. Simultaneously, however, the increase in stiffness causes ω^2 to be increased by the same factor λ, to leave the product $u_{max}\omega^2$, and the peak acceleration, unchanged.

However, a slight reduction in acceleration can be achieved by an increase in stiffness, since this leads to a reduction in the magnitude of the dynamic wind force $Pf(t)$, due to the available spectral wind energy generally decreasing with higher frequencies.

Equation 17.39 shows that increasing the mass produces a corresponding decrease in acceleration. However, controlling accelerations by increasing the building weight is an expensive and impractical solution, particularly since it may have other adverse effects such as increasing the loads on the foundations.

An obvious control on acceleration is by an increase in damping. Apart from air friction, energy is dissipated by hysteretic losses in the material of the structure and by friction between moving parts; consequently, materials with high hysteresis such as concrete, or joints with the possibility of large friction, such as bolted steel connections, are effective in reducing undesirable accelerations. If exceptionally severe problems of acceleration are encountered, it may be necessary to include devices such as tuned mass-dampers to dissipate energy [17.23].

Problems may arise due to the coupling between translational and rotational motions when their frequencies are close. If this is the case, it may be possible to separate the frequencies concerned by a change in the design configuration that would alter the way in which the bending and torsional resistances are mobilized in the structure.

Earthquake-induced motions, as far as human response is concerned, are quite different from those that result from wind forces. Earthquakes occur much less frequently than wind storms, and only in certain areas of the world. Their duration of vibration can be generally very short and transient in nature, and the motions induced are generally much more violent. In addition to the horizontal accelerations, significant vertical accelerations may also be present. People who experience earthquakes of any severity are simply thankful to have survived the ordeal; hence the design criterion is one of safety rather than comfort.

SUMMARY

Although for many tall buildings it is possible to treat the dynamic forces due to wind or seismic actions by equivalent static lateral loads, it may be necessary in certain cases to use a dynamic method of analysis to obtain an accurate representation of the peak dynamic forces and stresses, deflections, and accelerations. It is also important to ensure that the building motions do not affect the comfort and equanimity of the occupants. Guidelines are available to determine when a dynamic analysis is required.

Dynamic wind pressures produce sinusoidal or narrow-band vibration motions, and the building will generally undergo translations in both along-wind and cross-wind directions, and possibly rotations about a vertical axis. The vast amount of experimental and theoretical research of the past few decades has allowed the inclusion in Design Codes of simplified methods of dynamic analysis that do not require a formal solution of the equations of motion. The along-wind response can be analyzed by the gust factor method. Based on a number of environmental and structural parameters, a gust factor may be determined, such that, if the mean response to an equivalent static design factor is multiplied by the gust factor, it will provide a reasonable estimate of the peak dynamic response to the fluctuating wind forces. It has not yet proved possible to develop a corresponding technique for the cross-wind response, and the most accurate method is still to test an aeroelastic model in a wind tunnel. However, empirical formulas are available to allow a rough estimate of the cross-wind response to be determined.

The ground shaking that occurs in an earthquake can be described by a series of virtually random acceleration pulses, and the resulting building response will depend on its structural mass, stiffness, and amount of damping. If the input acceleration history is known, the equations of motion may be integrated numerically to give a time-history of the structural response to the earthquake. However, for design purposes it is necessary only to determine the peak response, and this may be obtained more directly from a design response spectrum analysis, using the modal superposition method of dynamic analysis. In this method, the equations of motion are uncoupled, and the response in each natural mode of vibration calculated independently of the others. The modal responses are then combined to give the total building response.

Although many other factors such as visual cues, body position and orientation, state of mind, and experience of previous happenings may play a part, it appears that the building acceleration is of paramount importance in determining the occupant's perception of motion. Threshold curves have been devised to provide limiting values of acceleration and frequency for both human motion perception and comfort, which apply to the vast majority of people. The predicted building accelerations can then be compared with the limiting criteria to establish whether problems are likely to occur.

REFERENCES

17.1 Simiu, E. and Scanlan, R. H. *Wind Effects on Structures*, 2nd ed. Wiley Interscience, New York, 1986.

17.2 Standard Association of Australia, Draft Australian Standard, Revision of AS 1170-Part II, 1983. 1987.

17.3 *National Building Code of Canada*. National Research Council of Canada, Ottawa, Canada, 1990.

17.4 Cook, N. J. *The Designer's Guide to Wind Loading of Building Structures*. Part 1. Building Research Establishment. Butterworths, London, 1985.

17.5 Davenport, A. G. "Gust Loading Factors." *J. Struct. Div.*, ASCE **93**, 1967, 11–34.

17.6 Vickery, B. J. "On the Assessment of Wind Effects on Elastic Structures." *Civil Engineer. Trans.*, Inst. of Engineers, Australia, 1966, 183–192.

17.7 Vickery, B. J. and Davenport, A. G. "A Comparison of Theoretical and Experimental Determination of the Response of Elastic Structures to Turbulent Flow." *Proc. Conf. on Wind Effects on Buildings and Structures*, Ottawa, University of Toronto Press, Toronto, 1967.

17.8 Gould, P. L. and Abu-Sitta, S. H. *Dynamic Response of Structures to Wind and Earthquake Loading*. Pentech Press, London, 1980.

17.9 Saunders, J. W. and Melbourne, W. H. "Tall Rectangular Building Response to Cross-Wind Excitation." *Proc. 4th Int. Conf. on Wind Effects on Buildings and Structures*, London 1975. University of Cambridge Press, Cambridge.

17.10 Kwok, K. C. S. and Melbourne, W. H. "Dynamic Analysis of Wind Sensitive Buildings and Structures to Wind Action—a Codified Approach." *Proc. 4th Int. Conf. on Tall Buildings*, Hong Kong and Shanghai, April/May, 1988, Vol. 1, 424–430.

17.11 *Tall Building Criteria and Loading*. Vol. CL. Monograph on Planning and Design of Tall Buildings, Chapter 3. ASCE, New York, 1980.

17.12 Clough, R. W. and Penzien, J. *Dynamics of Structures*. McGraw-Hill, New York, 1975.

17.13 Newmark, N. M. and Rosenblueth, E. *Fundamentals of Earthquake Engineering*. Prentice-Hall, Englewood Cliffs, NJ, 1971.

17.14 Chopra, A. K. and Newmark, N. M. "Analysis." In Chapter 2, *Design of Earthquake Resistant Structures*, E. Rosenblueth (Ed.). Pentech Press, London, 1980.

17.15 Newmark, N. M. and Hall, W. J. *Earthquake Spectra and Design*. Earthquake Engineering Research Institute, Berkeley, California, 1982.

17.16 Chopra, A. K. *Dynamics of Structures. A Primer*. Earthquake Engineering Research Institute, Berkeley, California, 1981.

17.17 Booth, E. D., Pappin, J. W., and Evans, J. J. B. "Computer Aided Analysis Methods for the Design of Earthquake Resistant Structures: A Review." *Proc. Inst. Civil Engineer. London* **84**(1), 1988, 671–691.

17.18 Ellis, B. R. "An Assessment of the Accuracy of Predicting the Fundamental Natural Frequencies of Buildings and the Implications Concerning the Dynamic Analysis of Structures." *Proc. Inst. Civil Engineer. London* **69**(2), 1980, 763–776.

17.19 Irwin, A. W. "Motion in Tall Buildings." *Proc. Conf. on Tall Buildings*, Second Century of the Skyscraper, Chicago, 1986, 759–778.

17.20 Yamada, M. and Goto, T. "The Criteria to Motions in Tall Buildings." *Proc. Pan-Pacific Tall Buildings Conference*, Hawaii, 1975, 233–244.

17.21 Chang, F.-K. "Human Response to Motions in Tall Buildings." *J. Struct. Div., A.S.C.E.* **99**, 1973, 1259–1272.

17.22 Irwin, A. W. "Human Response to Dynamic Motion of Structures." *Struct. Engineer, London* **56A**, 1978, 237–244.

17.23 *Planning and Environmental Criteria for Tall Buildings*. Vol. PC. Monograph on Planning and Design of Tall Buildings, Chapter 13. ASCE, New York, 1981.

CHAPTER 18

Creep, Shrinkage, and Temperature Effects

Until relatively recently, engineers paid little attention to the influence of creep and shrinkage on the design of concrete buildings. It was generally assumed that creep effects were beneficial in relieving overstressed sections by allowing a redistribution of load to less highly stressed regions. On the other hand, shrinkage was regarded as an inevitable fact that had to be tolerated. The structural movements caused by temperature changes were accommodated within relatively long members by a judicious arrangement of construction joints, in combination with appropriate reinforcement to reduce the effect and partially relieve the resulting stresses. However, with increasing heights of buildings, the time-dependent shortening of columns and shear walls becomes of much greater importance due to the cumulative nature of such deformations. An 80-story-high concrete column, for example, might be expected to shorten by about 4.6 cm (1.8 in.) due to creep and 6 cm (2.4 in.) due to shrinkage. In cases in which the building construction uses concrete pumping, which requires a more liquid mix than conventionally placed concrete, particular consideration should be given to the potentially greater creep and shrinkage movements.

18.1 EFFECTS OF DIFFERENTIAL MOVEMENTS

In concrete members, the strains due to creep and shrinkage depend on the percentage of reinforcement and the volume-to-surface ratio as well as on the basic material properties. When subjected to the same stress levels, the strains due to creep and shrinkage decrease with an increase in both percentage of reinforcement and volume-to-surface ratio.

In a high-rise building, adjacent vertical members may have different percentages of reinforcement due to different gravity loadings, which result from different tributary areas or from nonuniform stresses caused by lateral forces. Consequently, the differential shortening of adjacent columns will produce shear and moment in the connecting beams or slabs, due to the relative vertical displacement of the supports. A redistribution of load will result, with load transfer to the column that shortens less. Similarly, problems may arise if a large heavily reinforced column is located near a shear wall, for example. The column will attract additional loads from the shear wall, which suffers higher creep and shrinkage strains due to its lower volume-to-surface area and percentage of reinforcement. The differential movements are cumulative over the height of the building, being zero at ground level and reaching a maximum at roof level. Consequently, the effects become progressively more important as the height increases, and the cumulative distortions may cause damage to nonstructural elements such as partitions and windows as well as overstressing the slabs in the upper levels.

The influence of sequence and time of construction may also be important. For example, slipforming of service cores is frequently undertaken as the first step in the construction and, consequently, the bulk of the inelastic shortening of the core may have taken place before the adjacent columns and connecting beams and slabs have been cast. Significant differential movement of the slab supports may then occur as the columns are subjected to the full amount of creep and shrinkage relative to the virtually static core.

If a transfer girder is used to connect a shear core to exterior columns, any differential vertical movement between core and columns will be restrained by the transfer girder, and significant stresses may be induced in it.

One of the more critical elements affected by vertical movement is the facade. Problems have arisen in a number of masonry-faced buildings in which the interior concrete frame has crept, causing the exterior masonry to become load bearing, with a resultant buckling or spalling and crushing. In some cases in which this has happened, horizontal gaps have had to be cut in the facing to relieve the stresses.

In all cases, any restraint to the free movement will induce stresses in the structural elements.

18.2 DESIGNING FOR DIFFERENTIAL MOVEMENT

The characteristics of creep and shrinkage deformation are similar in that their strain rates are greatest initially, and decrease progressively with time. As a rough guide, some 40% of the inelastic strains that occur due to these phenomena take place in the first 28 days, while after 3 and 6 months, some 60 and 70% of the total deformation will have occurred, increasing to around 90% after 2 years. A structure will thus have experienced the majority of the eventual effects only a few years after construction.

The available data on shrinkage of concrete have generally been obtained from tests on small standard specimens stored in a controlled laboratory environment, while creep data have generally been obtained from specimens subjected to constant loads. The latter thus cannot be applied directly for the prediction of inelastic behavior of columns in tall buildings, since they are loaded progressively by increments arising from each story constructed above the level concerned. Although both creep and shrinkage produce shortening of the vertical elements, the time and sequence of construction have a pronounced effect on the total creep effect at any

18.2 DESIGNING FOR DIFFERENTIAL MOVEMENT

level, while shrinkage proceeds independently of construction time. Consequently, the creep and shrinkage strains must be estimated separately before their combined effects on the structure can be determined.

In the design process, the calculations are carried out in two stages:

1. The amount of creep and shrinkage movement occurring in the columns and shear walls must be determined, based on the loading history, member sizes and reinforcements, and the environment concerned.
2. The resulting member forces arising from the known amounts of differential elastic and inelastic shortenings of the vertical components must be calculated and assessed.

Temperature movements differ from the inelastic deformations caused by creep and shrinkage in that they are essentially elastic in nature, and vary continuously as they follow the seasonal and diurnal temperature fluctuations throughout the life of the structure. They can also produce lengthening as well as shortening of all members.

Under certain climatological conditions, significant temperature differences may occur between adjacent members of a building frame, particularly between the exterior and interior columns. The use of partially or fully exposed exterior columns in high-rise buildings, in combination with a heated or air-conditioned static internal temperature, can create further engineering problems due to the relative vertical elongation and contraction of the external facade. Although the exterior spandrel beams will also be generally at a different temperature from the interior parallel beams, the problem tends to be less acute since the total beam length across the building is usually less than the total height of an exterior column.

The members on the face of the building exposed to direct sunlight will be at a higher temperature than those on the opposite shaded side, and the building will thus tend to deflect away from the sun. It has been estimated that the solar deflection at the top of the building can in some situations be as much as 20% of the drift due to wind forces.

In both very long and very tall buildings, the cumulative effects of both time-dependent volume changes and temperature changes increase with both the length and the height of the building, and it may become necessary for the engineer to consider carefully the possible magnitudes of the various differential movements and develop suitable structural and architectural details for a satisfactory building performance [18.5].

Steps that may be taken to minimize differential movements include the design of a suitable concrete mix, close control of the curing procedures, and careful attention to the construction sequence, which involves the age of placing of the concrete in the different structural elements. If possible, adjacent columns and walls should be proportioned so that they contain similar percentages of reinforcement and are subjected to similar compressive stresses.

18.3 CREEP AND SHRINKAGE EFFECTS

This section describes the practical method developed by Fintel and Khan [18.1, 18.2] for estimating the effects of creep and shrinkage.

18.3.1 Factors Affecting Creep and Shrinkage Movements in Concrete

Creep consists of two components: true creep occurring under conditions of hygral equilibrium where no moisture movement occurs between the specimen and the surrounding medium, and drying creep, which results from an exchange of moisture between the stressed member and its environment, the latter being of importance only during the initial period of loading. Creep of concrete may be taken to be a linear function of stress up to values of around 40% of the ultimate strength; above this level creep becomes a nonlinear function of stress. A typical creep-time curve is shown in Fig. 18.1.

For a given concrete mix, the amount of creep depends to a considerable degree on the loading history as well as the applied stress. It has been found that a concrete specimen exhibits a much greater specific creep (that is, the ultimate creep strain per unit sustained stress) if it is first loaded at an early age rather than when it is older. This is of considerable practical importance, since creep decreases with the age of the concrete, and so each increment in a loading sequence will add a smaller component of specific creep to the final average creep of a column.

Provided that the stresses are not near the ultimate strength, the strains produced in a concrete member at any time by a load increment are virtually independent of the effects of any previous or subsequent applied loads. That is, each applied load produces a creep strain that corresponds to the strength–stress ratio at the time of application as if it were the only loading on the member.

Although accurate values of specific creep can be obtained only from a proper long-term comprehensive experimental investigation on the actual concrete mix to be used in a structure, this would clearly not be a very practical approach in the

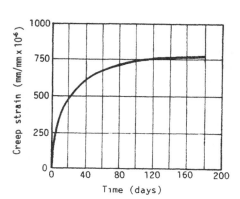

Fig. 18.1 Variation of creep strain with time.

18.3 CREEP AND SHRINKAGE EFFECTS   465

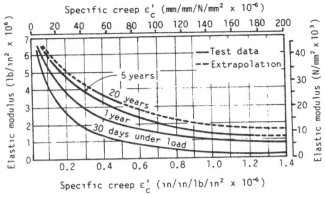

Fig. 18.2 Variation of specific creep with elastic modulus of concrete for different periods of loading.

design stage. Fortunately it has been shown that it is possible to predict basic creep values from the modulus of elasticity at the time of load application. For normal weight concrete, Fig. 18.2 shows the variation of specific creep values with the initial elastic modulus for different load durations [18.1], the extrapolated 20-year value being regarded as the ultimate creep for design purposes. If the 28-day concrete strength is specified, the basic specific creep for loading at 28 days may be determined, and the value then modified approximately to take account of construction sequence, member dimensions, and percentage of reinforcement, as described later.

Based on research data from numerous tests [18.3], a general curve showing the relationship between creep and age of concrete at loading has been established (Fig. 18.3), in which the creep has been expressed as a ratio λ_a of the unit datum

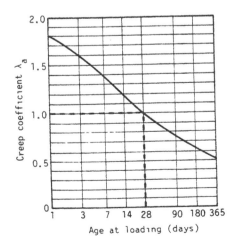

Fig. 18.3 Variation of creep ratio λ_a with age at loading.

creep value at an age of 28 days. The curve allows the effects of increments of loading on a column to be summed up to give the total creep effect produced by the application of all loads.

Because of the essential regularity in form of the floors, and the repetition involved in the construction of the majority of tall buildings, it may often be assumed that the columns or walls are loaded with equal load increments at equal time intervals. In that case, by taking the average specific creep, from Fig. 18.3, it is possible to convert the 28-day creep values into a curve showing the coefficient of average creep, λ_{av}, as a function of construction time (Fig. 18.4).

For example, if the rate of construction is $1\frac{1}{2}$ floors per week, and a particular column receives 54 equal load increments, it will take 252 days to apply all loads. From Fig. 18.4, the creep coefficient λ_{av} is 0.73, which can be used to factor the value of 28-day specific creep to give the average value to be used in a calculation. The curve in Fig. 18.4 illustrates graphically the influence of rate of load application on the resulting creep which ensues.

The true creep component is independent of the dimensions of a member, only the drying creep component being affected by the size and shape, and so creep is less sensitive than shrinkage to member size. Research [18.4] has indicated that drying creep is of importance only during the initial 3 months, and may be disregarded after about 4 months. The relationship between creep coefficient λ_c and volume-to-surface ratio is shown in Fig. 18.5.

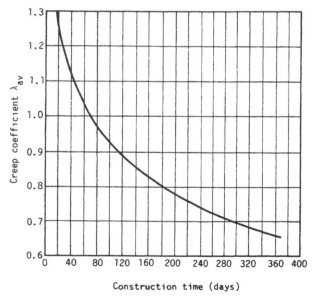

Fig. 18.4 Variation of creep coefficient λ_{av} with construction time.

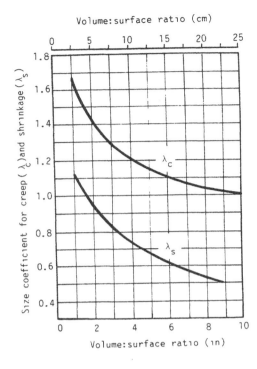

Fig. 18.5 Variation of creep coefficient λ_c and shrinkage coefficient λ_s with volume-to-surface ratio of column.

It must be expected that the creep of concrete will also be affected by variations in environmental humidity and temperature, but sufficient research data have not yet been gained to allow general conclusions to be reached. It appears likely, however, that the creep under conditions of varying temperature and humidity will be higher than that obtained in a controlled laboratory environment.

Since shrinkage of concrete is caused by moisture evaporation from the surface, the volume-to-surface ratio for the member will affect appreciably the resulting shrinkage. The variation of the magnitude of shrinkage coefficient λ_s with the volume-to-surface ratio [18.3, 18.4] is shown in Fig. 18.5. The shrinkage coefficient λ_s may be used to convert experimental data from small-scale specimens to that for full-scale columns. As with creep, it must be expected that higher rates of shrinkage will occur under fluctuating humidity conditions compared to those achieved in a controlled environment.

The rates of change of both creep and shrinkage with time follow an essentially similar curve of the form shown in Fig. 18.6 [18.3]. The curve allows the ultimate creep or shrinkage to be estimated from a laboratory test of a finite time duration, or, conversely, a value at a particular time to be estimated from a known ultimate value. If it is desired to evaluate the amount of creep or shrinkage that results after a particular time t_i, for example, the final value would have to be multiplied by a factor $(1 - \lambda_{ti})$ where λ_{ti} is the value of λ_t at the particular time t_i.

468 CREEP, SHRINKAGE, AND TEMPERATURE EFFECTS

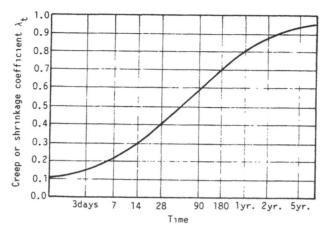

Fig. 18.6 Variation of creep or shrinkage coefficient λ_t with time.

18.3.2 Determination of Vertical Shortening of Walls and Columns

It is likely that rigorous estimates of the effects of elastic and inelastic shortening of the vertical members on the horizontal elements will be required only in the case of very tall buildings or in situations in which stiff slab systems connect vertical elements subjected to large differential shortening. An investigation may also be desirable where dissimilar elements, such as cores and columns, which shorten differentially, are closely spaced.

Figure 18.7 shows the idealized elevation of a building in which slabs have been placed to some level N. Since the supports for the slab from level 1 to level

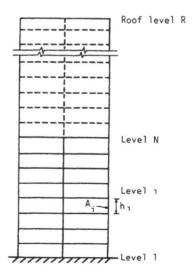

Fig. 18.7 Elevation of multistory building

18 3 CREEP AND SHRINKAGE EFFECTS

N have been subjected to both elastic and inelastic displacements prior to casting, these movements will have no influence on slab N. However, stresses will be induced in the slab due to differential movement of the supports that takes place after the slab has been cast, for all subsequent time (i.e., to time $t = \infty$).

The slab at level N will be subject to differential vertical movements of its supports occurring between levels 1 and N as a result of the following:

1. creep due to loads that were applied before the slab was cast (Δ_{c1}),
2. creep due to loads that are applied during the subsequent construction from level N up to roof level R (Δ_{c2}),
3. shrinkage (Δ_s), and
4. elastic shortening during the progress of construction caused by the additional loads added between levels N and R (Δ_e).

Shortening due to Creep. The creep strains can be determined from the effects of the individual floor loads summed to give the total shortening.

The creep of the ith story (Fig. 18.7), δ_{c1i}, due to loads that are applied before the slab at level N is placed, will be obtained by summing all the creep components that occur due to the loads applied to all stories from i to N. The total creep will occur over the time t that elapses after the slab N is placed at time t_N (i.e., $t = t_N$ to ∞). The creep strain will be obtained by the product of the specific creep $\epsilon'_c(28)$ (obtained either from a 28-day test on a specimen of the design mix, or estimated from the generalized curves of Fig. 18.2) and the particular stress σ_{ci} occurring on the transformed section due to the load increments between stories i and N. This value must be multiplied by the creep coefficients λ_c, to take account of the influence of the volume-to-surface area of the member, λ_a, to include the effect of the age of the column when each load increment is applied, and a factor $(1 - \lambda_t)$ to consider the component of creep that took place after the slab N was cast. The total vertical displacement of the story is then given by the product of the story height h_i and the creep strain, or

$$\delta_{c1i} = h_i \sum_i^N \epsilon'_c(28) \sigma_{ci} \lambda_c \lambda_a (1 - \lambda_t) \quad (18.1)$$

If the loads are applied in equal increments at equal time intervals, Eq. (18.1) can then be simplified to give

$$\delta_{c1i} = h_i \epsilon'_c(28) \sigma_c \lambda_c \lambda_{av} (1 - \lambda_t) \quad (18.2)$$

where σ_c is now the sum of the stresses on the column due to loads between levels i and N, and λ_{av} is the average creep coefficient to take account of the construction time from level i to level N (Fig. 18.4).

Hence, for all stories from the ground floor to level N, the total creep displacement Δ_{c1} due to loads applied prior to placing slab N will be obtained by summing

470 CREEP, SHRINKAGE, AND TEMPERATURE EFFECTS

the displacements in all stories involved. That is,

$$\Delta_{c1} = \sum_1^N \delta_{c1_i} \qquad (18.3)$$

The creep of the ith story, δ_{c2i}, due to the loads that are applied subsequent to the casting of slab N is the sum of the creep components caused by all loads applied between level N and the roof level R. Thus,

$$\delta_{c2i} = h_i \sum_N^R \epsilon_c'(28) \sigma_{ci} \lambda_c \lambda_a \qquad (18.4)$$

where σ_{ci} is the stress on the transformed section caused by the load increments between level N and the roof.

The total vertical shortening Δ_{c2} of all stories below level N due to loads applied subsequent to the casting of slab N is then

$$\Delta_{c2} = \sum_1^N \delta_{c2_i} \qquad (18.5)$$

Shortening Caused by Shrinkage. The columns and walls in the stories below the considered level N will contribute their remaining shrinkage δ_s, during the period subsequent to the time of casting of slab N. In a manner similar to that for creep deformations, the shrinkage deformation for each story will be given by the product of the story height h_i, the ultimate shrinkage strain ϵ_s, the size coefficient λ_s to take account of the influence of the volume-to-surface ratio of the member, and the coefficient $(1 - \lambda_t)$ to include only that part of the shrinkage that takes place after the slab N was cast. On summing all such shrinkage deformations of the stories below level N, the total shrinkage shortening Δ_s becomes

$$\Delta_s = \sum_1^N \delta_{s_i} = \sum_1^N h_i \epsilon_s \lambda_s (1 - \lambda_t) \qquad (18.6)$$

The ultimate shrinkage strain ϵ_s should ideally be obtained from a specimen of the concrete mix to be used in the structure and stored under site conditions.

Elastic Shortening. During the period of construction after slab N is cast, elastic shortening of the supports to the slab will be caused by the action of all load increments above level N. The total deformation Δ_e will be obtained by adding the individual deformations in each story, δ_{ei}. Thus,

$$\Delta_e = \sum_1^N \delta_{ei} = P \sum_1^N \frac{h_i}{E_i A_i} \qquad (18.7)$$

in which E_t is the elastic modulus and A_t is the transformed cross-sectional area of the member in story i, and P is the sum of all loads above the particular level N.

For regular buildings in which the column areas change uniformly, there should not be large variations in the differential elastic movement of the slab supports, as can be observed from Eq. (18.7). For the upper slabs, there are many stories below the slab but the loads above are small, while the position is reversed for the lower slabs.

18.3.3 Influence of Reinforcement on Column Stresses, Creep, and Shrinkage

When a reinforced concrete column is subjected to a sustained compressive force, there is a gradual transfer of stress from the concrete to the steel as a result of the shortening due to creep and shrinkage. In extreme cases of highly reinforced columns, it is even possible that all the axial load could be transferred to the steel, and tensile stresses induced in the concrete due to further shrinkage.

The transfer of stresses from the concrete may reduce significantly the shortening due to creep and shrinkage, and hence reduce the differential effects on the slabs. The influence of the reinforcing steel can be estimated by a separate calculation once the basic shortening has been determined.

By considering the conditions of equilibrium and compatibility between the steel and concrete, it can be shown [18.1] that the changes in steel stresses, $\delta \sigma_s$, and concrete stress, $\delta \sigma_c$, become

$$\delta \sigma_s = \frac{\sigma_c \epsilon_c' + \epsilon_s}{p \epsilon_c'} F$$

$$\delta \sigma_c = \left(\sigma_c + \frac{\epsilon_s}{\epsilon_c'} \right) F$$

(18.8)

where the function F is given by

$$F = 1 - \exp\left[-(pn/1 + pn)\epsilon_c' E_c\right]$$

In Eqs. (18.8)

σ_c = initial elastic stress in concrete
ϵ_s = total shrinkage strain of plain concrete adjusted for the volume-to-surface ratio
ϵ_c' = ultimate specific creep of plain concrete
E_c = modulus of elasticity of concrete
p = reinforcement ratio of cross section
n = modular ratio (E_s/E_c)

The residual creep and shrinkage strains $\delta\epsilon$ of a reinforced concrete column will then be equal to the additional steel strain and can be deduced directly from the change in steel stress.

$$\delta\epsilon = \delta(\epsilon_s + \epsilon_c) = \frac{\delta\sigma_s}{E_s} \qquad (18.9)$$

The total shortening will then be obtained by adding the elastic shortening and the modified (residual) inelastic creep and shrinkage effects, taking account of the reinforcement in the column or wall. Because the percentage of reinforcement, and therefore its influence, is usually much less in the case of walls, they tend to shorten more than columns. Conversely, the higher percentage of reinforcement in the lower story columns reduces considerably their creep and shrinkage deformations, as does the fact that loading takes place over a longer period of time.

18.3.4 Worked Example

To illustrate the use of the various curves and formulas, the ultimate residual creep and shrinkage strains, and the additional steel stresses, in a reinforced concrete column at a particular level will be considered.

The given data are as follows: Interior column, of dimensions 0.5×1.25 m (1.64×4.10 ft) situated 35 stories below roof level. The area of reinforcing steel is 262.5 cm^2 (40.69 in.^{2}) giving a percentage reinforcement of 4.2%.

For steel $E_s = 206.5 \times 10^3$ N/mm^2 (29.95×10^6 lb/in.^{2})

For concrete $E_c = 27.9 \times 10^3$ N/mm^2 (4.05×10^6 lb/in.^{2})

Modular ratio $n = E_s/E_c = 7.4$

Transformed column area $A_t = (0.5 \times 1.25) + (7.4 - 1) \times 262.5 \times 10^{-4}$

or $A_t = 0.793$ m^2 \quad (8.536 ft^2)

If construction is at the rate of one floor every 4 days, the total time to complete 35 floors (i.e., 35 load increments are applied) is $35 \times 4 = 140$ days. The imposed loading from each floor = 166 kN (32.32 kip). From Fig. 18.2, the 20-year specific creep $\epsilon_c'(28)$ for loading at 28 days is estimated to be 0.048×10^{-6} m/m/kN/m^2 (0.331×10^{-6} in./in./lb/in.^{2}) for the value of E_c adopted. It is estimated that the shrinkage strain for the first 90 days is 620×10^{-6} m/m (in./in.).

The required calculations are as follows:
From Fig. 18.4, at a time of 140 days, the coefficient $\lambda_{av} = 0.85$

For the column, volume-to-surface area = $\dfrac{0.5 \times 1.25}{2(0.5 + 1.25)} = 0.179$ m (0.587 ft)

$\therefore$ From Fig. 18.5, the size coefficient $\lambda_c = 1.06$

18.3 CREEP AND SHRINKAGE EFFECTS

Thus sustained stress on the concrete

$$\sigma_c = \frac{P}{A_t} = \frac{166 \times 35}{0.793} = 7327 \text{ kN/m}^2 \ (1063 \text{ lb/in.}^2)$$

$\therefore$ Total creep strain $\epsilon_c = \epsilon'_c(28) \times \sigma_c \times \lambda_c \times \lambda_{av}$

$$= 0.048 \times 10^{-6} \times 7327 \times 1.06 \times 0.85$$

$$= 316.8 \times 10^{-6} \text{ m/m (in./in.)}$$

From Fig. 18.6, the coefficient λ_t, which gives the ratio of shrinkage at 90 days to the ultimate shrinkage, is 0.60.

$$\therefore \epsilon_{\text{sult}} = \frac{620 \times 10^{-6}}{0.6} = 1033 \times 10^{-6} \text{ m/m (in./in.)}$$

From Fig. 18.5, the size coefficient $\lambda_s = 0.56$

$$\therefore \epsilon_s = 1033 \times 10^{-6} \times 0.56 = 578.5 \times 10^{-6} \text{ m/m (in./in.)}$$

$\therefore$ Total creep and shrinkage strain $= \epsilon_c + \epsilon_s = (316.8 + 578.5) \times 10^{-6}$

$$= 895.3 \times 10^{-6} \text{ m/m (in./in.)}$$

Converting the specific creep for loading at 28 days to consider incremental loading over a period of 140 days, and taking account of the volume-to-surface area effect, gives

Specific creep $\epsilon'_c = 0.048 \times 10^{-6} \times 1.06 \times 0.85 = 0.043 \times 10^{-6} \text{ m/m/kN/m}^2$

For a specific creep $\epsilon'_c = 0.043 \times 10^{-6} \text{ m/m/kN/m}^2 \ (0.296 \times 10^{-6} \text{ in./in./lb/in}^2)$

$$E_c = 27.9 \times 10^6 \text{ kN/m}^2 \ (4.05 \times 10^6 \text{ lb/in.}^2)$$

$$p = 0.042$$

$$n = 7.4$$

The function F in Eq. (18.8) is

$$F = 1 - e^{-0.285} = 0.248$$

Therefore, from Eq. 18.8, the additional stress in the steel becomes

$$\delta\sigma_s = \frac{7327 \times 0.043 \times 10^{-6} + 578.5 \times 10^{-6}}{0.042 \times 0.043 \times 10^{-6}} \times 0.248$$

$$= 122.7 \times 10^3 \text{ kN/m}^2 \ (17.8 \times 10^3 \text{ lb/in.}^2)$$

From Eq. (18.9), the residual creep and shrinkage strain is

$$\delta\epsilon = \frac{122.7 \times 10^3}{206.5 \times 10^6} = 594.2 \times 10^{-6} \text{ m/m (in./in.)}$$

$$\therefore \frac{\text{Residual strains}}{\text{Total strains}} = \frac{594.2 \times 10^{-6}}{895.3 \times 10^{-6}} = 0.663$$

This ratio indicates that the total strains have been reduced by 34% due to the presence of the reinforcement.

18.3.5 Influence of Vertical Shortening on Structural Actions in Horizontal Members

In view of the lack of experimental data available, precise details of the environmental conditions during and after construction, and the various simplifying assumptions that have had to be made to achieve a solution, the method of Khan and Fintel for determining the inelastic shortening must be regarded as approximate only. It is generally accepted that the accuracy in estimating creep strains is not high, and that errors of up to ±30% are possible.

The method assumes that the shortening of the vertical members can be calculated without reference to the horizontal connecting elements. However, the moments in the slabs caused by settlement of the supports will produce a redistribution of loads in the vertical members, and a different stress level for creep effects. The present state of knowledge makes it impractical to include such refinements in the analysis, and so, in view of the uncertainties in the estimation of differential shortening, an accurate model for the determination of stresses in the slabs is not warranted.

The effects of differential shortening between adjacent vertical elements may be estimated by replacing the floor slabs by beams of equivalent stiffness, as described in Appendix 1, to produce an analogous frame model. The analysis starts as usual with the known calculated differential vertical settlements Δ_i, unrestrained by any frame action, at the floor levels i. This movement produces equal end moments M of magnitude $6EI_i\Delta_i/L_i^2$, in which EI_i is the effective flexural rigidity of the equivalent beam of span L_i, with associated equal and opposite end shears of magnitude $2M/L_i$. The analysis of the equivalent frame subjected to those initial forces may then be achieved, using either a plane frame computer program or an approximate procedure using a simplified frame model (cf. Chapter 7). The analysis yields the distribution of the additional slab moments and axial forces in the columns throughout the structure.

When the columns shorten differentially over a period of time, at a progressively decreasing rate, the resulting slab and girder moments are relaxed by the creep in those members. Tests have indicated that the change in moments may be as much as 50% over a period of time, and it is suggested that the maximum value of the differential settlement moments should be taken to be half of the calculated values assuming no relaxation due to creep in the horizontal members.

The support that settles less will receive additional load from the other support, the transferred component V_i from beam element i being $(M_{i1} + M_{i2})/L_i$, where M_{i1} and M_{i2} are the end moments, reduced to allow for creep relaxation, on the beam element. The load transferred over the entire height is obtained by summing up all such values from the roof downward.

The stresses due to differential shortening should be treated as equivalent dead load stresses, modified by appropriate load factors before being combined with other loading conditions. When choosing an appropriate load factor, it must be borne in mind that the design relaxed moments occur only for a relatively short time during the life of the building and they continue to decrease after that.

18.4 TEMPERATURE EFFECTS

Until relatively recently, thermal effects on tall buildings were generally insignificant. The early buildings were well insulated by heavy cladding with low areas of fenestration, and any possible thermal movements were controlled by expansion joints. In recent years, however, many buildings, both steel and concrete, have been constructed with exterior columns either partially or fully exposed to the weather. When subjected to diurnal or seasonal temperature variations, exposed columns change their length relative to the interior columns, which remain essentially at a constant temperature in a controlled environment. Although the length changes within a single story are of secondary magnitude, the changes are cumulative, starting at ground level and increasing to a maximum at the top story. At heights above 20 stories the effects may be significant, and it becomes desirable to investigate the influence of thermal deformations on both the structural actions and on non-load-bearing elements such as internal partitions. If any restraint exists to prevent free deformations, loads will be induced in the members concerned. As well as vertical movements, horizontal deformations should be considered for buildings greater than about 60 m (200 ft) in breadth.

Thermal movements are elastic in nature, and a conventional linear elastic analysis may be employed to predict the structural actions produced by any temperature changes. Changes in length in the exterior columns will impose differential vertical displacements in the slabs and beams of the exterior bay. These in turn induce moments and shears in the slabs and beams, and moments and additional axial forces in the columns. The exterior columns will also be subjected to additional moments due to the temperature differentials that exist between their warm and cold sides, particularly if they are partially exposed only. The temperature differential between an exposed spandrel beam and the interior floor slab will produce tension in the beam in winter, and introduce compressive stresses into the floor slab near the corners.

The analysis of such structural actions requires four main steps:

1. Determination of a realistic design relative temperature for the building. Members in the interior of the building are maintained at a relatively constant temperature environment. Members on the outside of the building will be subject to the interior temperature on their inner faces, and to a variable external temperature on the opposite faces, which will be influenced both by the ambient air temperature and the effects of wind and solar radiation. Since steady-state heat conduction through a section is possible only if the boundary temperatures remain constant for a sufficient length of time, it is likely that true steady-state conditions will rarely occur, and an equivalent steady-state design temperature must then be chosen to allow temperature variations in the members to be determined. The mean highest and lowest effective seasonal temperatures for the specific locality in which the building is to be situated can generally be obtained from a local meteorological office.

The factors controlling heat transfer through the structural members are the time lag and the attenuation of the exterior temperature amplitudes, which depend on the frequency of temperature change and on the thermal properties of the members. Materials such as steel, with a high thermal conductivity, will respond more quickly to temperature changes and will require a consideration of changes over a shorter time period than for concrete. For the latter, it is considered adequate to use the minimum mean daily temperature with a frequency of recurrence of 40 years as the equivalent steady-state exterior design temperature. The design differential temperature is then the difference between exterior and interior mean values. Winter differentials are generally more severe than summer ones, and, in addition, the resulting contractions will be additive to creep and shrinkage movements, producing the most severe frame distortions.

2. Determination of heat flow through the members. After an equivalent steady-state design differential temperature has been selected, the interior temperatures may be calculated by a numerical solution of the heat flow equation, or by a graphic method using a flow net construction [18.6].

3. Evaluation of the thermal movements in the members. The two basic effects caused by temperature gradients through a column are bowing and an axial length change.

The axial length change of a member is equivalent to its being loaded by an axial force N, of magnitude

$$N = \int \alpha t E \, dA = \alpha E A t_{av} \qquad (18.10)$$

where A = the cross-sectional area of the member
t_{av} = the average temperature change
α = the coefficient of linear expansion of the material, which may be taken to be constant over the practical range of temperatures encountered in buildings.

Although the value of α varies with the concrete mix concerned, and is particularly affected by the aggregate used, an average value for concrete may be taken to be $\alpha = 9.9 \times 10^{-6}$ per °C. For steel, the value of α may be assumed to be $\alpha = 11.7 \times 10^{-6}$ per °C.

The temperature gradient will produce a corresponding initial effective moment M_y in the Y direction, given by

$$M_y = \int E\alpha t y \, dA \qquad (18.11)$$

The corresponding moment M_x in the X direction may be obtained in a similar manner if the column is subjected to biaxial bending. This will be of particular importance for corner columns.

Corresponding initial effective axial forces and bending moments may be obtained for the horizontal spandrel beams.

4. Elastic analysis of the frame for the imposed differential length changes. The initial forces and moments due to the thermal movements acting on the frame with all joints restrained in direction may be added directly to other load systems, using load factors appropriate to the National Code concerned, to give the design loads. The analysis of the complete building may then be undertaken using any acceptable elastic approach.

However, in multistory frames with partially or fully exposed exterior columns, the effects of temperature movements of the exterior columns will be significant only in the members of the exterior bay—that is, the exterior column, the exterior bay beam, and the first interior column—since all other interior columns will be subjected to the same uniform temperature and no relative deformations will occur. The exterior effects may be assumed not to extend far inward and will be negligible over most of the structure. Consequently, an analysis of sufficient accuracy to assess the relative importance of the thermal movements may be performed on an equivalent subframe, which will model the actions in the region of interest. The simplest model is a three-bay symmetrical frame shown in Fig. 18.8.

In the frame, the correct values of the axial and flexural stiffnesses will be prescribed for the exterior column 1, the interior column 2, and the connecting beam 1-2 at each story level. The assumption of the symmetrical three-bay frame ensures that the effects of thermal deformations of column 1 will not proceed beyond the line of symmetry, which is at the mid-span position of the second beam. Because of symmetry, only one-half of the frame need be analyzed.

With all joints initially restrained, the fixed end moments may be calculated for the beams at each floor level. The frame can then be analyzed using either a direct stiffness matrix approach or an iterative technique, such as the moment-distribution method [18.7]. The analysis will yield the moments in the beams and columns, and the axial forces in the columns. The shear in the interior beam will be zero because of the assumed symmetry.

The distortions of the topmost story panel may then be calculated to see whether the relative deflections involved may cause distress in the interior partitions.

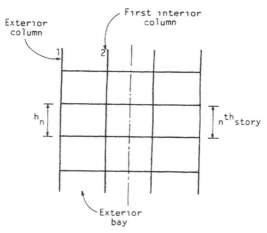

Fig. 18.8 Exterior bays of nth story in building.

To protect brickwork and plaster from cracking, it has been suggested that the thermal differential movement between exterior and interior columns should be limited to the smaller of 1/300 of the span, or 19 mm (0.75 in.). For buildings without masonry and plastered partitions, a thermal movement of 1/200 of the span should be acceptable.

SUMMARY

In very tall concrete buildings, the cumulative vertical movements due to creep and shrinkage may be sufficiently large to cause distress in nonstructural elements, and to induce significant structural actions in the horizontal elements, in the upper regions of the building. In assessing these long-term deformations, the influence of a number of significant factors must be considered, particularly the concrete properties, the loading history and age of the concrete at the time of load application, and the volume-to-surface ratio and amount of reinforcement in the members concerned. To simplify the calculations, it is assumed that the vertical deformations are initially independent of the horizontal member stiffnesses. The total shortening of all columns and walls is then equal to the sum of the elastic shortening under load and the combined effects of creep and shrinkage. The structural actions in the horizontal elements caused by the resulting relative vertical deflections of their supports can then be estimated. The differential movements due to creep and shrinkage must be considered structurally and accommodated as far as possible in the architectural details at the design stage.

In buildings with partially or fully exposed exterior columns, significant temperature differences may occur between the exterior and adjacent interior columns, and any restraint to their relative deformations will induce stresses in the members concerned. The analysis of such actions requires a knowledge of the differential temperatures that are likely to occur between the building and its exterior and the temperature gradient through the members. This will allow an evaluation of the free thermal length changes that would occur if no restraint existed, and, hence, using a standard elastic analysis, the resulting thermal stresses and deformations may be determined.

REFERENCES

18.1 Fintel, M. and Khan, F. R. "Effects of Column Creep and Shrinkage in Tall Structures—Prediction of Inelastic Column Shortening." *ACI J.* **66,** 1969. 957–967.

18.2 Fintel, M. and Khan, F. R. "Effects of Column Creep and Shrinkage in Tall Structures—Analysis for Differential Shortening of Columns and Field Observation of Structures." Paper SP 27-4. ACI Special Publication 27, 1971, 95–119.

18.3 *Recommendation for an International Code of Practice for Reinforced Concrete.* CEB, Published jointly by ACI and Cement and Concrete Association, London, 1970.

18.4 Hansen, T. C. and Mattock, A. H. "Influence of Size and Shape of Member on the Shrinkage and Creep of Concrete." *ACI J.* **63,** 1966. 267–290.

18.5 Khan, F. R. and Fintel, M. "Conceptual Details for Creep, Shrinkage and Temperature in Ultra High-Rise Buildings." Paper SP 27-9, ACI Special Publication 27, 1971, 215–228.

18.6 Fintel, M. and Khan, F. R. "Effects of Column Exposure in Tall Structures—Temperature Variations and Their Effects." *ACI J.* **62,** 1965, 1533–1555.

18.7 Khan, F. R. and Fintel, M. "Effect of Column Exposure in Tall Structures—Analysis for Length Changes of Exposed Columns." *ACI J.* **63,** 1966. 843–862.

APPENDIX 1

Formulas and Design Curves for Coupled Shear Walls

Detailed solutions and associated design curves were presented in Chapter 10 for coupled shear walls subjected to uniformly distributed lateral loading. This Appendix provides further information on coupled shear wall structures.

First, formulas and design curves are given for two other standard load cases, a triangularly distributed lateral loading and a concentrated top load. Second, solutions are given for coupled shear walls supported either on elastic foundations or on portal or column supports, subjected to the three standard forms of loading considered earlier. Finally, data are presented to enable the effective width and bending stiffness to be determined for floor slabs connecting shear walls of different cross-sectional shapes.

A1.1 FORMULAS AND DESIGN CURVES FOR ALTERNATIVE LOAD CASES

A1.1.1 Formulas for Top Concentrated Load and Triangularly Distributed Loading

Based on the continuum approach, solutions are now presented for the two other standard load cases, a concentrated load P at the top, and a triangularly distributed loading with a maximum intensity p at the top (Fig. A1.1). As noted in Chapter 3, a combination of these two static load cases can be used to simulate earthquake loading, and a combination of the triangularly distributed loading with uniformly distributed loading can be used to represent graduated wind loading.

The externally applied bending moment M in Eqs. (10.16) and (10.17) is

$$M = P(H - z) + \frac{1}{6}p(H - z)^2(2 - z/H) \tag{A1.1}$$

On following the same procedure as for the uniformly distributed loading in Section 10.3.3, the corresponding solutions become

A1.1 FORMULAS AND DESIGN CURVES FOR ALTERNATIVE LOAD CASES

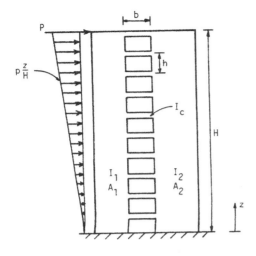

Fig. A1.1 Loading on coupled walls.

Concentrated Load P at Top

$$N = \frac{PH}{k^2 l}\left[\left(1 - \frac{z}{H}\right) - \frac{1}{k\alpha H \cosh k\alpha H}\sinh k\alpha(H-z)\right] \quad (A1.2)$$

$$q = \frac{P}{k^2 l}\left[1 - \frac{\cosh k\alpha(H-z)}{\cosh k\alpha H}\right] \quad (A1.3)$$

$$y = \frac{PH^3}{3EI}\left\{\frac{1}{2}\left(\frac{k^2-1}{k^2}\right)\left[3\left(\frac{z}{H}\right)^2 - \left(\frac{z}{H}\right)^3\right]\right.$$
$$\left. + \frac{3}{k^2}\left[\frac{1}{(k\alpha H)^2}\frac{z}{H} - \frac{\sinh k\alpha H - \sinh k\alpha(H-z)}{(k\alpha H)^3 \cosh k\alpha H}\right]\right\} \quad (A1.4)$$

Triangularly Distributed Loading of Intensity p (z/H)

$$N = \frac{pH^2}{k^2 l}\left\{\frac{[\sinh k\alpha H - (k\alpha H/2) + (1/k\alpha H)]}{(k\alpha H)^2 \cosh k\alpha H}\sinh k\alpha(H-z)\right.$$
$$- \frac{1}{(k\alpha H)^2}\cosh k\alpha(H-z)$$
$$\left. + \frac{1}{2}\left(1 - \frac{z}{H}\right)^2 - \frac{1}{6}\left(1 - \frac{z}{H}\right)^3 + \frac{1}{(k\alpha H)^2}\left(\frac{z}{H}\right)\right\} \quad (A1.5)$$

$$q = \frac{pH}{k^2 l}\left\{\frac{[\sinh k\alpha H - (k\alpha H/2) + (1/k\alpha H)]}{(k\alpha H)\cosh k\alpha H}\cosh k\alpha(H-z)\right.$$

$$-\frac{1}{k\alpha H}\sinh k\alpha(H-z)$$

$$\left.+\left(1-\frac{z}{H}\right) - \frac{1}{2}\left(1-\frac{z}{H}\right)^2 - \frac{1}{(k\alpha H)^2}\right\} \quad (A1.6)$$

$$y = \frac{1}{2}\frac{pH^4}{EI}\left\{\frac{1}{60}\left(\frac{k^2-1}{k^2}\right)\left(20\left(\frac{z}{H}\right)^2 - 10\left(\frac{z}{H}\right)^3 + \left(\frac{z}{H}\right)^5\right)\right.$$

$$+\frac{1}{k^2(k\alpha H)^2}\left\{\left[1-\frac{2}{(k\alpha H)^2}\right]\left(\frac{z}{H}\right) - \frac{1}{3}\left(\frac{z}{H}\right)^3\right.$$

$$+\frac{2}{(k\alpha H)^2\cosh k\alpha H}\left|\cosh k\alpha z - 1\right.$$

$$\left.\left.+\left(\frac{1}{k\alpha H}-\frac{k\alpha H}{2}\right)[\sinh k\alpha H - \sinh k\alpha(H-z)]\right]\right\}. \quad (A1.7)$$

A1.1.2 Design Curves

Using the above formulas, design curves corresponding to those presented in Sections 10.3.3 and 10.3.4 for a uniformly distributed load may be produced for these two standard load cases.

Concentrated Load P at Top. The percentage of composite cantilever action K_2 is given by

$$K_2 = 100\left|1 - \frac{\sinh k\alpha(H-z)}{k\alpha(H-z)\cosh k\alpha H}\right| \quad (A1.8)$$

and

$$K_1 = 100 - K_2$$

Figure A1.2 shows the variations of K_1 and K_2 with the relative height z/h and stiffness parameter $k\alpha H$. These may be used in a manner identical to that described in Section 10.3.4.

The shear flow q can be expressed as

$$q = P\frac{1}{k^2 l}F_2 \quad (A1.9)$$

A1.1 FORMULAS AND DESIGN CURVES FOR ALTERNATIVE LOAD CASES

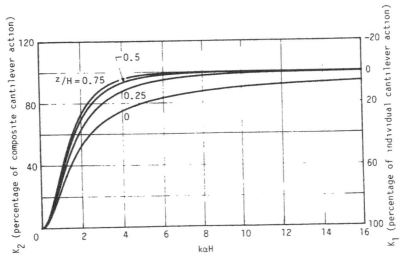

Fig. A1.2 Wall moment factors for concentrated load at top.

where the shear flow factor F_2 is

$$F_2 = 1 - \frac{\cosh k\alpha(H-z)}{\cosh k\alpha H}$$

The variation of F_2 with the two parameters z/H and $k\alpha H$ is shown in Fig. A1.3. It may be noted that in the case of a top concentrated load the maximum value always occurs at the top of the structure.

The maximum lateral deflection y_H at the top of the structure is

$$y_H = \frac{PH^3}{3EI} F_3 \tag{A1.10}$$

where

$$F_3 = 1 - \frac{3}{k^2}\left[\frac{1}{3} + \frac{\sinh k\alpha H}{(k\alpha H)^3 \cosh k\alpha H} - \frac{1}{(k\alpha H)^2}\right]$$

The variation of the deflection factor F_3 is shown in Fig. A1.4.

The factor indicates the reduction provided by the connecting beams to the top deflection of a pair of linked walls ($PH^3/3EI$).

Triangularly Distributed Loading. The percentage of composite cantilever action K_2 is

484 FORMULAS AND DESIGN CURVES FOR COUPLED SHEAR WALLS

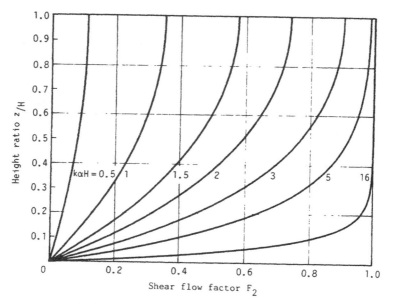

Fig. A1.3 Shear flow factor for concentrated load at top.

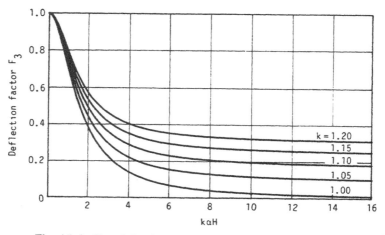

Fig. A1.4 Top deflection factor for concentrated load at top.

A1.1 FORMULAS AND DESIGN CURVES FOR ALTERNATIVE LOAD CASES

$$K_2 = \frac{100}{\frac{1}{6}(k\alpha H)^2[1-(z/H)]^2[2+(z/H)]} \left| \frac{z}{H} + \frac{1}{2}(k\alpha H)^2 \left(1-\frac{z}{H}\right)^2 \right.$$

$$- \frac{1}{6}(k\alpha H)^2 \left(1-\frac{z}{H}\right)^3 - \cosh k\alpha(H-z)$$

$$\left. + \frac{\sinh k\alpha H - (k\alpha H/2) + (1/k\alpha H)}{\cosh k\alpha H} \sinh k\alpha(H-z) \right| \quad (A1.11)$$

The variations of K_1 and K_2 are shown in Fig. A1.5.
The shear flow q in the connecting medium is

$$q = p\frac{H}{k^2 l} F_2 \quad (A1.12)$$

where the shear flow factor F_2 is

$$F_2 = \frac{\sinh k\alpha H - (k\alpha H/2) + (1/k\alpha H)}{k\alpha H \cosh k\alpha H} \cosh k\alpha(H-z)$$

$$- \frac{1}{k\alpha H} \sinh k\alpha(H-z)$$

$$+ \left(1-\frac{z}{H}\right) - \frac{1}{2}\left(1-\frac{z}{H}\right)^2 - \frac{1}{(k\alpha H)^2}$$

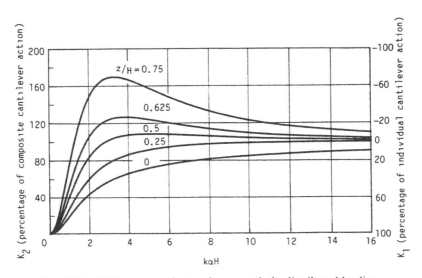

Fig. A1.5 Wall moment factors for triangularly distributed loading.

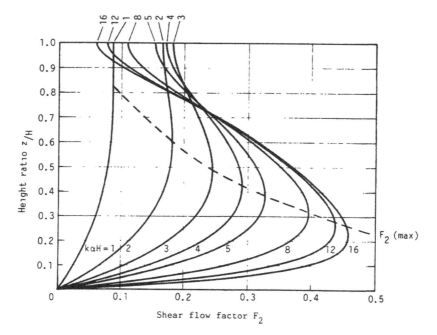

Fig. A1.6 Shear flow factor for triangularly distributed loading.

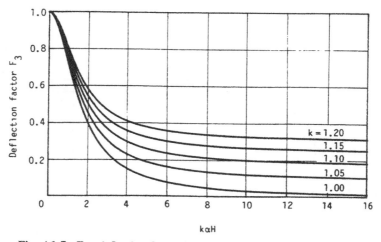

Fig. A1.7 Top deflection factor for triangularly distributed loading.

The variation of F_2 is shown in Fig. A1.6, on which the curve of the maximum value $F_2(\max)$ is shown by a broken line.

The maximum lateral deflection y_H may again be expressed in terms of the top deflection of a pair of free cantilevers by the deflection factor F_3:

$$y_H = \frac{11}{120} \frac{pH^4}{EI} F_3 \qquad (A1.13)$$

where

$$F_3 = 1 - \frac{1}{k^2} + \frac{120}{11} \frac{1}{k^2 (k\alpha H)^2} \left\{ \frac{1}{3} - \frac{1 + [(k\alpha H/2) - (1/k\alpha H)] \sinh k\alpha H}{(k\alpha H)^2 \cosh k\alpha H} \right\}$$

The variation of F_3 is shown in Fig. A1.7.

The factor again illustrates the reduction provided by the connecting beams to the top deflection of a pair of linked walls $[(11/120)(pH^4/EI)]$.

A1.2 FORMULAS FOR COUPLED SHEAR WALLS WITH DIFFERENT FLEXIBLE SUPPORT CONDITIONS

Section 10.3.7 outlined briefly how solutions could be achieved for coupled walls on flexible supports. This section presents formulas for walls supported on individual elastic foundations (Fig. A1.8) and on the range of portal frame or column

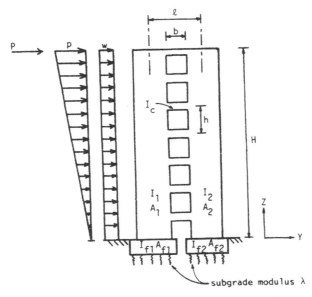

Fig. A1.8 Coupled shear walls on elastic foundations.

supports shown in Fig. A1.9. The symbol "x" in Fig. A1.9 indicates where hinges are either provided, or assumed to exist as a result of the large variation between the moments of inertia of the walls and columns. The necessary structural parameters r_I, r_L, or r_H used in the general formulas are defined in Fig. A1.9. Formulas are presented for the three standard load cases of a uniformly distributed loading

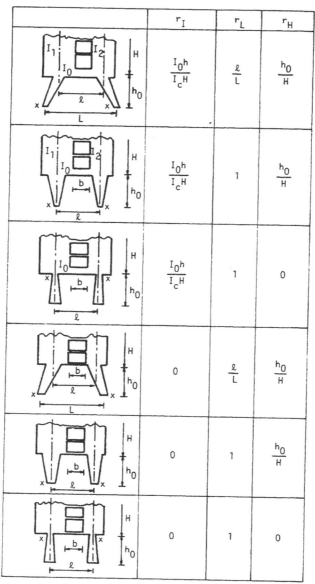

Fig. A1.9 Parameters for coupled walls supported on columns or portal frames.

A1.3 STIFFNESS OF FLOOR SLABS CONNECTING SHEAR WALLS

of intensity w per unit height, a concentrated load P at the top, and a triangularly distributed loading whose intensity varies linearly from zero at the base to a value of p at the top (Fig. A1.8) [A1.1].

For walls supported on separate bases, it is assumed that walls 1 and 2 are fixed to bases of cross-sectional areas A_{f_1} and A_{f_2} and second moments of area I_{f_1} and I_{f_2}, supported on soil with a modulus of subgrade reaction λ (Fig. A1.8). In the tables of results, the following additional parameters are used.

$$S = \frac{EI}{\lambda I_f H} \qquad I_f = I_{f_1} + I_{f_2}$$

$$m = \frac{E}{\lambda H} \qquad A_f = A_{f_1} + A_{f_2} \qquad (A1.14)$$

$$\alpha_i^2 = \frac{12 I_c l^2}{h a^3 I_f} \qquad k_i^2 = 1 + \frac{A_f I_f}{A_{f_1} A_{f_2} l^2}$$

Formulas for the lateral deflection y, the sum of the wall moments $(M_1 + M_2)$, and the axial forces N, all expressed in nondimensional form, are given in Tables A1.1, A1.2, and A1.3 for the three standard load cases.

Since the moment in a wall is proportional to its flexural rigidity, the moments in walls 1 and 2 are

$$M_1 = \frac{I_1}{I}(M_1 + M_2)$$
$$M_2 = \frac{I_2}{I}(M_1 + M_2) \qquad (A1.15)$$

In each table, the quantities are expressed in terms of the four integration constants C_1 to C_4, which are subsequently tabulated for each type of wall support system.

A1.3. STIFFNESS OF FLOOR SLABS CONNECTING SHEAR WALLS

In cross-wall structures, which are popular forms of construction for tall apartment blocks, the structural system consists of one-way slabs spanning between parallel assemblies of load-bearing walls that resist lateral as well as gravitational loads (Fig. A.1.10a). The analysis of slab-coupled walls can be conveniently performed using the techniques described in Sections 10.3 and 10.4 for beam-coupled wall systems, provided the effective bending stiffness of the slab, assumed to act as a wide connecting beam, can be established. Under the action of lateral forces, the free bending of a pair of shear walls is resisted by the floor slab, which is forced to rotate and bend out of its plane where it connects rigidly to the walls (Fig. A.10b

Table A1.1 Formulas for Concentrated Load at Top of Structure

Quantity	Expressions for Determining Deflections, Wall Moments, and Axial Forces
$\dfrac{EI}{PH^3}y$	$\dfrac{1}{6}\dfrac{k^2-1}{k^2}\left(1-\dfrac{z}{H}\right)^3 - \dfrac{1}{k^2(k\alpha H)^2}\left(1-\dfrac{z}{H}\right) + \dfrac{1}{k^2(k\alpha H)^3}\sinh k\alpha(H-z) + C_1 + C_2\dfrac{z}{H} + C_3 \sinh k\alpha z + C_4 \cosh k\alpha z$
$\dfrac{M_1+M_2}{PH}$	$\dfrac{k^2-1}{k^2}\left(1-\dfrac{z}{H}\right) - \dfrac{1}{k^2}\dfrac{1}{k\alpha H}\sinh k\alpha(H-z) + (k\alpha H)^2 (C_3 \sinh k\alpha z + C_4 \cosh k\alpha z)$
$\dfrac{Nl}{PH}$	$\dfrac{1}{k^2}\left[\left(1-\dfrac{z}{H}\right) - \dfrac{1}{k\alpha H}\sinh k\alpha(H-z)\right] - (k\alpha H)^2 (C_3 \sinh k\alpha z + C_4 \cosh k\alpha z)$

	Elastic Foundation	Portal or Column Base
C_1	$C_3 \tanh k\alpha H + \dfrac{1}{k^2}\dfrac{k^2-1}{(k\alpha H)^2} - \dfrac{1}{k^2}\dfrac{1}{(k\alpha H)^3}\sinh k\alpha H$	$\dfrac{1}{k^2}\dfrac{1}{(k\alpha H)^2}\left(1-\dfrac{\sinh k\alpha H}{k\alpha H}\right) + \dfrac{r_H r_L}{(k\alpha H)^2}(1-\cosh k\alpha H)$ $-\dfrac{1}{6}\dfrac{k^2-1}{k^2} + [\tanh k\alpha H + r_H r_L k^2(k\alpha H)]C_3$
C_2	$\dfrac{1}{2}\dfrac{k^2-1}{k^2} + m\dfrac{\alpha_f^2 k_i^2-1}{\alpha^2\,k_i^2} - \dfrac{1}{(k\alpha H)^2}\left(\dfrac{1}{k^3}-\dfrac{1}{k_i^2}\right)(1-\cosh k\alpha H)$ $-\left(1-\dfrac{k^2}{k_i^2}\right)k\alpha H C_3$	$\dfrac{1}{2}\dfrac{k^2-1}{k^2}\left(\dfrac{1}{k^3}-r_L\right)\left\|\dfrac{1-\cosh k\alpha H}{(k\alpha H)^2} + k^3(k\alpha H)C_3\right\|$
C_3	$\dfrac{1}{\Delta}\left\{\left\|S\dfrac{k^2-1}{k^2}(k_f\alpha H)^2 - m(\alpha_f H)^2(k_i^2-1)\right\| - \dfrac{1}{k^2}\left\|1-\cosh k\alpha H\right\|\right\}$ $-S\dfrac{k_i^2(\alpha H)}{k}\sinh k\alpha H$	$\dfrac{1}{\Delta}\left\{\dfrac{1}{k^2}\left\|\dfrac{\sinh k\alpha H}{k\alpha H} - (1-\cosh k\alpha H)r_l\right\| - \left(\dfrac{1}{k^3}-r_L+r_H r_L\right)\right\}$
C_4	$-C_3 \tanh k\alpha H$	$-C_3 \tanh k\alpha H$
Δ	$(k\alpha H)^3[S(k_i \alpha H)^2 \tanh k\alpha H + k\alpha H]$	$(k\alpha H)^3(k\alpha H r_l + \tanh k\alpha H)$

TABLE A1.2 Formulas for Uniformly Distributed Load

Quantity	Expressions for Determining Deflections, Wall Moments, and Axial Forces
$\dfrac{EI}{wH^4} y$	$\dfrac{1}{24}\dfrac{k^2-1}{k^2}\left\| \left(\dfrac{z}{H}\right)^4 - 4\left(\dfrac{z}{H}\right)^3 \right\| + \dfrac{1}{2(k\alpha H)^2}\left\| \dfrac{k^2-1}{k^2} + \dfrac{1}{2}(\alpha H)^2(k^2-1) - 1 \right\|\left(\dfrac{z}{H}\right)^2 + C_1 + C_2\left(\dfrac{z}{H}\right) + C_3 \sinh k\alpha z + C_4 \cosh k\alpha z$
$\dfrac{M_1 + M_2}{wH^2}$	$\dfrac{1}{2}\dfrac{k^2-1}{k^2}\left(1 - \dfrac{z}{H}\right)^2 - \dfrac{1}{k^2(k\alpha H)^2} + (k\alpha H)^2(C_3 \sinh k\alpha z + C_4 \cosh k\alpha z)$
$\dfrac{Nl}{wH^2}$	$\dfrac{1}{k^2}\left\|\dfrac{1}{2}\left(1 - \dfrac{z}{H}\right)^2 + \dfrac{1}{(k\alpha H)^2}\right\| - (k\alpha H)^2(C_3 \sinh k\alpha z + C_4 \cosh k\alpha z)$

	Elastic Foundation	Portal or Column Base
C_1	$C_3 \tanh k\alpha H - \dfrac{1}{k^2(k\alpha H)^4}\,\text{sech}\,k\alpha H$	$\dfrac{r_\text{II} r_\text{L}}{(k\alpha H)^2} - \dfrac{1}{k^2(k\alpha H)^4}\,\text{sech}\,k\alpha H + [\tanh k\alpha H + r_\text{H} r'_\text{L} k^2(k\alpha H)] C_3$
C_2	$\dfrac{1}{(k_1\alpha H)^2}\left\| \dfrac{1}{k^2} + \dfrac{m}{2}(\alpha_1 H)^2(k_1^2 - 1) \right\| - \left(1 - \dfrac{k^2}{k_1^2}\right) k\alpha H\, C_3$	$\dfrac{r_\text{L}}{(k\alpha H)^2} + (k^2 r_\text{L} - 1) k\alpha H\, C_3$
C_3	$\dfrac{1}{\Delta}\left\{ S\dfrac{k_1^2}{k^2}\left\| \dfrac{1}{k_1^2}(\text{sech}\,k\alpha H - 1) + \dfrac{1}{2}(\alpha H)^2(k^2-1) \right\| \right.$ $\left. - \dfrac{1}{k^2} - \dfrac{m}{2}(\alpha_1 H)^2(k_1^2 - 1) \right\}$	$\dfrac{1}{\Delta}\left\{ (k\alpha H)\left\| \dfrac{1}{k^2}(\text{sech}\,k\alpha H - 1) + \dfrac{1}{2}(\alpha H)^2(k^2 - 1) \right\| \right.$ $\left. - \dfrac{1}{k^2} r_\text{L} - \dfrac{1}{2}(1 - r_\text{L}) + r_\text{II} r'_\text{L} \right\}$
C_4	$-C_3$	$\dfrac{1}{k^2(k\alpha H)^4}\,\text{sech}\,k\alpha H - C_3 \tanh k\alpha H$
Δ	$(k\alpha H)^2[S(k_1\alpha H)^2 \tanh k\alpha H + k\alpha H]$	$(k\alpha H)^2(k\alpha H r_\text{L} + \tanh k\alpha H)$

TABLE A1.3 Formulas for Triangularly Distributed Load

Expressions for Determining Deflections, Wall Moments, and Axial Forces

Quantity	
$\dfrac{EI}{pH^4}\,y$	$\dfrac{1}{120}\dfrac{k^2-1}{k^2}\left(\dfrac{z}{H}\right)^5 - \dfrac{1}{6(k\alpha H)^2}\left[\dfrac{1}{k^2}+\dfrac{1}{2}(\alpha H)^2(k^2-1)\right]\left(\dfrac{z}{H}\right)^3 + \dfrac{1}{6}\dfrac{k^2-1}{k^2}\left(\dfrac{z}{H}\right)^2 + C_1 + C_2\left(\dfrac{z}{H}\right) + C_3 \sinh k\alpha z + C_4 \cosh k\alpha z$
$\dfrac{M_1+M_2}{pH^2}$	$\dfrac{k^2-1}{k^2}\left[\dfrac{1}{3}\left(\dfrac{1}{3}+\dfrac{1}{6}\dfrac{z}{H}\right)\left(1-\dfrac{z}{H}\right)^3 - \dfrac{1}{k^2(k\alpha H)^2}\left(\dfrac{z}{H}\right)\right] + (k\alpha H)^2 (C_3 \sinh k\alpha z + C_4 \cosh k\alpha z)$
$\dfrac{N I}{pH^3}$	$\dfrac{1}{k^2}\left[\left(\dfrac{1}{3}+\dfrac{1}{6}\dfrac{z}{H}\right)\left(1-\dfrac{z}{H}\right)^3 + \dfrac{1}{(k\alpha H)^2}\left(\dfrac{z}{H}\right)\right] - (k\alpha H)^2(C_3 \sinh k\alpha z + C_4 \cosh k\alpha z)$

	Elastic Foundation	Portal or Column Base
C_1	$C_3 \tanh k\alpha H - \dfrac{1}{k^2(k\alpha H)^4}\,\text{sech}\,k\alpha H$	$\dfrac{1}{(k\alpha H)^2}\left\|\dfrac{1}{2}-\dfrac{1}{(k\alpha H)^2}\right\|r_H r_1 - \dfrac{1}{k^2(k\alpha H)^4}\,\text{sech}\,k\alpha H\right\|C_1$
		$+ \left[\tanh k\alpha H + r_H r_1 k^2 (k\alpha H)\right]C_1$
C_2	$\dfrac{1}{(k_1\alpha H)^2}\left\|\left(\dfrac{1}{2}-\dfrac{1}{(k\alpha H)^2}\right)\dfrac{1}{k^2}+\dfrac{1}{3}m(\alpha_1 H)^2(k_1^2-1)\right\|$	$-\dfrac{1}{(k\alpha H)^2}\left\|\dfrac{1}{2}-\dfrac{1}{(k\alpha H)^2}\right\|r_1 + (k^2 r_1 - 1)k\alpha H\, C_1$
	$-\left(1-\dfrac{k^2}{k_1^2}\right)k\alpha H\,C_1$	
C_3	$\dfrac{1}{\Delta}\left\{\left\|S\dfrac{k_1^2}{k^2}\right\|\dfrac{1}{k^2}\,\text{sech}\,k\alpha H + \dfrac{1}{3}(\alpha H)^3(k^2-1)\right\|$	$\dfrac{1}{\Delta}\left\{\dfrac{1}{(k^2(k\alpha H)^2)}\left\|\,\text{sech}\,k\alpha H - \dfrac{1}{2}((k\alpha H)^2 - 2)\right\|r_1 \right.$
	$-\dfrac{1}{k^2}\left\|\dfrac{1}{2}-\dfrac{1}{(k\alpha H)^2}\right\|-\dfrac{1}{3}m(\alpha_1 H)^3(k_1^2-1)\right\}$	$\left. +\dfrac{1}{3}\dfrac{k^2-1}{k^2} - \dfrac{1}{3}(1-r_1) + \dfrac{1}{2}r_H r_1\right\}$
C_4	$-C_1$	$-\dfrac{1}{k^2(k\alpha H)^4}\,\text{sech}\,k\alpha H - C_1 \tanh k\alpha H$
Δ	$(k\alpha H)^3\left[S(k_1\alpha H)^3 \tanh k\alpha H + k\alpha H\right]$	$(k\alpha H)^2(k\alpha H r_1 + \tanh k\alpha H)$

A1.3 STIFFNESS OF FLOOR SLABS CONNECTING SHEAR WALLS

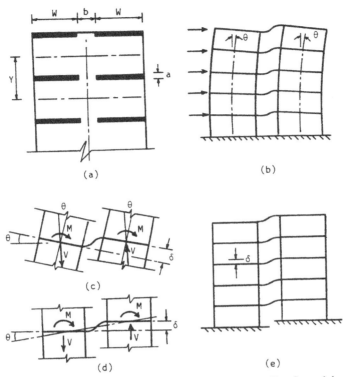

Fig. A1.10 Interaction between shear walls and coupling floor slabs.

and c). A similar mode of deformation of the slab occurs due to the relative vertical displacement of the walls caused by the induced axial forces, or to differential foundation settlement (Fig. A.1.10d and e).

A1.3.1 Effective Width of Floor Slab

The resistance of the floor slab against the displacements imposed by the shear walls is a measure of its coupling stiffness, which can be defined in terms of the displacements at its ends and the forces producing them. The stiffness can be determined from a finite element plate-bending analysis. The slab panel is discretized into an assembly of plate-bending elements using a suitable mesh graded so that the mesh used in the region close to the wall, where stress gradients are high, is finer than the mesh for other regions. Appropriate displacements are then prescribed for the wall nodes, either to give a unit wall rotation or a unit relative vertical wall movement, the slab being subjected to the same form of deformation, relative to the wall, in each case. The resulting solution then gives the displacement and stress values at all nodes, and also the slab reactions at the restrained nodes. The reactions at the set of wall nodes provide the static equivalent wall moment M and shear force V transferred from the wall to the slab when the wall undergoes the unit relative displacement involved.

To carry out an overall analysis of the coupled-wall structure, it is convenient to assume that a strip of slab acts effectively as a beam in coupling the pair of walls. The effective width of the slab may be established by equating the rotational or translational stiffness of the slab, obtained from the finite element analysis, to that of an equivalent beam. The results are normally given in terms of a width factor, expressed as the ratio of the effective width Y_e to the bay width, or longitudinal wall spacing Y (Fig. A1.10).

The procedure has been described in detail in Reference [A1.2].

Because of the architectural layouts involved, shear walls occur in many different cross-sectional shapes, and, in practice, plane, flanged, or box-shaped walls may be coupled together in cross-wall structures. Figure A1.11 shows a range of possible coupled systems.

Theoretical and experimental studies have shown that the main coupling actions take place in the corridor area and at the inner edges of the coupled walls. For walls with external facade flanges (Fig. A1.11c), the flange has a negligible effect on the coupling stiffness and the walls may be treated as plane walls (Fig. A1.11a). In the case of walls with internal flanges (Fig. A1.11d), very little bending of the slab occurs in the regions behind the flanges, and so the influence of the wall length may generally be disregarded.

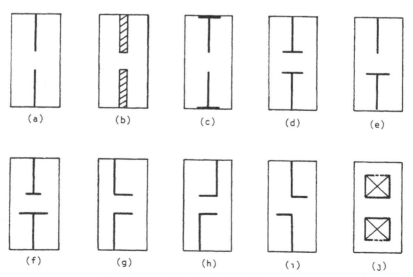

Fig. A1.11 Coupled wall configurations.

A1.3.2 Empirical Relationships for Effective Slab Width

For shear walls connected by flat slabs, the value of the characteristic relative stiffness parameter $k\alpha H$ is usually low, of the order of 1 to 3. It has been proved that errors in the effective width or stiffness of the slab produce much smaller errors in the important design parameters of wall stresses and lateral deflections and, consequently, it is not essential to evaluate the effective width with extreme accuracy. This is particularly the case since the effects of local stress concentrations, and a resultant cracking of the concrete, will produce a redistribution of forces and loss of stiffness.

Comprehensive sets of design curves have been presented in References [A1.2] and [A1.3]. However, it is possible to produce simple empirical relationships that fit the design curves fairly accurately, and that may be used for design calculations [A1.2]. These are considered for the various cross-sectional forms of shear walls commonly encountered in practice.

Plane Wall Configurations. For a slab coupling a pair of plane walls as shown in Fig. A1.12a the effective slab width ratio Y_e/Y may be taken to be

$$\frac{Y_e}{Y} = \frac{a}{Y} + \frac{b}{Y}\left(1 - 0.4\frac{b}{Y'}\right) \quad \text{for } 0 \leq \frac{b}{Y'} \leq 1$$

and
(A1.16)

$$\frac{Y_e}{Y} = \frac{a}{Y} + \frac{Y'}{Y}\left[1 - 0.4\left(\frac{b}{Y'}\right)^{-1}\right] \quad \text{for } 1 \leq \frac{b}{Y'} \leq \infty$$

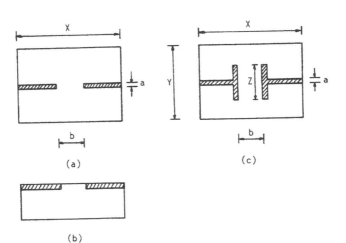

Fig. A1.12 Coupled plane and flanged walls.

in which

    a = wall thickness
    Y = bay width
    $Y' = Y - a$
    b = length of opening between walls.

Equations (A1.16) will generally yield results that are within a small percentage of the accurate values obtained directly from a finite element analysis.

The values from which the empirical relationships were derived were based on a Poisson's ratio for concrete of 0.15. The effective width is not sensitive to small differences in the value of Poisson's ratio, but, if desired, the value of Y_e/Y may be corrected approximately for the actual value of Poisson's ratio ν by multiplying by the factor $(1 - 0.15^2)/(1 - \nu^2)$.

If the wall thickness is neglected as being small, Eqs. (A1.16) reduce to the simpler expressions,

$$\frac{Y_e}{Y} = \frac{b}{Y}\left(1 - 0.4\frac{b}{Y}\right) \quad \text{for } 0 \le \frac{b}{Y} \le 1$$

$$\frac{Y_e}{Y} = 1 - 0.4\left(\frac{b}{Y}\right)^{-1} \quad \text{for } 1 \le \frac{b}{Y} \le \infty$$

(A1.17)

Effective Width for Slab in End Bay (Fig. A1.12b). End bays occur at the two gable ends of the building where the gable walls are coupled by the floor slab on one side of the wall only. With the asymmetric coupling of the slab, gable walls will generally undergo some out-of-plane bending that will depend on the relative stiffness of the wall. Since the gable edge of the slab is less restrained against transverse rotation than a continuous interior edge, the coupling stiffness of the end bay will be less than half the stiffness of an internal bay slab.

Corresponding finite element studies [A1.2] have shown that for a practical range of wall configurations, the effective width of an end bay varies between 44 and 47% of the value of an interior bay. As a convenient rule, the effective width of an end bay should be taken to be 45% of the corresponding interior value given by Eqs. (A1.16).

Flanged Wall Configurations—Equal Widths. It has been demonstrated both theoretically and experimentally that the main coupling actions in the slab take place in the corridor region between the internal edges of the walls. The coupling actions are dependent only on the flange dimensions, since the regions behind the flanges are essentially stress free. Consequently, the effective width is unaffected by the location of the web wall, and the results presented are equally appropriate for offset web walls provided the flanges are located opposite each other (Fig. A1.11g and h).

A1.3 STIFFNESS OF FLOOR SLABS CONNECTING SHEAR WALLS

For the configuration shown in Fig. A1.12c the effective width of a slab coupling two walls with equal flange widths may be taken to be

$$\frac{Y_e}{Y} = \frac{Z}{Y} + \frac{b}{Y'}\left(1 - 0.4\frac{b}{Y'}\right) \quad \text{for } 0 \leq \frac{b}{Y'} \leq 1$$

and (A1.18)

$$\frac{Y_e}{Y} = \frac{Z}{Y} + \frac{Y'}{Y}\left|1 - 0.4\left(\frac{b}{Y'}\right)^{-1}\right| \quad \text{for } 1 \leq \frac{b}{Y'} \leq \infty$$

in which Z = flange width and $Y' = Y - Z$.

Coupled Plane Walls and Tee-Shaped Flanged Walls. If a plane wall is coupled to a flanged wall as shown in Fig. A1.11e the rotational stiffnesses of the slab evaluated at its ends will not be equal because of the asymmetry of the walls, and, in order to replace it by a uniform equivalent beam, the average rotational stiffness must be taken to be the effective value.

Based on a range of parametric studies, a set of generalized design curves has been produced to give the variation of the effective width ratio Y_e/Y as a function of b/Y for various ratios Z/Y. These may be used to read off directly the value of Y_e/Y for a practical analysis. It has not yet proved possible to obtain a sufficiently accurate empirical relationship for this particular configuration, and practical assessments of the effective width must be obtained from Fig. A1.13.

A comparison of the results with those for a slab coupling two flanged walls shows that the omission of the flange from one wall results in a disproportionately large reduction in the effective width of the slab when the ratio b/Y is small.

The generalized design curves may be taken to be accurate unless the opening ratio b/X becomes relatively large, say greater than about 0.5, or when the ratio $b/X + Y/X$ exceeds unity, but in almost all practical situations the discrepancies are unlikely to exceed 10%.

Other Wall Configurations. Empirical formulas and associated design curves have also been presented in Reference [A1.2, A1.4] to enable the effective slab width to be determined for other wall configurations, such as flanged walls with unequal flanges (Fig. A1.11f), coupled ell-shaped walls (Fig. A1.11g-i), and coupled box cores (Fig. A1.11j).

Guidance on the magnitude of the peak bending stresses in the slab, and peak shear stresses at the slab-wall junction, in coupled plane and flanged wall configurations, has been given in Reference [A1.5, A1.6].

A1.3.3 Numerical Examples

1. Slab coupling two (thick) plane walls: it is required to evaluate the effective width of a slab in the following configuration (Fig. A1.12a).

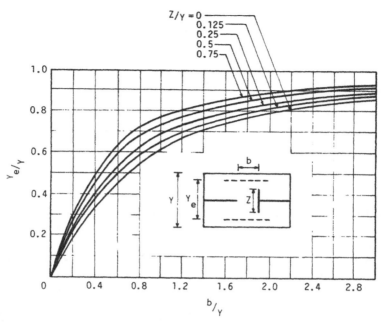

Fig. A1.13 Generalized design curves for effective slab width for planar-flanged wall configuration.

Bay width $Y = 6$ m (19.69 ft)
Building width $X = 12$ m (39.37 ft)
Corridor opening width $b = 2$ m (6.56 ft)
Wall thickness $a = 0.4$ m (1.31 ft)

Hence,

Aspect ratio $Y/X = 6/12 = 0.5$
Wall opening ratio $b/X = 2/12 = 0.167$
Wall thickness ratio $a/Y = 0.4/6 = 0.067$

The reduced width factor

$$Y'/Y = 1 - a/Y = 1 - 0.067 = 0.933$$

The ratio

$$\frac{b}{Y'} = \frac{b}{Y}\frac{Y}{Y'} = \frac{2}{6} \times \frac{1}{0.933} = 0.357$$

A1.3 STIFFNESS OF FLOOR SLABS CONNECTING SHEAR WALLS

Hence, using Eq. (A1.16), the effective width ratio is

$$\frac{Y_e}{Y} = 0.067 + \frac{2}{6}[1 - 0.4 \times 0.357]$$

$$= 0.353$$

∴ Effective slab width = $0.353 \times 6 = 2.116$ m (6.94 ft).

If the wall thickness is neglected, the effective width is given by Eq. (A1.17). In that case,

$$\frac{Y_e}{Y} = \frac{2}{6}\left(1 - 0.4 \times \frac{2}{6}\right) = 0.289$$

∴ Effective slab width = $0.289 \times 6 = 1.734$ m (5.69 ft)

2. Slab coupling two flanged walls: it is required to find the effective width of a slab for the configuration shown in Fig. A1.12c.

Bay width $Y = 7.5$ m (24.61 ft)
Building width $X = 12.5$ m (41.01 ft)
Flange width $Z = 5.6$ m (18.37 ft)
Corridor opening $b = 2.5$ m (8.20 ft)

The reduced slab width ratio

$$Y'/Y = 1 - Z/Y = 1 - 5.6/7.5 = 0.253$$

The reduced span/width ratio b/Y' is

$$\frac{b}{Y'} = \frac{b}{Y}\frac{Y}{Y'} = \frac{2.5}{7.5} \times \frac{1}{0.253} = 1.318$$

From Eq. (A1.18), the effective width ratio becomes

$$\frac{Y_e}{Y} = \frac{5.6}{7.5} + 0.253\left(1 - 0.4\frac{1}{1.318}\right)$$

$$= 0.923$$

∴ $Y_e = 0.923 \times 7.5 = 6.92$ m (22.70 ft)

If the slab thickness is t, then in each case the slab may be replaced by an effective second moment of area I_b given by

$$I_b = \frac{1}{12} Y_e t^3$$

and used directly in the earlier analyses using either the continuum or the wide-column frame method of analysis.

Loss of Slab Stiffness due to Cracking.

The results that have been presented are based on elastic analyses that assume a rigid joint between the shear wall and the slab. The analyses showed that practically all the shear transfer between slab and wall occurs around the inner corridor edges of the walls, and high stress concentrations occur in the slab at the inner edges of plane shear walls [A1.5]. The problem is less serious in the case of flanged walls, since the slab tends to be subjected to cylindrical bending along the length of the flange and stress concentrations occur only at the tips of the flanges [A1.6].

The high localized stresses tend to cause cracking of the concrete, with an attendant reduction in the slab stiffness. The true effective width and stiffness will then generally be less than the calculated elastic values, which must be considered as the maximum possible stiffness that can be developed.

The tabulated values should thus be factored to take account of concrete cracking, the greatest reduction being required for coupled plane walls and the least for coupled flanged walls.

Published research on this topic is limited, and it is not yet possible to give accurate recommendations for such reduction factors over the range of parameters concerned. However, the available evidence [A1.7, A1.8, A1.9] indicates that the elastic effective width of slabs coupling plane walls should be reduced by a factor of about 0.4–0.45 to take account of cracking.

If the slab is subjected to reversed cyclic loading, with increasing imposed inelastic deformations, considerable stiffness degradation may ensue [A1.8]. This additional loss of stiffness should be considered in earthquake-resistant design. It has been suggested [A1.9] that in the case of light earthquakes (i.e., ductility ratio ≤ 3), the initial cracked stiffness should be further reduced by a factor of 0.35, and for heavy earthquakes (ductility ratio ≥ 6), by a factor of 0.1. The last factor indicates that it is doubtful whether slab coupling could be used as the primary source of energy dissipation in earthquake-resisting ductile coupled shear walls.

No significant research has been reported on the effects of cracking on the stiffness of slabs coupling flanged walls. At the present time, it is suggested that the effective stiffness should be calculated from the inertia of the cracked section of the equivalent slab.

Fortunately, the overall behavior of the coupled wall system is not too sensitive to the value of the slab stiffness, and errors in the calculation of the effective width usually result in much smaller errors in the estimation of lateral deflections and wall stresses.

REFERENCES

A1.1 Coull, A. and Mukherjee, P. R. "Coupled Shear Walls with General Support Conditions." *Proc. Conference on Tall Buildings*, Kuala Lumpur, December 1974, pp. 4.24–4.31.

A1.2 Coull, A. and Wong, Y. C. "Bending Stiffness of Floor Slabs in Cross-Wall Structures." *Proc. ICE* **71**, Part 2, 1981, 17–35.

A1.3 Tso, W. K. and Mahmoud, A. A. "Effective Width of Coupling Slabs in Shear Wall Buildings." *J. Struct. Div., ASCE* **103**(ST3), 1977, 573–586.

A1.4 Coull, A. and Wong, Y. C. "Effective Stiffness of Floor Slabs Coupling Open Box Cores." *Proc. ICE* **83**, Part 2, 1987, 321–325.

A1.5 Coull, A. and Wong, Y. C. "Design of Floor Slabs Coupling Shear Walls." *J. Struct. Engineer. ASCE* **109**, 1983, 109–125.

A1.6 Coull, A. and Wong, Y. C. "Stresses in Slabs Coupling Flanged Shear Walls." *J. Struct. Engineer. ASCE* **110**, 1984, 105–119.

A1.7 Schwaighofer, J. and Collins, M. P. "Experimental Study of the Behavior of Reinforced Concrete Coupling Slabs." *J. ACI* **74**, 1977, 123–127.

A1.8 Paulay, T. and Taylor, R. G. "Slab Coupling of Earthquake-Resisting Shearwalls." *J. ACI* **78**, 1981, 130–140.

A1.9 Mirza, M. S. and Lim, A. K. W. "Behavior and Design of Coupled Slab-Shear Wall Systems." *Proc. Can. Soc. Civil Eng./Can. Prestressed Conc. Assoc. Structural Concrete Conference*, Montréal, Canada, 20–21 March 1989, 209–244.

APPENDIX 2

Formulas and Graphs for Wall–Frame and Core Structures

The differential equations for the deflection of wall–frame structures and the rotation of core structures are fully analogous, as shown in Chapters 11 and 13. Consequently, their solutions for displacements and forces are also analogous.

A complete set of formulas for the displacements and forces in wall–frame structures subjected to uniformly distributed loading, triangularly distributed loading, and a concentrated top loading is given here. Analogous terms for core structures are specified so that the formulas may be used also for the analysis of core structures.

Curves of coefficients K_1 to K_4 for the rapid solution of wall–frame and core problems, as explained in Chapters 11 and 13, are given. They also refer to the three loading cases: uniformly distributed load (Figs. A2.1–A2.4), triangularly distributed load (Figs. A2.5–A2.8), and a concentrated top load (Figs. A2.9–A2.12). Formulas for use with the coefficients are given with each graph.

A2.1 FORMULAS AND GRAPHS FOR DEFLECTIONS AND FORCES

A2.1.1 Uniformly Distributed Horizontal Loading

Loading Intensity w per Unit Height. Deflection y (or Rotation θ):

$$y(z) = \frac{wH^4}{EI} \left\{ \frac{1}{(\alpha H)^4} \left[\frac{\alpha H \sinh \alpha H + 1}{\cosh \alpha H} (\cosh \alpha z - 1) \right. \right.$$
$$\left. \left. - \alpha H \sinh \alpha z + (\alpha H)^2 \left[\frac{z}{H} - \frac{1}{2}\left(\frac{z}{H}\right)^2 \right] \right] \right\} \quad (A2.1)$$

where H is the height of the structure, EI is the flexural rigidity of the wall, and $\alpha = [(GA)/EI]^{1/2}$ in which (GA) is the effective shear rigidity of the frame.

For use with cores subjected to torsion, substitute rotation $\theta(z)$ for $y(z)$, torque per unit height m for w, sectorial moment of inertia I_ω for I, and take $\alpha = [GJ/EI_\omega]^{1/2}$ in which J is the torsion constant.

A2.1 FORMULAS AND GRAPHS FOR DEFLECTIONS AND FORCES

Inclination dy/dz (or Twist $d\theta/dz$):

$$\frac{dy}{dz}(z) = \frac{wH^3}{EI}\left\{\frac{1}{(\alpha H)^3}\left[\frac{\alpha H \sinh \alpha H + 1}{\cosh \alpha H}(\sinh \alpha z)\right.\right.$$

$$\left.\left. - \alpha H \cosh \alpha z + \alpha H\left(1 - \frac{z}{H}\right)\right]\right\} \quad (A2.2)$$

For cores subjected to torsion substitute $d\theta/dz$ for dy/dz, etc. Multiplying $d\theta/dz$ by $12EI_h\Omega/L^3$ gives the connecting beam shear Q_h [Eq. (13.55)], and hence the connecting beam maximum moment M_h ($M_h = Q_h L/2$), in which I_h is the moment of inertia of the beam, L is its span, and Ω is twice the area enclosed by the middle line of the core profile.

Wall Moment M_b (or Core Bimoment B):

$$M_b(z) = EI\frac{d^2y}{dz^2}(z)$$

$\therefore$

$$M_b(z) = wH^2\left\{\frac{1}{(\alpha H)^2}\left[\frac{\alpha H \sinh \alpha H + 1}{\cosh \alpha H}(\cosh \alpha z)\right.\right.$$

$$\left.\left. - \alpha H \sinh \alpha z - 1\right]\right\} \quad (A2.3)$$

For cores subjected to torsion, substitute negative bimoment $-B(z)$ for $M_b(z)$, etc.

Wall Shear Q_b:

$$Q_b(z) = -EI\frac{d^3y}{dz^3}(z)$$

$\therefore$

$$Q_b(z) = -wH\left\{\frac{1}{(\alpha H)}\left[\frac{\alpha H \sinh \alpha H + 1}{\cosh \alpha H}(\sinh \alpha z)\right.\right.$$

$$\left.\left. - \alpha H \cosh \alpha z\right]\right\} \quad (A2.4)$$

504 FORMULAS AND GRAPHS FOR WALL-FRAME AND CORE STRUCTURES

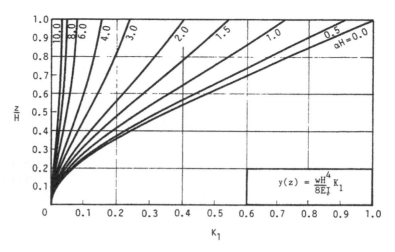

Fig. A2.1 K_1 factor, uniformly distributed horizontal loading.

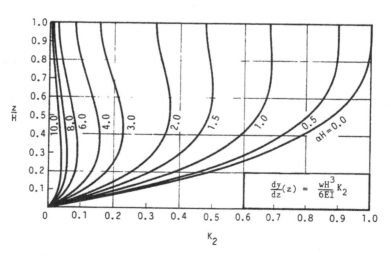

Fig. A2.2 K_2 factor, uniformly distributed horizontal loading.

A2.1 FORMULAS AND GRAPHS FOR DEFLECTIONS AND FORCES

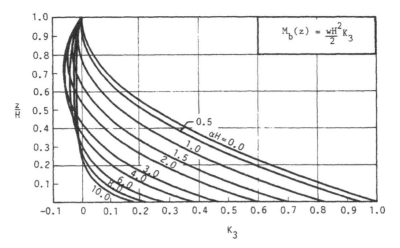

Fig. A2.3 K_3 factor, uniformly distributed horizontal loading.

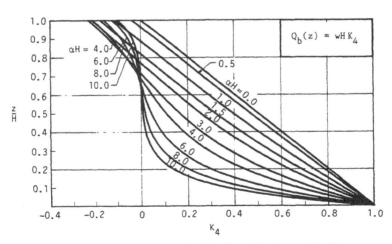

Fig. A2.4 K_4 factor, uniformly distributed horizontal loading.

A2.1.2 Triangularly Distributed Horizontal Loading

Maximum Intensity of Loading w_1 at the Top, Zero at the Base. Deflection y (or Rotation θ):

$$y(z) = \frac{w_1 H^4}{EI}\left\{\frac{1}{(\alpha H)^4}\left[\left(\frac{\alpha H \sinh \alpha H}{2} - \frac{\sinh \alpha H}{\alpha H} + 1\right)\left(\frac{\cosh \alpha z - 1}{\cosh \alpha H}\right)\right.\right.$$

$$\left.\left.+ \left(\frac{z}{H} - \frac{\sinh \alpha z}{\alpha H}\right)\left[\frac{(\alpha H)^2}{2} - 1\right] - \frac{(\alpha z)^2}{6}\left(\frac{z}{H}\right)\right]\right\} \quad (A2.5)$$

For cores subjected to triangularly distributed torque, substitute m_1 (maximum intensity of torque at the top) for w_1, etc.

Inclination dy/dz (or Twist $d\theta/dz$):

$$\frac{dy}{dz}(z) = \frac{w_1 H^3}{EI}\left\{\frac{1}{(\alpha H)^3}\left[\left(\frac{\alpha H \sinh \alpha H}{2} - \frac{\sinh \alpha H}{\alpha H} + 1\right)\left(\frac{\sinh \alpha z}{\cosh \alpha H}\right)\right.\right.$$

$$\left.\left.+ \left(\frac{1 - \cosh \alpha z}{\alpha H}\right)\left[\frac{(\alpha H)^2}{2} - 1\right] - \frac{\alpha z}{2}\frac{z}{H}\right]\right\} \quad (A2.6)$$

Wall Moment M_b (or Core Negative Bimoment $-B$):

$$M_b(z) = w_1 H^2\left\{\frac{1}{(\alpha H)^2}\left[\left(\frac{\alpha H \sinh \alpha H}{2} - \frac{\sinh \alpha H}{\alpha H} + 1\right)\left(\frac{\cosh \alpha z}{\cosh \alpha H}\right)\right.\right.$$

$$\left.\left.- \left(\frac{\sinh \alpha z}{\alpha H}\right)\left[\frac{(\alpha H)^2}{2} - 1\right] - \frac{z}{H}\right]\right\} \quad (A2.7)$$

Wall Shear Q_b:

$$Q_b(z) = -w_1 H\left\{\frac{1}{(\alpha H)}\left[\left(\frac{\alpha H \sinh \alpha H}{2} - \frac{\sinh \alpha H}{\alpha H} + 1\right)\left(\frac{\sinh \alpha z}{\cosh \alpha H}\right)\right.\right.$$

$$\left.\left.- \left(\frac{\cosh \alpha z}{\alpha H}\right)\left[\frac{(\alpha H)^2}{2} - 1\right] - \frac{1}{\alpha H}\right]\right\} \quad (A2.8)$$

A2.1 FORMULAS AND GRAPHS FOR DEFLECTIONS AND FORCES

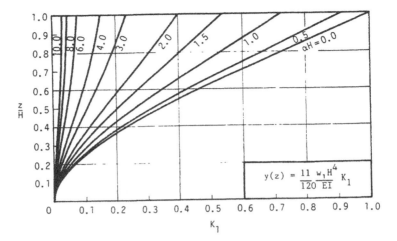

Fig. A2.5 K_1 factor, triangularly distributed horizontal loading.

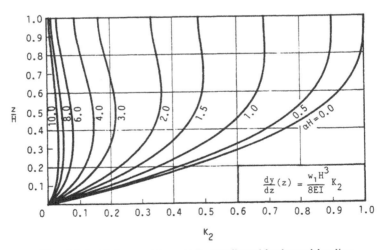

Fig. A2.6 K_2 factor, triangularly distributed horizontal loading.

508 FORMULAS AND GRAPHS FOR WALL-FRAME AND CORE STRUCTURES

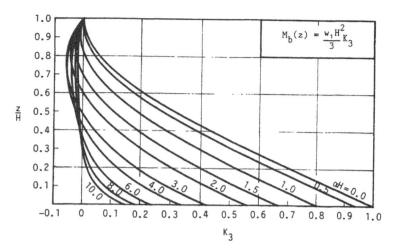

Fig. A2.7 K_3 factor, triangularly distributed horizontal loading.

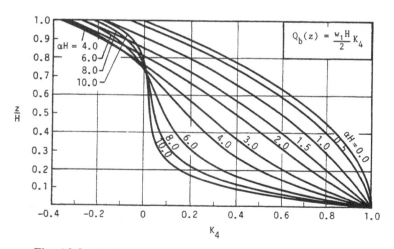

Fig. A2.8 K_4 factor, triangularly distributed horizontal loading.

A2.1.3 Concentrated Horizontal Load P at the Top

Deflection y (or Rotation θ):

$$y(z) = \frac{PH^3}{EI}\left\{\frac{1}{(\alpha H)^3}\left[\frac{\sinh \alpha H}{\cosh \alpha H}(\cosh \alpha z - 1) - \sinh \alpha z + \alpha z\right]\right\} \quad \text{(A2.9)}$$

Inclination dy/dz (or Twist $d\theta/dz$):

$$\frac{dy}{dz}(z) = \frac{PH^2}{EI}\left\{\frac{1}{(\alpha H)^2}\left[\frac{\sinh \alpha H}{\cosh \alpha H}(\sinh \alpha z) - \cosh \alpha z + 1\right]\right\} \quad \text{(A2.10)}$$

Wall Moment M_b (or Core Negative Bimoment $-B$):

$$M_b(z) = PH\left\{\frac{1}{(\alpha H)}\left[\frac{\sinh \alpha H}{\cosh \alpha H}(\cosh \alpha z) - \sinh \alpha z\right]\right\} \quad \text{(A2.11)}$$

Wall Shear Q_b:

$$Q_b(z) = -P\left[\frac{\sinh \alpha H}{\cosh \alpha H}(\sinh \alpha z) - \cosh \alpha z\right] \quad \text{(A2.12)}$$

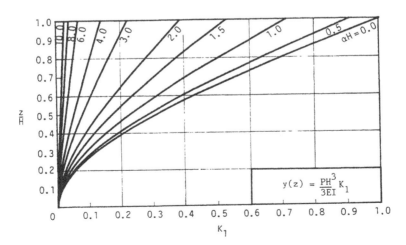

Fig. A2.9 K_1 factor, concentrated horizontal load at top.

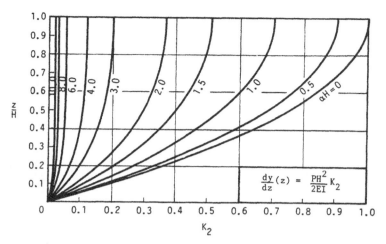

Fig. A2.10 K_2 factor, concentrated horizontal load at top.

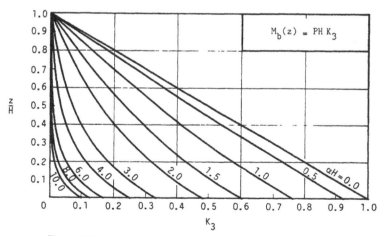

Fig. A2.11 K_3 factor, concentrated horizontal load at top.

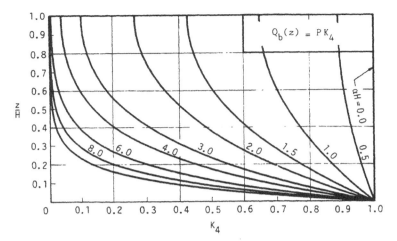

Fig. A2.12 K_4 factor, concentrated horizontal load at top.

Bibliography

CHAPTER 1

Beedle, L. (Ed.). *Second Century of the Skyscraper*. Council on Tall Buildings and Urban Habitat, Van Nostrand Reinhold, New York, 1988.

Beedle, L. (Ed.) *Developments in Tall Buildings*. Council on Tall Buildings and Urban Habitat. Hutchinson Ross, Stroudsburg, PA, 1983.

Billington, D. P. and Goldsmith M. (Eds.). *Techniques and Aesthetics in the Design of Tall Buildings*. Institute for the Study of the High-Rise Habitat, Lehigh University, Bethlehem, PA, 1986.

Condit, C. W. "The Two Centuries of Technical Evolution Underlying the Skyscraper." In *Second Century of the Skyscraper*, Council on Tall Buildings and Urban Habitat, Van Nostrand Reinhold, New York, 1988, pp. 11-24.

Coull, A. and Stafford Smith, B. "Analysis of Shear Wall Structures (A Review of Previous Research)." In *Tall Buildings*, A. Coull and B. Stafford Smith (Eds.). Pergamon Press, Oxford, 1967, pp. 139-155.

Coull, A. and Stafford Smith, B. "Recent Developments in Elastic Analysis of Tall Concrete Buildings." In *Developments in Tall Buildings*, L. S. Beedle (Ed.). Hutchinson Ross, Stroudsburg, PA, 1983, pp. 569-581.

Coull, A. and Stafford Smith, B. "Tall Buildings." Chapter 37, *Handbook of Structural Concrete*, F. K. Kong, R. H. Evans, E. Cohen, and F. Roll (Eds.). Pitman, London, 1983, pp. 37.1-37.46.

Fintel, M. "Multistory Structures." Chapter 10, *Handbook of Concrete Engineering*, M. Fintel (Ed.). Van Nostrand Reinhold, New York, 1974, pp. 287-344.

Grossman, J. S. "Slender Concrete Structures—The New Edge." *ACI Struct. J.* **87**(1), 1990, 39-52.

Hart, F., Henn, W., and Sontag, H. *Multi-Storey Buildings in Steel*, G. B. Godfrey (Ed.). Crosby Lockwood Staples, London, 1978.

Khan, F. R. "Current Trends in Concrete High-Rise Buildings." In *Tall Buildings*, A. Coull and B. Stafford Smith (Eds.). Pergamon Press, Oxford, 1967, pp. 571-590.

Peters, P. F. "The Rise of the Tall Building." *Civil Engineer.*, ASCE **56**(11), 1986, 46-49.

Schueller, W. *High-Rise Building Structures*. Wiley, New York, 1977.

Weisman, W. "A New View of Skyscraper History." Chapter 3, *The Rise of an American Architecture*, Kaufman (Ed.). Pall Mall Press, London, 1970, pp. 115-160.

"Philosophy of Tall Buildings." Chapter PC-1, Vol. PC, *Planning and Environmental*

Criteria for Tall Buildings, Monograph on Planning and Design of Tall Buildings. ASCE, New York, 1981, pp. 1–82.

"History of Tall Buildings." Chapter PC-2, Vol. PC, *Planning and Environmental Criteria for Tall Buildings, Monograph on Planning and Design of Tall Buildings.* ASCE, New York, 1981, pp. 83–166.

Monograph on Planning and Design of Tall Buildings, 5 Vols. ASCE, New York, 1978–81.

Taranath, B. S. *Structural Analysis and Design of Tall Buildings.* McGraw Hill, New York, 1988.

CHAPTER 2

Monograph on Planning and Design of Tall Buildings, ASCE, New York.
  Vol. CL. *Tall Building Criteria and Loading,* 1980 (Chapters 1, 2, 3, 4, 7).
  Vol. SC. *Tall Building Systems and Concepts,* 1980 (Chapters 1, 7, 8).
  Vol. SB. *Structural Design of Tall Steel Buildings,* 1979 (Chapters 1, 2, 4, 5, 8).
  Vol. CB. *Structural Design of Tall Concrete and Masonry Buildings,* 1978. (Chapters 2, 3, 9, 10, 11, 12, 13).

ACI Committee 442. "Response of Buildings to Lateral Forces." *J. Am. Conc. Inst.* **68**(2), 1971, 81–106.

ACI Committee 442. Abstract of "Response of Buildings to Lateral Forces." *J. Am. Conc. Inst.* **85**(4), 1988, 472–474.

ACI Committee 435, Subcommittee 1. "Allowable Deflections." *J. Am. Conc. Inst.* **65**(6), 1968, 433–444.

Robertson, L. E. "Design Criteria for Very Tall Buildings." *Proc. 12th National Conf. on Tall Buildings,* Sydney, 1973, 171–180.

Tregenza, P. "Association between Building Height and Cost." *Architects J. Inform. Library* **1**, November 1972.

CHAPTER 3

For bibliography, refer to collected references in Ref. [3.1], and to local building codes.

CHAPTER 4

ACI Committee 442. "Response of Buildings to Lateral Forces." *J. Am. Conc. Inst.* **68**(2), 1971, 81–106.

Anon. "Thinking Tall." *Progress. Architecture,* December 1980, 45–57.

Beedle, L. S. "Tall Buildings: A Look to the Future," *J. IABSE* 3(S-3/77), 1977, 1–20.

Colaco, J. P. "Mile High Dream." *Civil Engineer.,* ASCE **56**(4), April 1986, 76–78.

Coull, A. and Stafford Smith, B. "Tall Buildings," Chapter 37, *Handbook of Structural Concrete.* F. K. Kong, R. H. Evans, E. Cohen, and F. Roll (Eds.). Pitman, London, 1983, pp. 37.1–37.46.

Davis, D. "The Sky's the Limit." *Newsweek* November 8, 1982, 66-76.

Fairweather, V. "Record High-Rise, Record Low Steel." *Civil Engineer., ASCE* August 1986, 42-45.

Frischmann, W. W. and Prabhu, S. S. "Planning Concepts Using Shear Walls." In *Tall Buildings*, A. Coull and B. Stafford Smith (Eds.). Pergamon Press, Oxford, 1967, pp. 49-79.

Goldberger, P. "2nd Century of the Skyscraper: Architecture and Society." *IABSE Period. Bull.* **B-37/86**(1), February 1986, 2-9.

Goldsmith, M. "The Tall Building—Effects of Scale." *Quart. Column, Japan* No. 6, April 1963, 38-49.

Grossman, J. S. "780 Third Avenue: The First High-Rise Diagonally Braced Concrete Structure." *Conc. Int.* 7(2), 1985, 53-56.

Haryott, R. B. and Glover, M. J. "Developments in Multi-Storey Buildings." *Civil Engineer., London* March 1985, 40-42.

Hirschmann, E. W. "The Hull-Core Structure." *Architects J. Inform. Library* 29, December 1965, 1615-1623.

Iyengar, S. H. "Preliminary Design and Optimization of Steel Building Systems," Chapter SB-2, Vol. SB, *Structural Design of Tall Steel Buildings, Monograph on Planning and Design of Tall Buildings*. ASCE, New York, 1979, pp. 33-46.

Khan, F. R. "Current Trends in Concrete High-Rise Buildings." In *Tall Buildings*, A. Coull and B. Stafford Smith (Eds.). Pergamon Press, Oxford, 1967, pp. 571-590.

Khan, F. R. "Recent Structural Systems in Steel for High-Rise Buildings." *Proc. Conf. on Steel in Architecture*, British Constructional Steelwork Assoc., London, November 1969.

Khan, F. R. "The Future of High-Rise Structures," *Progress. Architecture* October 1972, 78-85.

Robertson, L. E. "On Tall Buildings." In *Tall Buildings*, A. Coull and B. Stafford Smith (Eds.). Pergamon Press, Oxford, 1967, pp. 591-607.

Tucker, J. B. "Superskyscrapers: Aiming for 200 Stories." *High Technol.* January 1985, pp. 50-63.

Zunz, G. J., Glover, M. J., and Fitzpatrick, A. J. "The Structure of the New Headquarters for the Hong Kong and Shanghai Banking Corp., Hong Kong." *Struct. Engineer* **63A**(9), 1985, 255-284.

CHAPTER 5

Boppana, R. R. and Naiem, F. "Modelling of Floor Diaphragms in Concrete Shear Wall Buildings," *Conc. Int.* 7(7), 1985, 44-46.

Coull, A. and Stafford Smith, B. "Tall Buildings." Chapter 37, *Handbook of Structural Concrete*, F. K. Kong, R. H. Evans, E. Cohen, and F. Roll (Eds.). Pitman, London, 1983, pp. 37.1-37.46.

Coull, A. and Stafford Smith, B. "Torsional Analysis of Symmetric Building Structures." *J. Struct. Div., ASCE* **99**(1), 1973, 229-233.

Goldberg, J. E. "Approximate Elastic Analysis." Chapter SB-2, Vol. SB, *Structural Design of Tall Steel Buildings, Monograph on Planning and Design of Tall Buildings*. ASCE, New York, 1979, pp. 46-61.

Macleod, I. A. "Lateral Stiffness of Shear Walls with Openings." In *Tall Buildings*, A. Coull and B. Stafford Smith (Eds.). Pergamon Press, Oxford, 1967, pp. 223-244.

Macleod, I. A. "Structural Analysis of Wall Systems." *Struct. Engineer* 55(11), 1977, 487-495.

Rutenberg, A. "Analysis of Tube Structures Using Plane Frame Programs." *Proc. of the Regional Conf. on Tall Buildings*, Bangkok, 1974 pp. 397-413.

Rutenberg, A. and Eisenberger, M. "Simple Planar Modeling of Asymmetric Shear Buildings for Lateral Forces." *Computers and Struct.* 24(6), 1986, 885-891.

Stafford Smith, B., Coull, A., and Cruvellier, M. "Planar Models for Analysis of Intersecting Bent Structures." *Computers and Struct.* 29(2), 1988, 257-263.

Stafford Smith, B. and Girgis, A. "Simple Analogous Frames for Shear Wall Analysis." *J. Struct. Engineer., ASCE* 110(11), 1984, 2655-2666.

CHAPTER 6

ASCE Sub-Committee 31. "Wind-Bracing in Steel Buildings."
First Progress Report, *Civil Engineer*. March 1931, 478-483.
Second Progress Report, *Proc. ASCE* 58, February 1932, 213-220
Third Progress Report, *Proc. ASCE* 59, December 1933, 1601-1631.
Fifth Progress Report, *Proc. ASCE* 62, March 1936, 1496-1498.
Sixth Progress Report, *Proc. ASCE* 65, June 1939, 969-1000.
Final Report, *Trans. ASCE*, No. 105, 1940, 1713-1739.

ASCE Sub-Committee 31. "Windforces on Structures." *Trans. ASCE* 126(II), 1961, 1124-1198.

Merovich, A. T., Nicoletti, J. P., and Hartle, E. "Eccentric Bracing in Tall Buildings." *J. Struct. Div., ASCE* 108(9), 1982, 2066-2080.

Norris, C. H., Wilbur, J. B., and Utku, S. *Elementary Structural Analysis*, 3rd ed. McGraw-Hill, New York, 1976.

Popov, E. P., and Roeder, C. W. "Design of an Eccentrically Braced Steel Frame," *J. Am. Inst. Steel Const.* 15(3), 1978, 77-81.

Spurr, H. V. *Wind Bracing, the Importance of Rigidity in High Towers*. McGraw-Hill, New York, 1930.

CHAPTER 7

Cheong-Siat-Moy, F. "Control of Deflections in Unbraced Steel Frames." *Proc. Inst. Civil Engineer., London* 57(2), 1974, 619-634.

Clough, R. W., King, I. P., and Wilson, E. L. "Structural Analysis of Multistory Buildings." *J. Struct. Div., ASCE* 90(3), 1964, 19-34.

Elias, Z. M. "Lateral Stiffness of Flat Plate Structures." *J. Am. Conc. Inst.* **80**(1), 1983, 50–54.

Fleming, R. *Wind Stresses*, Engineering News, New York, 1915.

Fraser, D. J. "Equivalent Frame Method for Beam-Slab Structures." *J. Am. Conc. Inst.* **74**(5), 1977, 223–228.

Pecknold, D. A. "Slab Effective Width for Equivalent Frame Analysis." *J. Am. Conc. Inst.* **72**(4), 1975, 135–137.

Rad, F. N. and Furlong, R. W. "Behavior of Unbraced Reinforced Concrete Frames." *J. Am. Conc. Inst.* **77**(4), 1980, 269–278.

Tall, L. *Structural Steel Design*. Ronald Press, New York, 1964.

Vanderbilt, M. D. "Equivalent Frame Analysis for Lateral Loads." *J. Struct. Div., ASCE* **105**(10), 1979, 1981–1998.

Vanderbilt, M. D. and Corley, W. G. "Frame Analysis of Concrete Buildings." *Conc. Int.* **5**(12), 1983, 33–43.

Weaver, W. and Nelson, M. F. "Three-Dimensional Analysis of Tier Buildings." *J. Struct. Div., ASCE* **92**(6), 1966, 385–404.

Williams, F. W. "Simple Design Procedures for Unbraced Multi-Storey Frames." *Proc. Inst. Civil Engineer., London* **63**(2), 1977, 475–479.

Wright, E. W. and Gaylord, E. H. "Analysis of Unbraced Multistory Steel Rigid Frames." *J. Struct. Div., ASCE* **94**(5), 1968, 1143–1163.

CHAPTER 8

Holmes, M. "Steel Frame with Brickwork and Concrete Infilling." *Proc. Inst. Civil Engineer., London* **19**, 1961, 473–478.

Irwin, A. W. and Afshar, A. B. "Performance of Reinforced Concrete Frames with Various Infills Subject to Cyclic Loading." *Proc. Inst. Civil Engineer., London* **67**(2), 1979, 509–520.

Kahn, L. F. and Hanson, R. D. "Infilled Walls for Earthquake Strengthening." *J. Struct. Div., ASCE* **105**(2), 1979, 283–296.

King, G. J. W. and Pandey, P. C. "The Analysis of Infilled Frames Using Finite Elements." *Proc. Inst. Civil Engineer., London* **65**(2), 1978, 749–760.

Klingner, R. E. and Bertero, V. V. "Earthquake Resistance of Infilled Frames." *J. Struct. Div., ASCE* **104**(6), 1978, 973–989.

Liauw, T. C. "Tests on Multistory Infilled Frames Subject to Dynamic Lateral Loading." *J. Am. Conc. Inst.* **76**(4), 1979, 551–563.

Liauw, T. C. and Kwan, K. H. "Plastic Theory of Infilled Frames with Finite Interface Shear Strength." *Proc. Inst. Civil Engineer., London* **75**(2), 1983, 379–396.

Liauw, T. C. and Kwan, K. H. "Plastic Theory of Non-Integral Infilled Frames." *Proc. Inst. Civil Engineer., London* **75**(2), 1983, 379–396.

Liauw, T. C. and Lee, S. W. "On the Behaviour and the Analysis of Multi-Storey Infilled Frames Subject to Lateral Loading." *Proc. Inst. Civil. Engineer., London* **63**(2), 1977, 641–656.

Mainstone, R. J. "On the Stiffness and Strength of Infilled Frames." *Proc. Inst. Civil Engineer.*, *London* **48**, 1971, 57–90.

Ockleston, A. J. "Tests on the Old Dental Hospital Johannesburg: The Effect of Floors and Walls on the Behavior of Reinforced Concrete Frameworks Subject to Horizontal Loading." The Concrete Assoc. of South Africa, Johannesburg, South Africa, Paper No. 3, November 1956.

Polyakov, S. V. "Calculation of the Horizontal Loads Acting in the Plane of the Walls of Framed Buildings." *Stroit. Mekh. Raschet Sooruzhenii, Moscow* **3**(2), 1961, 5–11, (in Russian).

Simms, L. G. "Behaviour of No-Fines Concrete Panels as the Infill in Reinforced Concrete Frames." *Civil Engineer. and Public Works Rev.* **62** November 1967, 1245–1250.

Stafford Smith, B., and Riddington, J. R. "The Design of Infilled Steel Frames for Bracing Structures." *Struct. Engineer* **56B**(1), 1978, 1–6.

Thiruvengadam, V. "On the Natural Frequencies of Infilled Frames." *J. Earthquake Engineer. and Struct. Dynam.* **13**(3), 1985, 401–419.

Wood, R. H. "Plasticity, Composite Action and Collapse Design of Unreinforced Shear Wall Panels in Frames." *Proc. Inst. Civil Engineer.*, *London* **65**(2), 1978, 381–411.

CHAPTER 9

Benjamin, J. R. and Williams, H. A. "Behavior of Reinforced Concrete Shear Walls." *Trans. ASCE* **124**, 1959, 669–708.

Biswas, J. K. and Tso, W. K. "Three-Dimensional Analysis of Shear Wall Buildings to Lateral Load." *J. Struct. Div., ASCE* **100**(5), 1974, 1019–1035.

Cardenas, A. E., Hanson, J. M., Corley, W. G., and Hognestad, E. "Design Provisions for Shear Walls." *J. Am. Conc. Inst.* **70**(3), 1973, 221–230.

Colaco, J. P. "Preliminary Design of High-Rise Buildings with Shear Walls." *J. Am. Conc. Inst.* **68**(1), 1971, 26–31.

Coull, A. and Stafford Smith, B. (Eds). *Tall Buildings*. Pergamon Press, Oxford, 1967.

Ghali, A. and Neville, A. M. "Three-Dimensional Analysis of Shear Walls." *Proc. Inst. Civil Engineer.*, *London* **51**, 1972, 347–357.

Green, N. B. "Rigid-Reinforced Walls for Earthquake Bracing of Tall Buildings." *Engineer. News-Record* Part II, July–December, 1931, 364–366.

Green, N. B. "Bracing Walls for Multi-Story Buildings." *J. Am. Conc. Inst.* **24**(3), 1952, 233–248.

Harman, D. J. and Johnson, A. A. "Analysis of Buildings with Interconnected Shear Walls." *Can. J. Civil Engineer.* **5**(2), 1978, 157–163.

MacLeod, I. A. "Structural Analysis of Wall Systems." *Struct. Engineer* **55**(11), 1977, 487–495.

Moudarres, F. R. and Coull, A. "Stiffening of Linked Shear Walls." *J. Engineer. Mech., ASCE* **112**(3), 1986, 223–237.

Rosenblueth, E. and Holtz, I. "Elastic Analysis of Shear Walls in Tall Buildings." *J. Am. Conc. Inst.* **31**(12), 1960, 1209–1222.

Rosman, R. "Analysis of Spatial Concrete Shear Wall Systems." *Proc. Inst. Civil Engineer., London* Paper 7266s, 1970, 131–152.

Rosman, R. "Statics of Non-Symmetric Shear Wall Structures." *Proc. Instn. Civ. Engineer., London* Paper 7393s, 1971, 211–244.

S.E.A.O.S.C. "Horizontal Bracing Systems in Buildings Having Masonry or Concrete Walls." Report of Special Committee, Structural Engineers Association of Southern California, 1949.

Taniguchi, T. et al. "Shear Walls." *Trans. Architect. Inst. Japan Toyko* No. 41, August 1950, (in Japanese).

Taranath, B. S. "Analysis and Design of Shear Wall Structures." *Struct. Engineer* **63A(7)**, 1985, 208–215.

Tomii, M. "Shear Walls." State-of-the-Art Report, Technical Committee 21, *Proc. 3rd Regional ASCE-IABSE Conf. on Planning and Design of Tall Buildings*, Tokyo, Japan, September 1971, pp. 212–215.

Tso, W. K. and Ast, P. F. "Special Considerations in Design of an Asymmetrical Shear Wall Building." *Can J. Civil Engineer.* 5(3), 1978, 403–413.

CHAPTER 10

Albiges, M. and Goulet, J. "Wind Bracing of Tall Buildings." *Ann. Inst. Batiment Travaux Publics* No. 149, May 1960, 473–500 (in French).

Beck, H. "Contribution to the Analysis of Coupled Shear Walls." *J. Am. Conc. Inst.* 59(8), 1962, 1055–1069.

Chan, H. C. and Cheung, Y. K. "Analysis of Shear Walls Using Higher Order Finite Elements." *Building and Environ. London* **14**, 1979, 217–224.

Cheung, Y. K. "Tall Buildings." Chapter 38, *Handbook of Structural Concrete*, F. K. Kong, R. H. Evans, E. Cohen, and F. Roll (Eds.). Pitman, London, 1983, pp. 38.1–38.52.

Chitty, L. and Wan, W.-J. "Tall Building Structures under Wind Load." *Proc. 7th Int. Conf. Appl. Mech.* **1**, Paper 22, 1948, 254–268.

Coull, A. "Stiffening of Coupled Shear Walls against Foundation Movement." *Struct. Engineer, London* **52**, 1974, 23–26.

Coull, A. "Contribution to the Continuum Analysis of Coupled Shear Walls." *Res. Mechanica* **26**, 1989, 353–370.

Coull, A., Boyce, D. W., and Wong, Y. C. "Design of Floor Slabs Coupling Shear Walls." Chapter 14, Vol. 1, *Structures, Civil Engineering Practice*, P. N. Chereminisoff, N. P. Chereminisoff, and S. L. Cheng (Eds.), Technomic, Lancaster, PA, 1987, pp. 357–399.

Coull, A. and Chantaksinopas, B. "Design Curves for Coupled Shear Walls on Flexible Bases." *Proc. Inst. Civil Engineer., London*, 57(2), 1974, 595–618.

Coull, A. and Choudhury, J. R. "Analysis of Coupled Shear Walls." *J. Am. Conc. Inst.* 64(9), 1967, 587–593.

Coull, A. and Irwin, A. W. "Analysis of Load Distribution in Multi-Storey Shear Wall Structures." *Struct. Engineer, London* **48**, 1970, 301–306.

Coull, A., Puri, R. D., and Tottenham, H. "Numerical Elastic Analysis of Coupled Shear Walls." *Proc. Inst. Civil Engineer., London* **55**, 1973, 109–128.

Gluck, J. "Elasto-Plastic Analysis of Coupled Shear Walls." *J. Struct. Div., ASCE* **99**(8), 1973, 1743–1760.

Irwin, A. W. "Design of Shear Wall Buildings." Report No. 102, Construction Industry Research and Information Association, London, 1984.

Kabaila, A. P. and Edwardes, R. J. "Hybrid Element Applied to Shear Wall Analysis." *J. Struct. Div., ASCE* **105**(12), 1979, 2753–2760.

Macleod, I. A. "New Rectangular Finite Element for Shear Wall Analysis." *J. Struct. Div., ASCE* **95**(3), 1969, 399–409.

Macleod, I. A. "Structural Analysis of Wall Systems." *Struct. Engineer, London* **55**, 1977, 487–495.

Michael, D. "The Effect of Local Wall Deformations on the Elastic Interaction of Cross Walls Coupled by Beams." In *Tall Buildings*, A. Coull and B. Stafford Smith (Eds.). Pergamon Press, Oxford, 1967, pp. 253–270.

Mukherjee, P. R. and Coull, A. "Free Vibrations of Coupled Shear Walls on Flexible Bases." *Proc. Inst. Civil Engineer., London* **57**(2), 1974, 493–511.

Paulay, T. "An Elasto-Plastic Analysis of Coupled Shear Walls." *J. Am. Conc. Inst.* **67**(11), 1970, 915–922.

Rosman, R. "Approximate Analysis of Shear Walls Subjected to Lateral Loads." *J. Am. Conc. Inst.* **61**(6), 1964, 717–732.

Rosman, R. *Tables for the Internal Forces of Pierced Shear Walls Subject to Lateral Forces*. Bauingenieur-Praxis, Heft 66, Verlag von Wilhelm Ernst and Sohn, Berlin, 1966.

Singh, G. and Schwaighofer, J. "A Bibliography on Shear Walls." Pub. No. 76-02, Department of Civil Engineering, University of Toronto, Toronto, 1976.

Stafford Smith B. "Modified Beam Method for Analyzing Symmetrical Interconnected Shear Walls." *J. Am. Conc. Inst.* **67**(12), 1970, 977–980.

Tso, W. K. and Biswas, J. K. "General Analysis of Non-Planar Coupled Shear Walls." *J. Struct. Div., ASCE* **99**(3), 1973, 365–380.

Tso, W. K. and Chan, H. B. "Dynamic Analysis of Plane Coupled Shear Walls." *J. Engineer. Mech. Div., ASCE* **97**(1), 1971, 33–48.

Tso, W. K. and Chan, P.C.K. "Flexible Foundation Effect on Coupled Shear Walls." *J. Am. Conc. Inst.* **69**(11), 1972, 678–683.

CHAPTER 11

Aktan, A. E., Bertero, V. V., and Sakino, K. "Lateral Stiffness Characteristics of R/C Frame-Wall Structures." In *Deflection of Concrete Structures*. ACI Pub. SP86-10, Detroit, 1985.

Arvidsson, K. "Non-Uniform Shear Wall-Frame Systems with Elastic Foundations." *Proc. Inst. Civil Engineer., London* **59**(2), 1975, 139–148.

Basu, A. K. and Nagpal, A. K. "Frame-Wall Systems with Rigidly Jointed Link Beams." *J. Struct. Div., ASCE* **106**(5), 1980, 1175–1190.

Basu, A. K., Nagpal, A. K., and Nagar, A. K. "Dynamic Characteristics of Frame-Wall Systems." *J. Struct. Div., ASCE* **108**(6), 1982, 1201-1218.

Cardan, B. "Concrete Shear Walls Combined with Rigid Frames in Multistory Buildings Subject to Lateral Loads." *J. Am. Conc. Inst.* **58**(3), 1961, 299-316.

Cheung, Y. K. and Kasemset, C. "Approximate Frequency Analysis of Shear Wall Frame Structures." *J. Earthquake Engineer. and Struct. Dynam.* **6**(2), 1978, 221-229.

Chowdhary, B., Nagpal, A. K., and Nayar, K. K. "Laterally Loaded Frame Wall Systems." *Int. J. Struct.* **6**(2), 1986, 57-70.

Coull, A. and Khachatoorian, H. "Analysis of Laterally Loaded Wall-Frame Structures." *J. Struct. Engineer., ASCE* **110**(6), 1984, 1396-1399.

Derecho, A. T. "Frames and Frame-Shear Wall Systems." In *Response of Multistory Concrete Structures to Lateral Forces*, ACI Publ. SP-36, Detroit, 1973, 13-37.

Frischmann, W. W., Prabhu, S. S., and Toppler, J. F. "Multi-Storey Frames and Interconnected Shear Walls Subjected to Lateral Loads." *Conc. and Construct. Engineer. London*, June 1963, 227-234.

Gould, P. L. "Interaction of Shear Wall-Frame System in Multistory Buildings." *J. Am. Conc. Inst.* **62**(1), 1965, 45-70.

MacLeod, I. A. "Simplified Analysis of Shear Wall-Frame Interaction." *Building Sci.* **7**(2), 1972, 121-125.

Majumdar, S. N. G. and Adams, P. F. "Test on Steel-Frame, Shear-Wall Structures." *J. Struct. Div., ASCE* **97**(4), 1971, 1097-1111.

Oakberg, R. G. and Weaver, W. "Analysis of Frames with Shear Walls by Finite Elements." *Proc. Symp. Application of Finite Element Methods in Civil Engineer.*, Vanderbilt University, November 1969, pp. 567-607.

Rutenberg, A. "Plane Frame Analysis of Laterally Loaded Asymmetric Buildings—An Uncoupled Solution." *Computers and Struct.* **10**(3), 1979, 553-555.

Rutenberg, A. and Heidebrecht, A. C. "Approximate Analysis of Asymmetric Wall-Frame Structures." *Building Sci.* **10**(1), 1975, 27-35.

Stamato, M. C. and Mancini, E. "Three-Dimensional Interaction of Walls and Frames." *J. Struct. Div., ASCE* **99**(12), 1973, 2375-2390.

Stamato, M. C. and Stafford Smith, B. "An Approximate Method for the Three-Dimensional Analysis of Tall Buildings." *Proc. Inst. Civil Engineer., London* **43**, 1969, 361-379.

Weaver, W., Brandow, G. E., and Manning, T. A. "Tier Buildings with Shear Cores, Bracing and Setbacks." *Computers and Struct.* **1**(1/2), 1971, 57-83.

Winokur, A. and Gluck, J. "Lateral Loads in Asymmetric Multistory Structures." *J. Struct. Div., ASCE* **94**(3), 1968, 645-656.

CHAPTER 12

Chang, P. C. and Foutch, D. A. "Static and Dynamic Modeling and Analysis of Tube Frames." *J. Struct. Engineer., ASCE* **110**(12), 1984, 2955-2975.

Chang, P. C. and Foutch, D. "Numerical Solution of Tube Buildings Modeled as Continuum." *Computers and Struct.* **21**, 1985, 771-776.

Chang, P. C. "Analytical Modeling of Tube-in-Tube Structures." *J. Struct. Engineer., ASCE* **111**(6), 1985, 1326-1337.

Coull, A. and Subedi, N. K. "Framed-Tube Structures for High-Rise Buildings." *J. Struct. Div., ASCE* **97**(8), 1971, 2097-2105.

Coull, A. and Bose, B. "Simplified Analysis of Framed-Tube Structures." *J. Struct. Div., ASCE* **101**(11), 1975, 2223-2240.

Coull, A. and Ahmed, A. K. "Deflections of Framed-Tube Structures." *J. Struct. Div., ASCE* **104**(5), 1978, 857-862.

Coull, A., Bose, B., and Ahmed, A. K. "Simplified Analysis of Bundled-Tube Structures." *J. Struct. Div., ASCE* **108**(5), 1982, 1140-1153.

Feld, L. S. "Superstructure for 1350 ft. World Trade Center." *Civil Engineer., ASCE* **41**, 1971, 66-70.

Grossman, J. S. "780 Third Avenue: The First High Rise Diagonally Braced Concrete Structure." *Conc. Int., Design and Construct.* **7**, 1985, 53-56.

Iyengar, H., Amin, N. R., and Carpenter, L. "Computerized Design of World's Tallest Building." *Computers and Struct.* **2**, 1972, 771-783.

Khan, F. R. "Current Trends in Concrete High Rise Buildings." In *Tall Buildings,* A. Coull and B. Stafford Smith (Eds.). Pergamon Press, Oxford, 1967, pp. 571-590.

Khan, F. R. "Tubular Structures for Tall Buildings." Chapter 11, *Handbook of Concrete Engineering,* M. Fintel (Ed.). Van Nostrand Reinhold, New York, 1974, pp. 345-355.

Khan, F. R. "100-Story John Hancock Center in Chicago—A Case Study of the Design Process." *J. I.A.B.S.E.* 3(J-16/82), 1982, 27-34.

Khan, F. R. and Amin, N. R. "Analysis and Design of Framed Tube Structures for Tall Concrete Buildings." In *Response of Multistory Concrete Structures to Lateral Forces.* ACI, Pub. SP-36, 1973, 39-60.

Liauw, T. C. and Luk, W. K. "Unified Method for Torsional Analysis of Shear Core, Framed-Tube, and Tube in Tube Structures." *Proc. Inst. Civil Engineer., London* **71**(2), 1981, 463-477.

Mancini, E. "Analysis of Framed Tube Structures by Continuous Medium Technique." Prelim. Pub. IABSE, *Seminar on Tall Structures and Use of Prestressed Concrete in Hydraulic Structures,* Srinagar, India, 1984.

Mazzeo, L. A. and De Vries, A. "Perimetral Tube for 37-Story Steel Building." *J. Struct. Div., ASCE* **98**(6), 1972, 1255-1272.

Nair, R. S. "Modified Tube Concept for Tall Buildings." *J. Am. Inst. Steel Construct.* **23**, 1986, 126-130.

Notch, J. S. "Octagonal Framed Tube Office Building: A Case History." *J. Struct. Engineer., ASCE* **109**(12), 1983, 2872-2892.

Notch, J. S. "Circu-Rectangular Bundled Tube Office Tower—A Case History." *J. Struct. Engineer., ASCE* **110**(7), 1984, 1598-1612.

Patel, K. S., Logcher, R. D., Harmon, T. G., and Hansen, R. J. "Computer-Aided Design—Standard Oil of Indiana Building." *J. Struct. Div., ASCE* **99**(4), 1973, 621-635.

Picardi, E. A. "Structural System—Standard Oil of Indiana Building." *J. Struct. Div.*, ASCE **99**(4), 1973, 605–620.

Roesset, J. M., Harmon, T. G., Efimba, R. E., and Hansen, R. J. "Some Structural Problems—Standard Oil of Indiana Building." *J. Struct. Div.*, ASCE **99**(4), 1973, 637–654.

Schwaighofer, J. and Ast, P. F. "Tables for the Analysis of Framed-Tube Buildings." Pub. 72-01, Department of Civil Engineering, University of Toronto, Toronto, 1972.

CHAPTER 13

Coull, A. and Tawfik, S. Y. "Analysis of Core Structures Subjected to Torsion." *Building and Environ.* **16**(3), 1981, 221–228.

Gluck, J. "Lateral Load Analysis of Asymmetric Multistory Structures." *J. Struct. Div.*, ASCE **96**(2), 1970, 317–333.

Ho, D. and Liu, C. H. "Analysis of Shear-Wall and Shear-Core Assembly Subjected to Lateral and Torsional Loading." *Proc. Inst. Civil Engineer.*, London. **79**(2), 1985, 119–133.

Irwin, A. W. and Bolton, C. J. "Torsion of Tall Building Cores." *Proc. Inst. Civil Engineer.*, London **63**(2), 1977, 579–591.

Khan, M. A. H. and Stafford Smith, B. "Restraining Action of Bracing in Thin-Walled Open Section Beams." *Proc. Inst. Civil Engineer.*, London **59**(2), 1975, 67–78.

Khan, M. A. H. and Tottenham, H. "Methods of Bimoment Distribution for the Analysis of Continuous Thin-Walled Structures Subject to Torsion." *Proc. Inst. Civil Engineer.*, London **63**(2), 1977, 843–863.

Liauw, T. C. "Torsion of Multi-Storey Spatial Core Walls." *Proc. Inst. Civil Engineer.*, London **65**(2), 1978, 601–609.

Liauw, T. C. and Leung, K. W. "Torsion Analysis of Core Wall Structures by Transfer Matrix Method." *Struct. Engineer* **53**(4), 1975, 187–194.

Liauw, T. C. and Luk, W. K. "Torsion of Core Walls of Nonuniform Section." *J. Struct. Div.*, ASCE **106**(9), 1980, 1921–1931.

MacLeod, I. A. and Hosny, H. M. "Frame Analysis of Shear Wall Cores." *J. Struct. Div.*, ASCE **103**(10), 1977, 2037–2047.

Mallick, D. V. and Dungar, R. "Dynamic Characteristics of Core Wall Structures Subjected to Torsion and Bending." *Struct. Engineer* **55**(6), 1977, 251–261.

Michael, D. "Torsional Coupling of Core Walls in Tall Buildings." *Struct. Engineer* **47**(2), 1969, 67–71.

Mukherjee, P. R. and Coull, A. "Free Vibrations of Open-Section Shear Walls." *J. Earthquake Engineer. and Struct. Dynam.* **5**(1), 1977, 81–101.

Osawa, Y. "Seismic Analysis of Core-Wall Buildings." *Proc. 3rd World Conf. on Earthquake Engineer.* New Zealand, 1965, pp. II-458–II-475.

Rosman, R. "Torsion of Perforated Concrete Shafts." *J. Struct. Div.*, ASCE **95**(5), 1969, 991–1010.

Rutenberg, A., Shtarkman, M., and Eisenberger, M. "Torsional Analysis Methods for Perforated Cores." *J. Struct. Engineer.*, ASCE **112**(6), 1986, 1207–1227.

Rutenberg, A. and Tso, W. K. "Torsional Analysis of Perforated Core Structures." *J. Struct. Div.*, *ASCE* **101**(3), 1975, 539–550.

Stafford Smith, B. and Girgis, A. M. "The Torsion Analysis of Tall Building Cores Partially Closed by Beams." *Proc. Symp. on Behavior of Building Systems and Building Components*, Vanderbilt University, March 1979, pp. 211–228.

Tso, W. K. and Biswas, J. K. "Analysis of Core Wall Structures Subject to Applied Torque." *Building Sci.* **8**(3), 1973, 251–257.

Vasquez, J. and Riddell, R. "Thin-Walled Core Element for Multistory Buildings." *J. Struct. Engineer.*, *ASCE* **110**(5), 1984, 1021–1034.

CHAPTER 14

Boggs, P. C. and Gasparini, D. A. "Lateral Stiffness of Core/Outrigger Systems." *J. Am. Inst. Steel Const.* **20**(4), 1983, 172–180.

Moudarres, F. R. "Outrigger-Braced Coupled Shear Walls." *J. Struct. Engineer.*, *ASCE* **110**(12), 1984, 2876–2890.

Moudarres, F. R. and Coull, A. "Free Vibrations of Outrigger-Braced Structures." *Proc. Inst. Civil Engineer.*, *London*, **79**(2), 1985, 105–117.

Rutenberg, A. "Earthquake Analysis of Belted High-Rise Building Structures." *Engineer. Struct.* **1**(3), 1979, 191–196.

Stafford Smith, B. and Salim, I. "Formulae for Optimum Drift Resistance of Outrigger-Braced Tall Building Structures." *Computers and Struct.* **17**(1), 1983, 45–50.

Taranath, B. S. "Optimum Belt Truss Location for High-Rise Structures." *Struct. Engineer* **53**(8), 1975, 345–347.

CHAPTER 15

Hoenderkamp, J. C. D. and Stafford Smith, B. "Approximate Rotation Analysis for Plan-Symmetric High-Rise Structures." *Proc. Inst. Civil Engineer.*, *London* **83**(2), 1987, 755–767.

Hoenderkamp, J. C. D. and Stafford Smith, B. "Simplified Torsion Analysis for High-Rise Structures." *Building and Environ.* **23**(2), 1988, 153–158.

Murashev, V., Sigalov, E. Y., and Baikov, V. N. "Members of Multi-Story Framed and Panel Residential and Civil Buildings." Section 19-2, *Design of Reinforced Concrete Structures*, MIR Publishers, Moscow, 1971, pp. 542–587 (in English).

Stafford Smith, B. and Hoenderkamp, J. C. D. "Simple Deflection Analysis of Planar Wall Frames." *Proc. Asian Regional Conf. on Tall Buildings and Urban Habitat*, Kuala Lumpur, Malaysia, August 1982, pp. 3-33–3-41.

Stafford Smith, B., Hoenderkamp, J. C. D., and Kuster, M. "A Graphical Method of Comparing the Sway Resistance of Tall Building Structures." *Proc. Inst. Civil Engineer.*, *London* **73**(2), 1982, 713–729.

CHAPTER 16

Adams, P. F. "The Design of Steel Beam-Columns." Canadian Steel Industries Construction Council, Willowdale, Ontario, 1974.

Beaulieu, D. and Adams, P. F. "Significance of Structural Out-of-Plumb Forces and Recommendations for Design." *Can. J. Civ. Engineer.* **7**(1), 1980, 105–113.

Beaulieu, D. and Adams, P. F. "The Results of a Survey on Structural Out-of-Plumbs." *Can. J. Civ. Engineer.* **5**(4), 1978, 462–470.

Birnstiel, C. and Iffland, J. S. B. "Factors Influencing Frame Stability." *J. Struct. Div., ASCE* **106**(2), 1980, 491–504.

Gluck, J. "The Buckling of Tier Buildings with Variable Lateral Stiffness Properties." *Building Sci.* **9**(1), 1974, 39–43.

Goldberg, J. E. "Buckling of Multi-Story Buildings." *J. Engineer. Mech., ASCE* **91**(1), 1965, 51–70.

Horne, M. R. "Approximate Method for Calculating the Elastic Critical Loads of Multi-Storey Plane Frames." *Struct. Engineer* **53**(6), 1975, 242–248.

LeMessurier, W. J. "A Practical Method of Second Order Analysis." *J. Am. Inst. Steel Const.* **14**(2), 1977, 49–67.

MacGregor, J. G. "Out-of-Plumb Columns in Concrete Structures." *Conc. Int.* **1**(6), 1979, 26–31.

MacGregor, J. G. and Hage, S. E. "Stability Analysis and Design of Concrete Frames." *J. Struct. Div., ASCE* **103**(10), 1977, 1953–1970.

Nair, R. S. "Overall Elastic Stability of Multistory Buildings." *J. Struct. Div., ASCE* **101**(12), 1975, 2487–2503.

Razzaq, Z. and Naim, M. M. "Elastic Instability of Unbraced Space Frames." *J. Struct. Div., ASCE* **106**(7), 1980, 1389–1400.

Rosenblueth, E. "Slenderness Effects in Buildings." *J. Struct. Div., ASCE* **91**(1), 1965, 229–252.

Stafford Smith, B. and Gaiotti, R. "Iterative Gravity Load Method for P-Delta Analysis." Structural Engrg. Report No. 88-4, Department of Civil Engineering and Applied Mechanics, McGill University, Montreal, Quebec, 1988.

CHAPTER 17

Berg, G. V. *Seismic Design Codes and Procedures*. Earthquake Engineering Research Institute, Berkeley, California, 1983.

Blume, J. A., Newmark, N. M., and Corning, L. H. *Design of Multistory Reinforced Concrete Buildings for Earthquake Motions*. Portland Cement Association, Skokie, Illinois, 1961.

Cermak, J. E. (Ed.). "Wind Engineering." *Proc. of Fifth Int. Conf. on Wind Engineering*, Fort Collins. Pergamon Press, Oxford, 1980.

Chen, P. W. and Robertson, L. E. "Human Perception Thresholds of Horizontal Motion." *J. Struct. Div., ASCE* **98**(8), 1972, 1681–1695.

Davenport, A. G. "The Treatment of Wind Loading on Tall Buildings." In *Tall Buildings*, A. Coull and B. Stafford Smith (Eds.). Pergamon Press, Oxford, 1967, 3–44.

Derecho, A. T. "New Developments in Earthquake Loading and Response." In *Advances in Tall Buildings*, L. S. Beedle (Ed.). Van Nostrand Reinhold, New York, 1986, 241–255.

Derecho, A. T. and Fintel, M. "Earthquake-Resistant Structures." Chapter 12, *Handbook of Concrete Engineering*, M. Fintel (Ed.), Van Nostrand Reinhold, New York, 1974, pp. 356–432.

Dowrick, D. J. *Earthquake Resistant Design*. Wiley, New York, 1977.

Green, N. B. *Earthquake Resistant Building Design and Construction*. Van Nostrand Reinhold, New York, 1978.

Hansen, R. J., Reed, J. W., and Vanmarcke, E. H. "Human Response to Wind Induced Motion of Buildings." *J. Struct. Div., ASCE* **99**(7), 1973, 1589–1605.

Housner, G. W. and Jennings, P. C. *Earthquake Design Criteria*. Earthquake Engineering Research Institute, Berkeley, California, 1982.

Hurty, W. C. and Rubinstein, M. F. *Dynamics of Structures*. Prentice-Hall, Englewood Cliffs, NJ, 1964.

Kwok, K. C. S. "Cross-Wind Response of Tall Buildings." *Engineer. Struct.* **4**, 1982, 256–262.

Melbourne, W. H. and Cheung, J. C. K. "Designing for Serviceable Accelerations in Tall Buildings." *Proc. 4th Int. Conf. on Tall Buildings*, Hong Kong and Shanghai, Y. K. Cheung and P. K. K. Lee (Eds.), Vol. 1, Hong Kong, 1988, pp. 148–155.

Muto, K. *Aseismic Design Analysis of Buildings*. Maruzen, Tokyo, 1974.

Paz, M. *Structural Dynamics, Theory and Computation*, 2nd ed. Van Nostrand Reinhold, New York, 1985.

Rathbun, C. J. "Wind Forces on a Tall Building." *Trans. ASCE* **105**, Paper No. 2056, 1938, 1–41.

Rosenblueth, E. (Ed.). *Design of Earthquake Resistant Structures*. Pentech Press, London, 1980.

Stafford Smith, B. and Crowe, E. "Estimating Periods of Vibration of Tall Buildings." *J. Struct. Engineer., ASCE* **112**(5), 1986, pp. 1005–1019.

Thomson, W. T. *Vibration Theory with Applications*. Prentice-Hall, Englewood Cliffs, NJ, 1980.

Wakabayashi, M. *Design of Earthquake-Resistant Buildings*. McGraw-Hill, New York, 1986.

Warburton, G. B. *The Dynamical Behaviour of Structures*. Pergamon Press, Oxford, 1976.

Weigel, R. (Ed.). *Earthquake Engineering*. Prentice-Hall, Englewood Cliffs, NJ, 1970.

Wiss, J. F. and Parmelee, R. A. "Human Perception of Transient Vibrations." *J. Struct. Div., ASCE* **100**(4), 1974, 773–787.

Yamada, M. and Goto, T. "Human Response to Tall Building Motion." In *Human Response to Tall Buildings*, Conway (Ed.). Dowden, Hutchison and Ross, Stroudsburg, PA, 1977, pp. 58–71.

"Earthquake Loading and Response." Chapter 2, Vol CL. *Tall Building Criteria and*

Loading, Monograph on Planning and Design of Tall Buildings. ASCE, New York, 1980, pp. 47–141.

CHAPTER 18

Bljuger, F. "Temperature Effects in Buildings with Panel Walls." *Building and Environ.* **17**, 1982, 17–21.

Bazant, Z. P. and Wittman, F. H. (Eds.). *Creep and Shrinkage in Concrete Structures.* Wiley, New York, 1982.

Fintel, M. and Ghosh, S. K. "High Rise Design: Accounting for Column Length Changes." *Civil Engineer., ASCE* **54**, 1984, 55–59.

Fintel, M. and Ghosh, S. K. "Column Length Changes in Ultra-High-Rise Buildings." In *Advances in Tall Buildings.* Van Nostrand Reinhold, New York, 1986, pp. 503–515.

Gardner, N. J. and Fu, H. C. "Effects of High Construction Loads on the Long-Term Deflections of Flat Slabs." *ACI Struct. J.* **84**(4), 1987, 349–360.

Karp, J. "Temperature Effects in Tall Reinforced Concrete Buildings." *Planning and Design of Tall Buildings, Proc. 1972 ASCE-IABSE Int. Conf. ASCE,* **III**, 1973, 793–814.

Khan, F. R. and Fintel, M. "Effects of Column Exposure in Tall Structures—Design Considerations and Field Observations of Buildings." *J. Am. Conc. Inst.* **65**, 1968, 99–110.

Liu, X. and Chen, W. F. "Effect of Creep on Load Distribution in Multistory Reinforced Concrete Buildings during Construction." *ACI Struct. J.* **84**(3), 1987, 192–199.

Neville, A. M. *Creep of Concrete: Plain, Reinforced and Prestressed.* North-Holland, Amsterdam, 1970.

Neville, A. M., Dilger, W. H., and Brooks, J. J. *Creep of Plain and Structural Concrete.* Construction Press, London, 1983.

Rusch, H., Jungwirth, D., and Hilsdorf, H. K. *Creep and Shrinkage: Their Effect on the Behavior of Concrete Structures.* Springer-Verlag, Berlin, 1982.

Russell, H. G. and Corley, W. G. "Time-Dependent Deformations of Vertical Members in Ultra-High Concrete Buildings." Technical Report, Portland Cement Association, Skokie, IL, 1976.

Weidlinger, P. "Temperature Stresses in Tall Reinforced Concrete Buildings." *Civil Engineer., ASCE* **34**, 1964, 58–61.

"Creep, Shrinkage and Temperature Effects." Chapter CB-10, Vol. CB, *Structural Design of Tall Concrete and Masonry Buildings, Monograph on Planning and Design of Tall Buildings.* ASCE, New York, 1978, 427–500.

"Prediction of Creep, Shrinkage and Temperature Effects in Concrete Structures." In ACI Pub. SP-27, Detroit, 1971.

INDEX

Page numbers in *italic* refer to the main discussion of that topic.

Acceleration, 13, 419, 429, 431, 441
  along-wind, 429
  cross-wind, 429
  ground, 419, 431, 435, 447
  limits of, 453, 454
  pseudo, 436, 441
  seismic, 420
  spectral, 438, 441, 442, 443
Accelerogram, 419, 425, 433, 437
Accelerograph, 433
Admittance:
  aerodynamic, 425
  mechanical, 425
Alcan Building, San Francisco, 124
Amplification factor, 251, 389, 395, 399
Amplitude, modal, 439
Analogy:
  beam on elastic foundation, 173
  braced-frame, 207, 210, 344
  deep-beam, 103
  frame, 207, 344
  single-column, 349
  two-column, 345
  wide-column, 103
Analysis:
  accurate, 6, 7, 65, 78
  anologous frame, 207, 246
  approximate, 70, 71, 117, 138, 141, 150, 355, 356
  assumptions for, 67, 216, 260, 356, 383, 386
  cantilever method, 146, 147, 150
  computer, 113, 130, 161, 199, 206, 341
  continuum, 77, 216, 260, 480
  deflection, 115, 116
  dynamic, 24, 29, 163, *419*
  final, 7, 66
  graphic, 231, 268, 332
  hybrid, 65, 67, 283
  intermediate, 66
  modal, 26, 29, 433, 438
  modeling for, *65*
  P-delta, 130, 161
  portal method, 141–144, 150
  preliminary, 6, 7, 65, 66, 71, 141, 161, 358
  static, 21, 25, 420
  stiffness matrix, 66, 77, 78, 161, 349
  three-dimensional, 66, 67, 83, 296, 305, 433
  time-history, 433
Analysis program:
  frame, 113, 190, 206
  general purpose, 6, 215, 247, 277
  structural, 78, 81, 88, 206, 344, 347, 353
Antisymmetry, 89, 278
Apartment building, 12, 202, 213
Architecture, 4, 9, 34, 50, 52, 106, 107, 113
  post-modern, 4
Area:
  shear, 343
  negative, 86, 87
  tributary, 5, 18, 20, 69, 111, 163, 461
Arm, rigid, 71, 73, 216, 247, 250

Bank of China Building, Hong Kong, 53
Base discontinuity, 202
Base flexibility, 220, 487
Base shear, 25, 26, 27, 433, 434
Bay, end, 496
Beam:
  auxiliary, 82, 83, 86, 88, 341, 353
  connecting, 36, 75, 78, 213, 250, 267, 308, 329, 331, 344, 372, 380, 489, 509
  equivalent uniform, 250, 493
  coupling, 213, 489
  deep, 67, 103, 249, 306
  equivalent, 87, 158, 250, 474, 493
  fictitious, 86, 94, 161, 253, 299
  haunched, 250
  one-way system, 60

527

Beam (*Continued*)
  rigid-ended, 248
  shear, 509
  spandrel, 127, 284, 285, 286, 306, 463
  three-way system, 61
  transfer, 44
  two-way system, 61
Beam and slab, 59
Beam on elastic foundation analogy, 173
Beam element, 66, 78, 79, 81, 88, 161, 206
Beam floor, steel, 60
Behavior:
  bracing, 109
  coupled shear walls, 213
  infilled frame, 169
  nonlinear, 68, 447, 463
  nonproportionate structures, 189
  rigid frame, 131
  shear wall, 184
  static, 419, 420
  structural, 55, 67–69, 109, 111, 137, 169, 184, 213, 257, 283, 290, 293
  tubular structures, 283
  wall-frames, 257
  warping, 308, 310, 345
Belt-braced structure, 355
Belt girder, 49, 355
Belt truss, 49
Bending:
  double-curvature, 102, 131, 151, 158, 165, 171, 208, 214, 343, 372, 380
  reverse, 214
  transverse, 75
Bending stiffness, 13, 39, 208
Bent:
  braced, 106, 107, 109, 255
  mixed, 381, 382, 384
Bimoment, 314, 323, 327, 328, 331, 333, 337, 509, 510, 511
Blockwork, 40, 168
Boundary conditions, 222, 262
Boundary layer:
  turbulence, 421
  wind tunnel, 23
Box core, 65, 66
Box section, 308
Braced:
  frame analogy, 207
  frame structure, 37, *106*
  outrigger-, structure, *355*
  -tube structure, 36, 46, 127, 283, 289, 305
Bracing, 69, 106, 108, 109, 168, 184
  diagonal, 2, 36, 37, 46, 106, 107, 113, 124, 127, 169, 178, 289, 290, 380
  eccentric, 109

K-, 107, 109, 113, 124, 127, 380
knee, 380
large-scale, 124
offset diagonal, 380
outrigger, 355, 356
Brickwork, 40, 168, 478
Buckling, 388, 462
  combined modes of, 388, 391, 392
  overall, 388, 389, 392
  P-Delta, 398–413
  torsional, 388, 394
  transverse, 388, 394
  transverse-torsional, 388, 411
  wall-frames, 392
Building(s):
  apartment, 202
  commercial, 1, 9
  concrete, 285, 305, 461
  function, 34
  hotel, 1, 12, 35, 41
  office, 34, 35, 55
  residential, 4, 35, 41, 56, 213, 285
Building Code, 5
  of Canada, National, 24, 30, 422, 425, 429, 437, 456
  Uniform, 133, 176, 318, 321
Bundled-tube structures, 45, 46, 283, 288, 305

Canadian Building Code, 24, 30, 422, 425, 429, 437, 456
Cantilever, 66, 67, 257
  composite, 42, 213, 232, 482, 483
  equivalent, 67
  factor, 233, 482, 485
  flexural, 118, 146, 154, 257, 260, 378, 392
  independent, action, 232, 233
  shear, 257, 260, 378, 392
  shear-flexure, 372, 382
  vertical, 36, 41, 69, 111, 184, 213, 308, 355, 388, 399
Cantilever method, 146, 147, 150
Cantilever structure, 372
Center, shear, 73, 188, 310, 317, 345, 346, 349
Center of mass, 447
Center of rigidity, 140, 141
Center of resistance, 140, 447
Center of rotation, 70, 395
Center of twist, 187, 188
Chicago:
  Home Insurance Building, 2
  John Hancock Building, 46, 127
  Masonic Temple, 2
  Monadnock Building, 2
  Rand-McNally Building, 2
  Sears Tower, 45, 46

INDEX **529**

Chord, 36, 37, 111, 372
Chrysler Building, New York, 106
Citicorp Building, New York, 127
Cladding, 4, 12, 65, 68, 422, 462, 475
  masonry, 462
Classification, structural, 422
Code, 5, 12, 21, 22, 24, 26, 30, 32, 414, 424
  Australian Wind, 422
  National Building, of Canada, 24, 30, 422, 425, 429, 437, 456
  Uniform Building, 133, 176, 318, 321
Code of practice, 168, 176
Coefficient:
  creep, 466
  factor, exposure, 22
  seismic, 26
  zone, 27
  shrinkage, 467
  thermal expansion, 476
Coefficient of thermal expansion, 476
Column(s):
  analogous, 71, 81
  auxiliary, 342
  bending, 162, 165, 298
  equivalent, 246
  exposed, 463, 475
  fictitious, 86, 301, 403, 406, 408
  shortening, 461, 475
  supports, 244, 480, 487, 488
  wide, 73, 102, 103, 159, 216, 246, 306, 360
Column model:
  single, 72
  single, warping, 349
Combination(s), load, 29, 130
Comfort criteria, human, 6, 12, 13, 452, 455, 457
Compatibility, vertical, 220, 299, 305
Component(s):
  flexural, 117, 120, 388
  participating, 65, 68
Computer analysis, 113, 130, 161, 199, 206, 341
Computer program, 78, 81, 88, 190, 215, 247
Concrete, 14, 40, 41, 161, 414, 432, 461
  cast in place, 40
  reinforced, 36, 37, 39, 44, 56, 161, 168, 308, 355
Concrete building, 285, 305, 461
Concrete cracking, 12, 69, 161, 414, 432, 500
Concrete creep, 461, 464
Concrete pumping, 4, 21, 461
Concrete shrinkage, 461, 464
Configuration:
  flexural, 42, 131, 372
  shear, 44, 131, 372

Connection method:
  continuous, 77
  shear, 378
Constant:
  shear torsion, 321
  torsion, 73, 84, 86, 313, 321, 345
  warping, 313
Constraint, internodal, 277, 299, 303, 305
Construction, 2, 3, 20, 21, 285, 419, 462, 463, 466
  loads, 10, 20, 21
  materials, 2, 419
  prefabricated, 36
Continuum, 7, 66, 71, 77, 215, 216, 260, 265, 382–384, 480
  method, 7, 77, 216, 260, 382, 385, 480
Contraflexure (inflexion):
  line of, 217
  point of, 131, 141, 142, 145, 146, 152, 217, 257, 280
Coordinate(s):
  generalized, 439
  principle sectorial, 318, 322–323
  sectorial, 314, 315, 317, 318
Core, 36, 45, 50, 52, 65, 66, 73, 84, 127, 130, 184, 257, 271, 301, 308, 355, 379, 462, 468, *502*, 509
  box, 65, 66
  elevator, 73, 257, 308
  hull-, structure, 45
Council on Tall Buildings and Urban Habitat, 7
Coupled (shear) wall, 7, 42, 78, 159, *213*, 356, 373, 380, 384, 386, *480*, 487
Coupling:
  beam, 213, 489
  lateral-torsional, 447, 458
  slab, 480, 489, 493
  stiffness, 489
Cracking:
  concrete, 12, 69, 161, 414, 432, 500
  diagonal, 168, 171
Crane, climbing, 4, 21
Creep, 6, 11, 12, 14, *461*, 464
  coefficient, 466
  drying, 46
  specific, 464
  strain, 461, 469
  ultimate, 464, 465
Criteria:
  design, 9, 176
  human comfort, 6, 12, 13, 452, 455, 457
Cross wall, 4, *34*, 213, 489

Crushing:
    diagonal, 172, 173
    failure, 172, 173, 177
    infill, 172, 173
Curtain wall, glass, 4
Curves, design, 231, 268, 332, 333, 480, 482, 502

Damper, tuned mass, 458
Damping, 29, 420, 422, 423, 431, 451
    critical, 425, 436
    matrix, 431
    ratio, 422, 425, 432, 439, 441, 442, 451
Deflection:
    factor, 227, 483, 487
    horizontal, 6, 91, 226, 362, 363, 481, 502, 510
    lateral, 11, 12, 226, 483, 487
    shear, 69, 218, 251
    solar, 463
Deflection analysis, 115, 116
Deflection factor, 227, 483, 487
Deformations, negligible, 69
Dependent node option, 297
Design:
    Code, 12, 18, 388, 414, 423
    criteria, 9, 176
    curves, 231, 268, 332, 333, 480, 482, 502
    formulas, 174, 231, 268, 480, 482, 489, 502
    foundation, 15
    graphic method of, 231, 268
    limit states, 10, 30
    method, 168, 179
    philosophy, 9
    plastic, 30
    procedure (process), 5, 174, 177
    response spectrum, 435, 437, 440
    wall-frames, 279
    wind speed, 22, 422
    working stress, 9, 30
Diaphragm:
    rigid, 287
    shear, 331
    vertical, 69
Discontinuity, base, 202
Displacement:
    modal, 439, 441
    spectral, 441
Drift, 11, 38, 109, 115, 124, 130, 150, 154, 158, 257, 269, 356, 381, 383, 399
    components, 117–123, 152, 164
    correction, 156
    excessive, 156
    index, 12, 268, 269
    (inter)story, 12, 116, 117, 119, 131, 154, 164, 263, 268, 269, 390, 401, 405, 441
    limits, 11, 116
Drying creep, 464
Ductility, 27, 41, 109
    factor, 448
Dynamic:
    analysis, 6, 13, 21, *419*
    motions, 13, 419
    response, 14, 18, 420, 431, 449
    wind pressures, 419

Earthquake, 5, 109, 419, 431, 500
    design response spectrum, 435, 437, 440
    El Centro, 437
    loading, 18, 25, 69, 431, 480
    response, 431
    response spectrum, 420, 433, 440
Element(s):
    beam, 66, 78, 79, 80, 81, 88, 161, 206, 248
    finite, 80, 382, 384, 386
        method, 172, 206
        model, 66, 78
    incompatible mode, 342
    line finite, 216, 253
    membrane, 66, 78, 79, 80, 81, 82, 84, 88, 102, 206, 252, 306, 341
        finite, 66, 78, 79, 81, 82, 84, 88, 102, 172, 206, 341
    plane stress, 216, 342
Elevator, 2, 5, 12, 20, 34, 40, 41, 107, 168, 308
    core, 73, 257, 308, 329
    shaft, 41, 107
Empire State Building, New York, 3, 106
Equitable Life Building, New York, 2
Exchange Building, Seattle, 3
Expansion, coefficient of thermal, 476
Exposed column, 463, 475
Exposure coefficient factor, 22

Facade, 2, 213, 462, 463
Factor:
    amplification, 251, 389, 395, 399
    axial force, 224
    composite cantilever, 233, 482, 485
    deflection, 227, 483, 487
    exposure coefficient, 22
    gust, 24, 424
    importance, 27, 433
    independent cantilever action, 233
    live load reduction, 19
    load, 10, 475
    modal participation, 439
    partial safety, 10
    peak, 423, 425, 428
    safety, 176
    shear flow, 225, 483, 485

INDEX **531**

size reduction, 427
Failure:
  compressive, 176, 177, 179
  crushing, 172, 173, 177, 462
  diagonal tension, 176
  limit state, 10
  shear, 171, 174–176, 179
Fire, 2, 4, 14, 36
  compartment, 15
  insulator, 41
  protection, 14, 36
  resistance, 15
Flange frame, 44, 46, 127, 285, 288, 293, 298, 303
Flat plate structure, 41, 58, 158
  slab structure, 41, 59
Flexibility:
  foundation, 220, 243, 425, 449, 489
  outrigger, 368
  shear, 388
Floor, 5, 285
  composite, 61
  loading, 19
  slab, 36, 68, 190, 199, 213, 308, 489
  slab effective stiffness (width), 158, 489, 493
  steel beam, 60
  system, 37, 55–63
Fluctuation rate, 427
Flying form, 4
Force:
  equivalent lateral, 7, 21, 26, 433, 439
  inertia, 18, 169, 420
Force factor, axial, 224
Form(ing):
  gang, 285
  slip, 4, 462
  structural, *34*
Foundation, 7, 15
  design, 15
  elastic, 220, 243, 480, 487
  flexibility, 415, 419, 487
  movement, 220
  rotation, 16, 220, 415
  settlement, 15, 220
Frame:
  analogous, 81, 158, 172, 174, 206, 207, 210, 211, 247, 344
  analogy, 246, 344
    braced, 207, 210, 344
  analysis program, 113, 190, 206, 215, 247, 277
  braced, 4, 7, 37, 44, 65, 79, *106*, 126, 174, 207, 210, 355, 372, 378–380, 384, 386
  equivalent, 71, 72, 216, 246, 297, 303, 474
  equivalent single-bay, 71, 72, 163

flange, 44, 46, 127, 285, 288, 293, 298, 303
infilled, 4, 7, 40, 168, 172
multi-bay, 71
plane, 79
racking, 372
rigid, 7, 36, 38, 66, 72, 75, 124, *130*, 255, 266, 372, 380, 384, 386, 405
single-bay, 71, 72, 111, 163
space, 36
substitute, 162, 163
three-bay, 477
two-dimensional equivalent, 277, 297, 303
web, 44, 46, 127, 285, 288, 293, 298, 303
wall-, structure, *255*
Framed tube structure, 44
Frequency:
  fundamental, 422, 427, 449
  natural, 420, 427, 435, 436, 439, 441
Frequency of vibration, 421, 427, 435, 436, 439, 441

Generalized theory, *372*
Girder, 20, 37, 65, 69, 106, 109, 111, 130, 132, 133, 156, 161, 168, 171, 260, 380
  belt, 49, 355
  connecting, 372
  lumped, 100
  spandrel, 44, 48, 306, 355, 463
  transfer, 127
Glass curtain wall, 4
Gradient, temperature, 14, 475, 476
Gravity load, 10, 18, 20, 34, 40, 69, 87, 106, 109, 110, 127, 130, 133, 166, 168, 180, 184, 213, 279, 290, 295, 308, 389, 398
Ground acceleration, 419, 431, 435, 447
Ground motion, vertical component of, 447
Gust, 21, 421, 424
Gust energy, 425, 428
Gust factor, 24, 424
Gust loading, 21

Heating, solar, 476
Home Insurance Building, Chicago, 2
Hong Kong, Bank of China Building, 53
Hotel building, 1, 12, 35, 41
Hull-core structures, 4, 45, 301
Human comfort criteria, 6, 12, 13, 452, 455, 457
Hybrid structures, 4, 54, 282

Impact gravity loading, 20
Importance factor, 27, 433
Index, drift, 12, 268, 269

Inertia:
  load, 18, 25, 169, 420
  negative, 87, 406
  sectorial moment of, 320
  warping moment of, 313, 345
Inertia force, 169
Infill, 12, 13, 68, 168
  crushing, 172, 173
  design, 175, 179
  masonry, 41, 68
  stresses, 172
Infilled frame, 40, *168*
Inflexion (contraflexure), point of, 131, 141, 142, 145, 146, 152, 217, 257, 280
Inplane rigidity (floor slabs), 36, 68, 75, 84, 86, 91, 138, 169, 190, 199, 255, 277, 287, 296, 303, 382
Inplane stiffness, 184
Instability, 388
  aerodynamic, 421
  galloping, 421
Interaction, 84, 168, 190, 202, 378, 387
  soil-structure, 15, 415, 420, 449, 487
Interstory drift, 116, 117, 119, 131, 154, 164
Iron, wrought, 2

John Hancock Building, Chicago, 46, 127
Joist, pan, 57

K-bracing, 107, 109, 113, 124, 127, 380
Knee bracing, 380

Lag, shear, 46, 127, 342
Limit(s):
  ambulatory, 13, 453
  drift, 116
Limit state:
  serviceability, 10
  ultimate, 10
Line of contraflexure, 217
Line finite element, 216, 253
Link, 70, 78, 91, 94, 109, 190, 199, 207, 209, 213, 257, 277, 301
  connecting, 78, 91, 213, 257
Load(ing), 10, *18*
  combinations, 29, 130
  concentrated top, 230, 260, 265, 480, 482, 502–511
  construction, 11, 20
  critical, 389, 390, 391, 392, 394
  critical buckling, 389, 390, 391, 392, 394
  dead, 10, 18, 19, 20, 29
  dynamic, 419
  earthquake, 3, 10, 25, 69, 419, 431, 441, 480
  equivalent lateral, 7, 21, 25, 26, 433, 439

  equivalent static, 7, 22, 25, 26, 420, 421, 433
  floor, 19
  gravity, 10, 18, 20, 34, 40, 69, 87, 106, 109, 110, 127, 130, 133, 180, 184, 279, 290, 295, 389, 398
  gust, 21
  horizontal, 34, 109, 130
  impact gravity, 20
  inertia, 18, 25, 169, 420
  inset reduction, 19, 20
  lateral, 4, 10, 21, 25, 109, 213, 216, 255, 293, 433, 439
  live, 6, 10, 18, 19
  pattern live, 19, 132
  permanent, 10
  redistribution of, 70, 461, 471, 474
  seismic, 3, 10, 25, 419, 431, 441, 480
  sequential, 10
  standard, cases, 223, 262, 266, 480, 502
  static, 21, 25
  triangularly distributed, 266, 480, 483, 502–511
  uniformly distributed, 359, 502–511
  wind, 2, 18, 21, 34, 69, 419, 480
    dynamic, 23
    equivalent static, 7, 22, 25, 420
Lobby, 4, 202
Location of outriggers, optimum, 355, 364, 366
Lumping, 67, 99
  lateral, 100
  vertical, 100, 102

Masonic Temple Building, Chicago, 2
Masonry, 2, 4, 7, 41, 68, 168, 169, 462
  cladding, 462
  infill, 41, 68
Mass, lumped, 439
  damper, 458
  matrix, 431
Master-slave technique, 297
Material(s), 2, 3, 34, 68, 419, 461
Matrix:
  second order, 406
  stiffness, 66, 215, 248, 277, 296, 406, 431
  stiffness, analysis, 7, 66, 77, 78, 161, 215, 247, 248, 277, 296, 349, 447
Medium:
  connecting, 260, 265, 392
  continuous, 66, 71, 77, 216, 229, 260, 265, 392
Member:
  fictitious, 94, 297, 301
  haunched, 250
Member stiffness, 414, 431

Membrane element, 66, 78, 79, 81, 82, 84, 88, 102, 206, 341
Membrane finite element, 66, 67, 78, 79, 81, 82, 84, 88, 102, 172, 206, 252, 306, 341
Mercantile Tower, St. Louis, 124
Method(s):
   cantilever, 146, 147, 150
   continuous connection, 77, 216, 260
   continuum, 7, 77, 216, 260, 382, 385, 480
   design, 168, 179
   finite element, 7, 16, 172, 206, 216, 252, 493
   graphic design, 231, 268
   generalized, 379, 384, 385, 386
   moment-area, 117, 359-362
   portal, 141-144, 150
   of sections, 115
   shear connection, 215, 378
Modal analysis, 26, 29, 433, 438
Mode(s):
   flexural, 42, 257, 372, 388, 391
   fundamental (natural), 423, 449
   shear, 42, 257, 372, 388, 390
   shear-flexure, 372, 391
Model:
   analytical, 65
   finite element, 66, 78
   planar, 91, 94, 95, 100, 202, 255, 260, 277, 297, 303
   single-column, 72
   single warping-column, 349
   two-dimensional, 67, 91, 94, 277
Mode element, incompatible, 342
Modeling, 65
   structural, 65, 66-105
Modulus of sub-grade reaction, 220, 243, 489
Moment:
   connection, 213
   distribution, 133, 135, 136, 190, 191, 477
   external, 35, 36, 69, 111
   fixed-end, 135
   girder, gravity load, 134
   overturning, 11, 433
   restraining, 215, 218, 257, 359, 362, 363, 368
   reverse, 215, 218
Moment-area method, 117, 359-362
Moment of inertia:
   sectorial, 320, 502
   warping, 313, 345
Moment transfer, 86, 88
Monadnock Building, Chicago, 2
Motion(s):
   along-wind, 429
   cross-wind, 419, 423, 429, 431
   discomfort, 452

dynamic, 13, 419
ground, 419, 431, 437
perception, 13, 452
sickness, 452
Movement:
   differential, 6, 11, 14, 461, 462, 474, 475
   foundation, 220
   thermal, 475

National Building Code of Canada (NBCC), 24, 29, 30, 422, 425, 429, 437, 456
New York:
   Chrysler Building, 106
   Citicorp Building, 127
   Empire State Building, 106
   Equitable Life Building, 2
   780 Third Avenue Building, 46
   Statue of Liberty, 106
   Woolworth Building, 106
   World Trade Center, 3
Node, dependent, 91, 297

Office, open plan, 36
Office building, 3, 12, 34, 35, 55, 285
One-bay frame, equivalent, 71, 72, 111, 163
One-way slab, 56, 58
Open plan office, 34
Out-of-plumb, 414
Outrigger(s), 36, 355
   -braced structure, 4, 48, *355*, 356
   efficiency, 368
   flexibility, 368
   optimum location of, 355, 364, 366

Pan joist, 57
Partial safety factor, 10
Participation factor, modal, 439
Partition, 12, 13, 34, 40, 65, 168, 169, 462
Pattern live loading, 19
P-Delta analysis, 11, 130, 161, 398, 399, 401, 403, 406
P-Delta effect, 6, 12, 16, 86, 398
Percentage of reinforcement, 461, 462, 463
Perception:
   human, 452, 453, 454
   limits, 453
   motion, 453
   thresholds, 454
Period, fundamental (natural), 12, 13, 25, 27, 433, 436
Planning, 34, 38, 285
Plan structure:
   asymmetric, 88
   symmetric, 88, 138, 255, 297
Plate, flat, 37, 41, 58, 68, 75, 130, 158, 166

Plumb, out of, 414
Portal frame support, 223, 244, 480, 487
Portal method, 141–144, 150
Procedure:
  design, 174, 177
  lumping, 67, 99, 100, 102
Program:
  computer, 78, 81, 88, 190, 215, 247
  frame analysis, 113, 190, 206
  general purpose analysis, 6, 215, 247, 277
  structural analysis, 78, 81, 88, 206, 215, 247, 277, 344, 347, 353
Properties:
  sectorial, 315, 349, 353
  warping, 318
Pumping, concrete, 4, 21, 461

Racking, 131, 141, 150, 285
  frame, 372
  rigidity, 71, 72, 266, 376, 377, 379, 380
Racking shear, 71, 72, 285, 378–380
Rand-McNally Building, Chicago, 2
Ratio:
  damping, 422, 425, 432, 439, 441, 442, 451
  modular, 471
  slenderness, 388
Redistribution of load, 70
Reduction (structural) techniques, 88, 161–165
Reinforcement, percentage of, 14, 461, 462, 463
Resistance, center of, 140
Resonance, 421
Response:
  earthquake, 431
  human, 452
  peak, 423, 428, 429, 431, 433
  seismic, 431
  spectrum, inelastic, 448
  time-history, 433, 449
Response spectrum, 420, 433, 435, 440, 447
  analysis, 435
  inelastic, 448
Rigid frame structures, 130
Rigidity:
  axial, 131, 360
  axial flexural, 379
  center of, 140, 141
  effective shear, 157
  flexural, 110, 184, 186, 187, 216, 247, 314, 359, 360, 373, 376, 377, 379, 383, 393, 474, 502
  in-plane, 36, 68, 75, 84, 86, 91, 138, 169, 190, 199, 217, 255, 277, 287, 296, 303, 382
  parameters, 379
  racking, 71, 72, 376, 377, 379, 380
  shear, 71, 72, 157, 220, 262, 266, 314, 376, 379, 380, 393
  torsional, 394
Rotation, 68, 308, 356–359, 502, 509, 510
  center of, 70, 394
  foundation, 16, 220, 243, 415

St. Louis, Mercentile Tower, 124
St. Venant principle, 173
St. Venant torsion, 345, 347, 349
San Francisco, Alcan Building, 124
Sears Tower, Chicago, 3, 45, 46
Second-order effect(s), 388–398
Second-order stiffness matrix, 406
Section(s):
  box, 308
  method of, 115
Sectorial properties, 315–321
Serviceability limit state, 10
Service shaft, 34, 41, 70
Service systems, 3, 4, 9, 34, 35
Setback, 4, 113, 150
Settlement:
  differential, 15, 220
  foundation, 15, 220, 243, 487
780 Third Avenue Building, New York, 46
Shaft:
  elevator, 41, 107
  service, 34, 41, 70
  stair, 34, 41, 65, 70, 107, 168, 184
Shape:
  flexural, 111, 178, 257, 372
  shear, 44, 111
Shear:
  area, 251, 408, 412
  area, negative, 408, 412
  base, 25, 433
  cantilever, 378
  center, 73, 188, 310, 317, 345, 346, 349
  configuration, 44, 131, 372
  connection method, 69
  core, 5, 7
  deflection, 69, 218, 251
  drift formula, 117, 120
  external, 35, 36, 39, 69
  failure, 15, 171, 174–176, 179
  flexibility, 388
  flow, 224, 482, 485
  flow factor, 225, 483, 485

horizontal, 37, 69, 109–111
lag, 46, 127, 283, 285, 293, 342
mode, 257, 388, 390
racking, 71, 72, 378–380
rigidity, 71, 72, 157, 220, 262, 266, 314, 376, 379, 380, 393, 502
shape, 44, 111, 257
stiffness, 208, 342
story, 442
stress analysis, 206
torsion constant, 321
transfer, 70
vertical, 70, 286
wall, 2, 4, 5, 35–37, 41, 45, 66, 71, 73, 79, 130, 184, 257, 308, 372, 382–384, 462
walls, coupled, 42, 78, 159, *213*, 356, 372, 373, 380, 386, *480*, 487
wall structure perforated, 41, *184*, 213
Shortening, 461, 462, 468, 474
  differential, 11, 461, 468
  elastic, 463
  inelastic, 462, 463
  vertical, 461, 474
Shrinkage, 6, 11, 14, *461*, 464, 470
  coefficient, 467
  strain, 461, 470
Sickness, motion, 452
Size reduction factor, 427
Skew symmetry, 296, 299, 305
Slab, 35, 65, 68, 69, 70, 73, 130, 213, 372, 462, 480
  connecting, 36, 308, 329, 372, 480, 489
  effective width (stiffness), 158, 159, 480, 493, 495
  flat, 37, 41, 58
  floor, 36, 68, 190, 199, 217, 255, 308, 480, 489
  one-way, 56, 58
  two-way, 59
  waffle, 59
Slab and beam, 59
Slenderness ratio, 388
Slip form(ing), 4, 462
Soil–structure interaction, 15, 415, 420, 449, 487
Space:
  frame, 36
  structure, 53
  truss, 127
Spandrel beam, 127, 284, 285, 286, 306, 463
Spandrel girder, 44, 48, 306, 355, 463
Spectrum, response, 420, 433, 435, 440, 447
Stability, 11, 163, *388*
Stair shaft(s) (case or well), 34, 41, 65, 70, 107, 168, 184
Statue of Liberty, New York, 106

Steel structure, 2, 36, 37, 42, 44, 109, 169, 255, 355, 476
Stiffness:
  axial, 209
  bending, 39, 208
  coupling, 489
  equivalent, 87, 216, 474, 493
  flexural, 158, 403, 410
  in-plane, 184
  lateral, 9, 11, 390
  limitations, 11
  member, 414, 431
  negative flexural, 87, 410, 412
  negligible, 68
  reduction (due to cracking), 69, 414, 432, 500
  shear, 208, 342
  slab, 158, 159
  torsional, 86, 161, 308, 394
Story drift, 12, 116, 117, 119, 131, 154, 164, 263, 268, 390, 401, 405, 441
Story shear, 442
Strain, creep, 461, 469
Strain, shrinkage, 461, 470
Strength, *11*
Stress:
  diagonal tensile, 173
  infill, 172
  warping, 308, 326
Stress design, working, 9, 30
Stress element, plane, 342
Structural form, *34*
Structure:
  asymmetric plan, 88, 255
  belt-braced, 355
  braced frame, 37, *106*
  bundled-tube, 45, 303
  cantilever, 372
  core, *308*
  coupled shear wall, 42, *213*
  doubly-symmetric, 91, 318
  framed-tube, 44, 305
  hybrid, 4, 54, 282
  hull-core, 45, 301
  infilled frame, *168*
  non-load-bearing, 13
  nonproportionate, 184, 190
  nontwisting, 186, 187, 190, 260
  outrigger-braced, 4, 48, *355*, 356
  proportionate, 184
  rigid frame, 38, *130*
  shear wall, 41, *184*
  space, 53
  steel, 2, 36, 37, 42, 44, 109, 169, 255, 355, 476

Structure (*Continued*)
 suspended, 36, 50
 symmetric plan, 88, 138, 255, 297, 392
 top hat, 355
 tube-in-tube, 45, 301
 tubular, *283*
 twisting, 66, 70, 186, 187, 199
 wall-frame, 36, 42, 78, *255*, 356, 378, 384, 385, 386, 392, 502
Strut, diagonal, 40, 169, 174
Subgrade reaction, modulus of, 220, 243, 489
Support(s):
 column, 244
 flexible, 243, 415, 487
 portal frame, 223, 244, 480, 487
Sway, 390
Symmetry, 66, 68, 235, 255, 277, 288, 296, 305, 477
 skew, 296, 299, 305
System(s):
 floor, 37, 55
 nonproportionate, 186, 190, 199
 proportionate, 184, 186
 service, 3, 4, 9, 34, 35

Tall building, definition of, 1
Technique, reduction, 88
Temperature:
 differential, 6, 463
 effect, 11, 14, *461*, 475
 gradient, 14, 475, 476
Tension failure, diagonal, 176
Theory:
 generalized, 372
 Vlasov's, 323
Time-history analysis, 433
 base acceleration, 420, 432
 response, 433, 499
Top hat structure, 355
Top load, concentrated, 230, 260, 265, 480, 482, 502
Torque, 99, 141
 external, 69
 uniformly distributed, 325
Torsion, 13, 23, 28, 69, 70, 141, 303, 308, 393, 502, 509
 warping, 308, 346
Torsional buckling, 388, 394
 rigidity (stiffness), 394
Torsion constant, 73, 84, 86, 313, 321, 345, 502
 shear, 321
Transfer:
 moment, 86, 88
 shear, 70

Transfer beam, 44
Transfer girder, 127, 285, 462
Transfer truss, 127
Truss, 2, 4, 79
 belt, 49
 space, 127
 transfer, 127
 vertical, 2, 37, 69, 106, 124, 372
Tube, 36, 283
 braced, 36, 46, 127, 283, 289, 305
 bundled, 45, 46, 283, 288, 305
 framed, 3, 7, 44, 103, 283, 285, 296, 297, 303, 305
 modular, 288
Tube-in-tube, 45, 283, 301
Tubular structure, *283*
Tunnel, wind, 5, 23, 423, 425, 429, 457
Turbulence, 421, 423
 background, 425, 426
 boundary layer, 421
 building generated, 421
 resonant, 425
Twist(ing), 66, 70, 186, 255, 308
 center of, 187, 188

Uniform Building Code, 21, 22, 26, 29, 30, 133, 176, 318, 321
Uniformly distributed load, 223, 237, 262, 488, 502
Uplift, 11, 36, 111

Velocity:
 pseudo, 436
 spectral, 436
Vibration:
 free, 435
 frequency of, 421, 427, 435, 439, 441
 natural mode of, 423, 449
 period of, 12, 13, 25, 27, 433, 436
Viscous damping, 431
Vlasov's theory, 323
Volume–surface ratio, 14, 461, 467
Vortex shedding, 13, 423, 429

Waffle slab, 59
Wall(s), 2, 4, 34, 35, 65, 69, 184
 bearing, 2, 213
 coupled (shear), 42, 78, 159, 213, 356, 372, 373, 380, 384, 386, *480*
 cross, 4, 42, 213, 489
 flanged, 494, 496, 497
 –frame structures, 36, 42, 78, 255, 356, 378, 384, 385, 386, 392, 502
 glass curtain, 4

linked, 184, 213, 483, 487
nonproportionate, 190
plane, 41, 79, 495
proportionate, 186–187
shear, 2, 5, 35–37, 41, 45, 66, 71, 73, 130, 184, 257, 308, 372, 382–384, 462
Wall-frame structure, 36, 42, 78, 255, 356, 378, 384–386, 392, 502
Warping, 70, 308, 349, 353
  restrained, 308, 315, 323
    behavior, 308, 310, 345
Warping column model, single, 349
Warping constant, 313
Warping moment of inertia, 313, 345
Warping properties, 318
Warping stress, 308, 326
Warping torsion, 308, 346
Wavelength, inverse, 427

Web frame, 44, 46, 127, 285, 288, 293, 298, 303
Wide column, 73, 102, 103, 159, 216, 246, 306, 360
Width, slab effective, 159
Wind, 2, 10, 18, 109, 419, 420
Wind Code, Australian, 422
Wind load, 2, 10, 21, 34, 69, 420, 422, 480
  dynamic, 419
  equivalent static, 7, 22, 25, 420
  pressure, 22, 24, 420, 422
Wind speed, design, 22, 422
Wind tunnel (test), 5, 23, 423, 425, 429, 457
  boundary layer, 23
Woolworth Building, New York, 2, 106
Working stress design, 9, 30
World Trade Center, New York, 3

Zone coefficient, seismic, 27